SPACES OF CONTENTION

Spaces of Contention
Spatialities and Social Movements

Edited by

WALTER NICHOLLS
University of Amsterdam, The Netherlands

BYRON MILLER
University of Calgary, Canada

JUSTIN BEAUMONT
University of Groningen, The Netherlands

Routledge
Taylor & Francis Group

LONDON AND NEW YORK

First published in paperback 2024

First published 2013 by Ashgate Publishing

Published 2016
by Routledge
4 Park Square, Milton Park, Abingdon, Oxon OX14 4RN

and by Routledge
605 Third Avenue, New York, NY 10158

Routledge is an imprint of the Taylor & Francis Group, an informa business

British Library Cataloguing in Publication Data
Spaces of contention : spatialities and social movements.
1. Human geography – Political aspects. 2. Spatial
behavior – Political aspects. 3. Social movements.
I. Nicholls, Walter II. Miller, Byron A., 1957–
III. Beaumont, Justin.
304.2'3–dc23

The Library of Congress has cataloged the printed edition as follows:
Spaces of contention : spatialities and social movements / [edited] by
Walter Nicholls, Byron Miller, and Justin Beaumont.
pages cm
Includes bibliographical references and index.
ISBN 978-0-7546-7778-9 (hardback)
1. Social movements. 2. Spatial
behavior. 3. Human geography. I. Nicholls, Walter. II. Beaumont, Justin.
III. Miller, Byron A., 1957–
HM881.S69 2013
303.48'4—dc23

2012047423

ISBN: 978-0-7546-7778-9 (hbk)
ISBN: 978-1-03-292226-3 (pbk)
ISBN: 978-1-315-61019-1 (ebk)

DOI: 10.4324/9781315610191

Contents

List of Figures and Tables

Figures

Tables

List of Contributors

John Agnew, Professor, Department of Geography, University of California, Los Angeles

Justin Beaumont, Assistant Professor, Department of Planning, University of Groningen

Andrew Cumbers, Professor, School of Geographical and Earth Sciences, University of Glasgow

Andrew D. Davies, Lecturer, Department of Geography, University of Liverpool

Donatella della Porta, Professor, Department of Political and Social Sciences, European University Institute

Jan Willem Duyvendak, Professor, Department of Sociology and Anthropology, University of Amsterdam

Maria Fabbri, PhD Candidate, University of Trento

David Featherstone, Senior Lecturer, School of Geographical and Earth Sciences, University of Glasgow

Martin Jones, Professor, Institute of Geography and Earth Sciences, University of Wales, Aberystwyth

Deborah G. Martin, Associate Professor, Department of Geography, Clark University

Margit Mayer, Professor, John F. Kennedy Institute, Free University Berlin

Byron Miller, Associate Professor, Department of Geography, University of Calgary

Don Mitchell, Distinguished Professor, Maxwell School, Syracuse University

Johan Moyersoen, co-founder and partner of i-propeller

Corinne Nativel, Senior Lecturer, Social Sciences, University of Paris-Est Créteil

Walter Nicholls, Assistant Professor, Department of Sociology and Anthropology, University of Amsterdam

Ulrich Oslender, Assistant Professor, Department of Global and Sociocultural Studies, Florida International University

Gianni Piazza, Assistant Professor, Department of Political and Social Sciences, University of Catania

Paul Routledge, Reader, Department of Geographical and Earth Sciences, University of Glasgow

Ted Rutland, Assistant Professor, Department of Geography, Planning, and Environment, Concordia University

Erik Swyngedouw, Professor, Department of Geography, University of Manchester

Loes Verplanke, Post-Doctoral Researcher, Department of Sociology and Anthropology, University of Amsterdam

Dingxin Zhao, Professor, Department of Sociology, University of Chicago

Introduction

Conceptualizing the Spatialities of Social Movements

Walter Nicholls, Byron Miller and Justin Beaumont

Virtually every day we witness a range of powerful movements challenging the privileges and power of governing elites. Early in the second decade of the new millennium we have seen the successful mobilization of democratic movements in North Africa and Burma, and anti-neoliberal movements worldwide, most notably the Occupy movement. These movements demonstrate the continued importance of challenging established orders, as well as the ability of movements to topple long-standing power structures. Forged in the context of particular political opportunities, organizational resources, and discursive frames, the dynamics of these movements reveal the centrality of spatial relations. In a great many cases streets and public squares became the vortices through which radically different groups connected and assembled to protest against, or wrest control from, those holding the reins of power. Many of the people who flooded into public spaces to protest did so spontaneously, prompted only by a tweet, a Facebook post, a phone call, a flyer, or word of mouth. But behind much of the seemingly spontaneous mobilization lay the work of long-standing communities and organizations that developed strategies and tactics to activate pre-existing networks, prompting members, allies, acquaintances and strangers to come out in support. In some cases apparently local mobilizations had transnational organizational connections, such as the links between the Egyptian revolution and the Serbian Centre for Applied Nonviolent Action and Strategies (Rosenberg 2011).

The processes of assembling thousands of different individuals and groups – with their own distinctive worldviews and political imaginaries – for defiant acts of spatial appropriation not only affirmed activists' commitment to horizontal and democratic modes of political action, it also demonstrated very clever spatial strategies and tactics. They reminded people and organizations across countries and continents that another world is indeed possible, instilling courage and changing the calculus of future acts of collective action. The multitude's expressions of dissent and solidarity through the occupation of highly symbolic public spaces were critical, not only because they breached the order of things and revealed the frailties and blind spots of state power. Images of mass public demonstrations – and knowledge of the strategies and tactics employed – were diffused through personal and organizational networks and mass media, with actors close and far discovering possibilities where none had previously been perceived. States responded with a range of techniques in attempts to repress mobilization and assert their claims to territorial sovereignty – from blocking the diffusion of information by censoring and shutting down the internet to forcibly removing protesters from protest sites. Yet, as some of the spatial channels of mobilization were disrupted, others were created. Networked and mobile activists have shown a remarkable capacity to overcome state barriers, though of course there are limits and exceptions to this, as evidenced by the repression of dissent in Iran and North Korea, among other countries.

Shared outrage, solidarity, and hope are fundamental prerequisites for virtually all mass mobilizations. But before people will flood the public squares they must understand that they are not alone. Feelings of isolation must be vanquished and fear must be replaced with hope. Hope and solidarity require the building of relationships that assure protestors that their views are shared, that their fellow activists can be trusted, that resources – however meagre – will be available, and that the collective force to change society can be rallied. These relationships – and all social relationships – are fundamentally, inextricably spatial. That social relations are spatial relations has long been recognized (Hagerstrand 1970; Lefebvre 1974 [1991 translation]; Soja 1980; Giddens 1984), but the way in which this ontological fact has been conceptualized and operationalized has frequently fallen short. Conceptualizations of space as simply a container of social action, such as a nation-state or city, or as a distance variable to be considered among other variables, can still be found in the contemporary literature.

But more commonly a particular spatiality, e.g., place, scale, networks, mobility, is selected as the epistemological lens through which analysis proceeds. Such analyses are often sharp and insightful, yet provide only partial accounts of the spatial constitution of the social processes under consideration. The discipline of Geography has undergone a sequence of spatial emphases: regions in the 1930s, '40s and '50s; space in the 1960s and '70s; place in the 1980s; scale in the 1990s and 2000s; and networks and mobility more recently. From time to time major debates erupt among scholars emphasizing different spatialities, e.g., the 'scale debate' in which some scholars sought to expunge the notion of scale from geographic research and replace it with a flat ontology focusing on networks (Marston et al. 2005; Leitner and Miller 2007; Jonas 2006; Leitner, Sheppard and Sziarto 2008). From our perspective these debates are both enlightening and frustrating. Enlightening, because they point out the weaknesses and blind spots in dominant epistemologies and research agendas. Frustrating, because they often present us with an either/or zero-sum intellectual choice: if we are to consider a new spatial epistemology we must discard what we have learned in the past – or so we are told. While recognizing that some spatial ontologies and epistemologies may be irreconcilable, we contend that there is actually considerable complementarity among the spatial ontologies and epistemologies employed in the spatial research of recent years (Lefebvre 1991/1974; Soja 1989, 1996; Harvey 2006; Jessop et al. 2008). All spatialities, properly conceived, are relational. These relational spatialities, e.g., place, space, scale, territory, networks, mobility, play distinctive yet interlocking roles in shaping the structures, strategies, dynamics and power of social movements.

The aim of this edited volume is to provide readers with a state-of-the-art analysis of how space plays a constituting role in social movement mobilization. The contributing authors identify a variety of ways in which this occurs, stressing that multiple spatialities are implicated and, indeed, co-implicated in contentious politics (Sheppard 2002; Leitner and Miller 2007; Jessop et al. 2008; Leitner et al. 2008; Leitner and Sheppard 2009; Nicholls 2009; Jones and Jessop 2010). Multiple spatialities intersect and shape social movements, but which spatialities are most relevant is a context-dependent question, dependent upon the positionality of movement actors as well as of the researcher. Most scale-focused research, for example, tends to concentrate on questions of relational state structures and political economy, while network-focused work tends to concentrate on actors and the relations they build. While the questions asked and the positionalities of researchers may be very different, to represent these bodies of research as irreconcilable is to pose a false antithesis (Castree 2002). Each spatiality has implications for the constitution of social movements (e.g., mobilization capacities, internal cohesion, frames, internal conflicts, etc.). In some cases particular spatialities may not be immediately evident or relevant, such as the role of a multi-scalar state in a highly localized movement against the exploitive practices of a local business. But social movements are always dynamic: the significance of scalar relations may not be immediately present, but emerge as events unfold and tactics and strategies evolve.

Our intention is not to make a broad statement that all social movements reflect an endlessly complex entanglement of spatialities and end it there. Rather, our analytical aim is to first disentangle the spatial complexity of social movements by identifying some of the distinctive roles spatialities may have – while acknowledging that complete disentanglement is impossible. Then, in keeping with current debates around an assemblage perspective on social movements (see Davies 2012; Legg 2009; McFarlane 2009; McCann and Ward 2010), as well as long-standing critical realist perspectives that combine ontological realism with epistemological relativism (e.g., Archer et al. 1998; Sayer 2010), we assess some of the ways in which spatialities may be implicated and co-implicated in the dynamics and pathways of social movements.

(Dis)entangling the spatialities of social movements

Place: enabling – disabling social movements

John Agnew (1987, 2002) has forcefully argued that places are sites where wider economic and political process are played out (*locations*), social and organizational relations develop to mediate micro responses to macro level processes (*locale*), and spatial imaginaries form to give people a sense of meaning in their particular worlds (*sense of place*) (Agnew 1987: 28). These qualities of place overlap within one another and form the contexts through which people experience and live out their everyday worlds. Agnew maintains that general social processes (e.g., formation of class, gender, 'race', etc.) intersect with one another in concrete places, serving as fields through which ongoing processes become inscribed into social habitus, identities, relations of trust and belonging, and political dispositions. People do not become familiar with their social positions by situating their locations in abstract structures, but through everyday interactions with other people, things and images that make up their *locale*. Significantly, place is not to be equated with 'the local'. As nodes of social interaction, places may be locations where geographically extensive processes, even global processes, intersect and play out. Indeed, Doreen Massey (1994) explicitly addresses myriad ways in which global movement and intermixing can give rise to specific local characteristics, including 'a global sense of place'. Places are where social relations are bundled or 'condensed', regardless of the territorial extent of those relations.

Places shape political subjectivities of people and provide them with frameworks for interpreting whether injustices have been done and whether collective and contentious responses are merited. Within locales people form 'epistemic communities' to interpret whether the abuse they face amounts to a violation of the 'social contract' and merits a forceful and collective response. For example, in France all nineteenth-century artisans faced similar threats by the forces of increased industrialization, but their interpretations of the nature of these threats varied sharply according to the cultures and relations found in different places across the country (Tilly 1964, 1986). While artisans in Paris interpreted structural changes as meriting a forceful collective response in 1848, artisans in the Vendée region had an entirely different interpretation of their changing structural position which precipitated a counter-mobilization to support the old regime.

While place has been shown to play an important role in shaping the basic political dispositions of people, scholars have also shown that place plays a vital role in helping disparate actors form into a cohesive political force. Sustained and proximate interaction over time can create strong trusting relations among actors, which can then be drawn on to enable collective action (Granovetter 1983; Coleman 1988; Diani 1997; Routledge 1997; Miller 2000; Tarrow and McAdam 2005; Tilly 2005; Nicholls 2008). Relations of trust are critical because they provide certainty that when actors contribute their scarce resources to high risk political enterprises, their contributions will not be squandered because of the malfeasance or incompetence of others (Tilly 2005). Trust and knowledge lowers uncertainties and increases the willingness of actors to risk their lives, resources, and freedom for different political enterprises. Strong relations not only help generate greater trust in others but also a sense of obligation to contribute to the struggles of fellow comrades. Strong emotional bonds knit actors together while obligations are enforced through collective surveillance (Coleman 1988).

In his now classic discussion of the Paris Commune, Roger Gould (1993; 1995) demonstrates that strong ties developed in working class neighbourhoods provided the social solidarity that permitted residents to risk their lives to protect the Commune. 'What tied workers from different occupations together in the Commune were the *tangible bonds they experienced as neighbors, not the abstract bonds of joint structural position in the capitalist mode of production*' (1993: 751, emphasis added). Although proximity is not the only condition for securing strong ties (kin networks, religious affiliations, and common history are also important), geographic stability increases the likelihood of repeated contact and bonding experiences among people, which in turn favours stronger ties (Coleman 1988; Collins 2004).

Place therefore helps generate strong relations among activists within a social movement network. These types of relations ultimately lower the uncertainties of high risk mobilizations, helping compel people to contribute their scarce resources to these mobilizations in spite of the risks. Thus, place plays a crucial role in producing the kinds of intensive relations needed to facilitate the mobilization of high end resources to high stakes struggles.

The relational interdependencies discussed here are also constituted through what Deborah Martin has called 'place frames', which function to provide activists with a common sense of who they are, their difference from adversaries, and the merits of their cause. Activists draw on the common symbolic repertoires found in places to assemble mobilizing frames and harness collective emotions (Miller 2000; Martin 2003, Bosco 2006). By contributing to the production of solidarity networks and these symbolic frames, 'place' makes it possible for marginalized people to contribute scarce resources to high risk political movements.

While place can enhance the mobilization powers of activists by strengthening relations and building common mobilizing frames and identities, states may attempt to short-circuit and disrupt movements by enacting a range of place-based strategies (Castells 1983; for a historical account, see Mann 1993). Through the development and enhancement of local political and administrative institutions, states can channel political grievances away from the central state and other oppressive institutions, toward more proximate and readily identifiable targets (Katznelson 1981; Castells 1983; Purcell 2006; Miller 2007). They can also channel aggrieved activists into thousands of particularistic battles. Moreover, each of these administrative and political units provides public officials with information and means to monitor and discipline the conduct of the activist communities within them. For example, David Harvey (2003) recounts that one of the most important measures to ensure social and political control by the French state in the years after the 1848 revolution was to create administrative districts (*arrondissements*) in Paris to monitor and disrupt activities within micro-insurgent spaces. The rollout of countless new local political and administrative spaces has only been accelerated in recent years with decentralized and 'participatory' urban development projects (Raco and Imrie 2000; Beaumont 2003; Nicholls 2006; Dikec 2007; Beaumont and Nicholls 2008; Uitermark 2010). While some of these new administrative spaces may actually provide new opportunities for empowered citizenship (Fung 2006), it appears that most are consistent with a long and well established state tradition of penetrating the places where insurgent networks arise and disrupting them before they evolve into anti-systemic movements.

Place matters for social movements because it empowers activists, but also it ensnares them in thousands of local traps (Purcell 2006). David Harvey (2001) cuts to the heart of the modalities of place through the concepts of 'place in itself' and 'place for itself'. Place in itself refers to the enabling effects of place-based relations which are, for him, the essential building blocks for assembling scattered individuals into a cohesive force for social and political change. Place for itself, by contrast, is exemplified by efforts to turn the geographic defence of places and territories against all outsiders, undermining the ability of such mobilizations to connect to distant others and present a unified front in the face of new geometries of capitalist power.

Territory and region: inclusion/exclusion

If place can be understood primarily as a locus of social interaction, meaning construction, and collective action calculation, the spatialities of 'territory' and 'region' address the areal constitution of social and political life. Territory and region are difficult to define, with numerous definitions and strands of analysis found in the geographical literature. Indeed, in his comprehensive overview of the concept of territory, Elden (2010) refuses to provide a definition. Likewise 'region' is difficult to pin down, and the two concepts are frequently used interchangeably (MacLeod and Jones 2007; Jonas 2012). If 'region' is to be differentiated from 'territory', it is likely to be on the basis of the mechanisms by which geographical areas are constructed and claimed. The notion of region is less likely to be directly associated with state action and, arguably, more likely to be associated with the geographical patterns of everyday life and the claims made in praise and defence of such patterns. Nonetheless, states may also promote the construction of regions and regionalism. Both concepts represent claims to geographical areas, claims of coherence within those areas, and claims of authority over the populations and resources of those areas, including the ability to include or exclude actors based on how the territory/region is defined.

While problematic, a common definition of territory can be offered: a 'bounded space' associated with the constitution of the modern state. The concept of territory typically implies state sovereignty over a bounded area, including the population and resources within. John Agnew, in his work on the 'territorial trap' (1994, 2009a), has pointed out the flaws in this conception: state sovereignty need not be inextricably linked to exclusive bounded territory, there is no clear distinction between foreign and domestic relations, and modern societies should not be viewed as territorially contained. Much of Agnew's argument resonates in the work of Massey (1994, 2005) and many poststructuralists who go a step further to argue that coherent spatial 'containers' are implausible and indefensible in the postmodern globalized world. Indeed, some poststructuralists have argued that scalar and territorial concepts be jettisoned in favour of attention to 'bordering practices' (Marston et al. 2005). Yet the fluidity and permeability of borders can neither be equated with, nor entirely account for, social and political relations within a geographical territory. As Elden (2010) explains, 'Focusing on the determination of space that makes boundaries possible, and in particular the role of calculation, opens up the idea of seeing boundaries not as a primary distinction that separates territory from other ways of understanding political control of land, but as a second-order problem founded upon a particular sense of calculation and concomitant grasp of space' (2010: 811). For Elden territory is not something fixed and ahistorical, but rather a 'political technology' employed in the exercise of power. Elden's notion of territory as political technology is consistent with Agnew's (2009b) explanation of the way modern territorial states function, i.e., through exclusion (of entities that cannot claim territorial sovereignty) and mutual recognition (of entities that effectively exercise territorial sovereignty).

The ability to exclude others and exercise control over populations and resources within an area is the essence of territory as political technology. While globalization and compromised bordering practices may undermine this technology, they by no means eliminate it. According to Agnew (2009b):

> Territories and networks exist relationally rather than mutually exclusively. If territorial regulation is all about tying flows to places, territories have never been zero-sum entities in which the sharing of power or the existence of external linkages totally undermines their capacity to regulate. If at one time territorial states did severely limit the local powers of trans-territorial agencies, that this is no longer the case does not signify that the states have lost all of their powers (2009b: 747).

Territorial (or regional) regulation has clear implications for collective action. The ability to control and marshal resources within a territory, as well as ability to regulate the flow of resources across borders, is of critical importance to social movements and states alike. 'Regions and territorial organization [operate] at the nexus of tensions between fixity and flow' (Jonas 2012: 266) and these tensions are resolved, if temporarily, through struggle.

The construction of territory, moreover, can provide the basis for territorial identity, an extremely powerful form of identity and solidarity in collective action (Paasi 2009). When effective, the state territorial discourses are capable of legitimating the existing order. When they fail to achieve legitimation, however, aggrieved actors may mobilize against the institutions of the central state. Or they may mobilize but nonetheless target local state institutions, seeking to advance particularistic concerns. Interests may be couched in terms of the protection of local regions from forces that threaten a group's status, privileges, or way of life. Many secessionist mobilizations, gated communities, Nimbyisms, and regionalist movements are rooted in such dynamics (Boudreau and Keil 2001). In these instances, mobilization is not only channelled away from the central organs of state power, but may also target marginalized social groups rather than structural forces or political and economic elites.

Castells (1997) adds that localizing state discourses and strategies, under conditions of advanced globalization, have intensified the disjuncture between global elites and local activists. According to Castells, as power has concentrated in global networks ('spaces of flows'), associational and representational institutions are increasingly circumscribed in localities ('spaces of places'). Global elites have been able to position themselves at the intersection of these global and local spaces, developing 'reflexive dispositions' that allow them to maximize their power in both. Non-elites, by contrast, may lack access to global networks and have few options but to use local associational and representational institutions in defensive manoeuvres. In other words, networks are themselves regionalized, in ways that frequently favour global elites. Massey (2007) adds that political imaginaries fuelled by nostalgia for a past undisrupted by 'outsiders' may energize such localized struggles.

Scaling the geometries of power

Territorial power is unevenly articulated across a range of relational geographic scales. Social movements often unfold at the intersection of a series of overlapping and hierarchical state spaces (municipality, regions, nation state, international agencies), each providing a complex yet malleable mix of opportunities and constraints (Miller 1994, 2004; Sikkink 2005). In her discussion of transnational social movements, Sikkink (2005) maintains that political opportunities and constraints can vary sharply between national and international scales depending on countries, international institutions, and the nature of political issues. In certain instances, movements may have strong political alliances at the national scale but regulatory levers concerning an issue reside in international institutions with few openings. In other instances the situation may be reversed, with national institutions charged with a particular issue remaining closed and international institutions displaying greater degrees of openness. Miller (2000) makes a similar observation regarding municipal, regional, and nation-state institutions.

Others have shown how elites employ a range of techniques to transmit the rationalities of neoliberalism across the scalar landscape of the socio-political body (Rose 1999; Larner 2000; Raco and Imrie 2000; Sparke 2006; Huxley 2008; McCann and Ward 2011). Capital, states, and intellectuals have succeeded in producing discourses that not only advocate neoliberalism as an economic strategy, but also as a moral system whose categories, codes, and values are superior to all others (Larner 2000; Ong 2006; Peck 2010). These discourses have been disseminated through variously scaled circuits (e.g., media, schools, public institutions, etc.). Quasi-state international institutions like the World Trade Organization and the International Monetary Fund play strategic roles in supporting neoliberal understandings through policy recommendations that shape the disciplining and development of nation-states, regions, localities, and bodies

(Ong 2006; Sparke 2006). Other governing institutions, like nation-states, operate in similar ways and diffuse their own neoliberal discourses through various institutional circuits. A new set of constraints, as well as some opportunities, exist for activists operating within this mesh of overlapping neoliberal discourses and disciplinary powers, with neoliberal discourses often becoming incorporated into 'normal' ways of thinking and addressing social problems. For example, Ong (2006) suggests that many non-governmental organizations in Asia have embraced conceptions of 'human rights' based on a particularly neoliberal understanding of humanity, values, and rights. In these instances, rather than NGOs contesting the existing order and proposing radical alternatives, they become important vehicles for reinforcing neoliberal values and producing new neoliberal subjectivities in Asia and elsewhere.

The uneven scaling of power, resources, and opportunities has important implications for the geographical strategies of activists and elites in the pursuit of their respective goals. It must be stressed that scale is not a spatiality simply to be 'found' in the political landscape, but rather is actively and relationally produced through struggle. Proponents of the state, capital, social movements, and other collectivities continually reshape scalar relations in an ongoing process of asserting and contesting power (Peck and Tickell 1994; Brenner 2004; Miller 2007). As Smith (1992: 74) puts it, 'the scale of struggle and the struggle over scale are two sides of the same coin'. States often rescale policy and decision-making processes to those scales where popular pressure can be muted or diffused, and where state resources are insufficient to implement democratic decisions (Peck and Tickell 1994; Miller 2007). When central state officials transfer responsibility for labour regulations to regional or international institutions, they reduce the ability of national labour unions to influence labour policy (Herod 2001; Keck and Sikkink 1998). Devolving welfare policies to local scales of government diffuses targets by requiring social movements to make claims in countless local bodies rather than a single national one. While politically astute elites have shown great ability to rescale administrative and policy making process to evade pressures from below, social movements have also effectively pursued multi-scalar strategies. Sikkink (2005), for instance, shows that movements exploit opportunities on one scale to open up opportunities on others. In the campaign to have Pinochet tried in Chile, activists used opportunities and alliances developed on the international scale to pressure national institutions to bring the former dictator to justice – an example of one of several possible multi-scalar strategies employed by social movements to achieve their objectives. To counter the efforts of elites to spatially contain or fragment activists, the latter can respond by rescaling conflict and engaging in complex multi-scalar strategies.

Rescaling and multi-scalar strategies strongly affect the relational dynamics of social movements. Employment of such strategies is by necessity relational, requiring the development and reconfiguration of social networks and power relations across geographical and institutional boundaries (Routledge 2003; Sikkink 2005; Cumbers et al. 2008; Nicholls 2009). Tarrow and McAdam (2005) have identified two general mechanisms by which social movements shift scales. 'Relational diffusion' refers to the extension of a movement through pre-existing relational ties, i.e., social networks. The existence of trust and shared identities contained in existing relational ties not only facilitates the spread of social movements, it also provides a durable relational base for sustainable mobilizations. The limitation of this mechanism is its dependence upon *existing* relational ties: the social and geographical reach of this scale shift tends to be limited and localized, which reduces the impact of these types of mobilizations on wider political spheres of influence. The second of these mechanisms is 'brokerage': the spread of mobilization resulting from linking two or more actors who were previously unconnected (Tarrow and McAdam 2005: 127). Brokerage results in new relations across traditional geographical and social boundaries, enhancing the potential reach and effect of collective actors. Though brokerage can expand the scope of social movements, alliances resulting from brokerage are, according to Tarrow and McAdam, more fragile because they are comprised of many different groups and possess weak mechanisms of social integration.

Scale-shifting strategies not only complicate the networking structures and dynamics of social movements, they also affect their framing and messaging strategies. Activists modulate their framing strategies according to the circumstances present on different geographic scales. For example, the recent immigrant rights movement in the United States employs different messages depending on the geographic scale of their audience (Nicholls forthcoming). When activists are pitching their message in localities where they have a strong organizational base, they stress frames and messages that resonate with immigrants. The aim in these instances is not necessarily to build broad support for their cause. Instead, the aim of discourse production is to strengthen internal identity, cohesion, and commitment of core activists. Consequently, lead discursive producers will stress themes, values, and symbols with great meaning for these core activists. However, when mobilization shifts to the national scale, the primary aim is to win over the support of large segments of the public. On this scale, activists must employ frames and messages that resonate with the values of a rather hostile public (flags, immigrants as extensions of the American Dream, etc.) rather than speak to the emotions and imaginaries of their core constituencies.

Scale-shifting strategies strongly affect who participates in social movements, participants' relations to one another, their capacities to achieve goals, and the ways in which they frame their struggles. Rather than this particular spatiality operating as an exogenous variable affecting social movements, scalar strategies play a direct and constituting role in their growth, development, and decline.

Networks and flows

Some sociologists and geographers have argued that heavy emphasis on the institutional forces that structure social movements has come at the expense of attention to the networks that permit the flow of information, ideas, and emotions among activists in different places (Diani 2005; Cumbers et al. 2008; Featherstone 2003; 2008). The geographical literature on networks is distinct from the sociological literature and begins from a very different problematic. Some geographers have questioned the assumptions underlying the traditional conceptions of place, territory, and scale (Amin and Thrift 2002; Amin 2004; Massey 1994, 2004, 2005, 2007; Marston et al. 2005). It is argued that people who reside within a common location have very different sociological attributes, histories, geographical ties and mobilities. Cohabitation and proximity do not by necessity produce strong ties, similar political dispositions, or feelings of solidarity. Moreover, structured places and scales are constantly undone and re-made by global flows of people, capital, and ideas. Traditional conceptions of place, territory, and scale may overemphasize their stability and underemphasize their mobility and flux (Amin 2004: 33). Lastly, traditional conceptions of place, territory, and scale may feed into imaginaries of nostalgia rather than those of progressive change (Massey 1994, 2004; also see Duyvendak 2011). Indeed, Massey argues that territorial conceptions of place fuel '... localist or nationalist claims to place based on eternal essential, and in consequence exclusive, characteristics of belonging' (2004: 6).

Featherstone (2003, 2005, 2008) has applied these general critiques to the study of social movements. He suggests that geographers are bound by binaries that 'counterpose local and global, of space and place' (2003: 405). These binaries are problematic because they privilege local relations over distant ones, with the former assumed to be more *authentic* and therefore more politically legitimate than distant relations and forces. Such views may reinforce reactionary, localist, and nationalist claims to power. Moreover, the use of binaries project fixed interests and identities on the actors we study, with certain actors being essentially 'local', 'global', or bound to particularistic territories. This results in representations of actors as essentially different from one another and engaged in zero-sum negotiations with others. Featherstone's critique echoes Eric Wolff's warning not to 'create a model of the world as a global pool hall in which the entities spin off each other like so many hard and round billiard balls' (Wolf 1982: 6). Identities and interests are not established prior to power struggles with multiple others (near and far) but through these struggles, with relational exchanges through struggles becoming the driving force for shaping the dispositions and outlooks of different subjects. A 'global sense of place' (Massey 1994) for Featherstone means that locations are traversed by a wide range of power networks, with actors in different sites engaging with one another through multiple relations. Employing a more fluid and networked understanding of place allows us to open up the study of activism and better understand the complex mechanisms that shape the identities and subjectivities of actors.

Attempting to reconcile these different positions in human geography, Nicholls (2009, 2011) has argued that urban places are unique sites for networking in social movements because they favour the formation of strong-tie relations *and* they permit fleeting and contingent weak ties among mobile actors. Fleeting interactions allow actors in strong-tie relations to regularly interact and share ideas with the diverse others that make up a local activist milieu. The qualities of place therefore favour a network structure that is both internally well structured and open to contacts with multiple others in the vicinity. When activists in these messy places connect to activists elsewhere, they form a distinctive 'social movement space' constituted by and through uneven networks. Because certain places within this networked space are more powerful than others in terms of their material and symbolic power, they become a structuring and driving force (i.e. hub) within the broader social movement network. While these places may provide the movement with a degree of cohesion and focus that enhances its powers to achieve collective goals, they also introduce powerful cleavages between centralizing hubs and multiple peripheries. Thus, while Nicholls largely agrees with the general theoretical spirit of the 'relational approach' to geography, he also argues that networks come together and form distinctive sociospatial structures with their own logics of inequality and conflict.

Entangling spatialities

The growing prominence of network and relational approaches to geography has resulted in efforts to identify the ways in which networks are co-implicated with other spatialities like place, territory, and scale (Jessop et al. 2008; Leitner et al. 2008). Most geographers implicitly recognize that multiple spatialities produce overlapping yet distinctive effects on politics and social movements. However, these same geographers often employ one spatiality over others because it provides a useful entry point to analyse complex geopolitical processes (Jessop et al. 2008).

Leitner and her colleagues (2008) argue that human beings are positioned simultaneously in multiple spatialities. By identifying these spatialities and their distinctive yet overlapping effects on social movements, we are in a stronger position to examine how space in general plays a constituting role in social movements. Using the case of the Immigrant Workers' Freedom Ride in the United States, they show how mobility and trans-local networks were instrumental in circulating alternative imaginaries and catalyzing a series of locally situated but nationally connected immigrant rights coalitions. The theoretical task at hand is therefore not to demonstrate which spatiality is more important than another but rather to identify the various roles of different spatialities in social movements and how they intersect with one another to affect social movement dynamics.

While in broad agreement with this line of analysis, we wish to also stress that different spatialities may matter in different ways, that different spatialities will be of greater or lesser importance at different times and in different places, and that how different spatialities become co-implicated and entangled affects the mobilization capacities, resources, frames, and internal conflicts found in movements. We therefore agree that all spatialities are important, but they are not always equally important at all times and in all kinds of conflict.

Structure of the book

Three major conceptual affinities form the organizational framework of the book.

Place and Space: Sites of Mobilization

The first section examines how the characteristics of place and space influence the abilities of insurgents to mount collective challenges to their political opponents. This section illustrates how political actors nourish solidarities, deepen bonds, and forge collective identities with recruits in their various *places* of socialization, all through socially regulated practices of mobility, assembly, and interaction. Place-based ties provide the relational and cultural building blocks that make robust systemic challenges possible. As insurgents build their powers unevenly across different places, states frequently respond by employing a range of spatially sensitive techniques to survey and repress these efforts.

The first chapter on *Place and Space*, Chapter 1 by Donatella della Porta, Maria Fabbri and Gianni Piazza, presents results from their research on three infrastructure protest campaigns – No Tav, No Bridge, No Dal Molin – in Italy. After an introduction to the three campaigns, the authors show how these protests around the construction of large infrastructures not only demonstrate contestation of the use of specific spaces, with the elaboration of an alternative conception of those spaces, but also how new spaces were created as terrains of resistance. The symbolic contestation of the conception of space interacted with the physical occupation of particular places that not only acquired high symbolic meaning but also had strong effects on the protests, allowing for the development of intense relations up to the formation of shared (territorially based) identities. If the sense of place influenced the protests, the protests produced new definitions of places as collective identities developed in 'liberated' as well as 'contested' spaces.

In Chapter 2 Don Mitchell addresses the paradox of free speech in liberal democracies and how regulatory practices in the latter have, in effect, rendered politically dissident speech ineffectual. These practices, Mitchell argues, are primarily spatial in nature: they govern the regulation of space, not the regulation of political speech or action *per se*. The nature of this regulatory regime – and its effects – are profoundly important for contemporary social movements, especially those who seek to radically transform the established political and economic order. It is increasingly apparent that spatial regulation is the means by which dissident speech and dissident groups are silenced and suppressed. Through an examination of significant Supreme Court decisions and three case studies, Mitchell shows that the regulation-cum-suppression of dissident speech relies less and less on *what* is said than *where* it is said. Silencing speech – silencing protest – is a function of geography. And free speech law in the United States, the case studies show, is increasingly geographically astute.

In Chapter 3 Jan Willem Duyvendak and Loes Verplanke examine social movements explicitly aiming at a sense of belonging or a new sense of 'home', in particular the movement fighting against 'total institutions' for people with psychiatric and intellectual disabilities and the parallel story of the gay and lesbian movement. Distinguishing conceptually between 'heaven', 'haven' and 'hell', the authors argue that these two movements demonstrate quite different struggles for new practices of home making in place. The new home making practices at the community level were far less successful for people with disabilities than for gays. In practice, the former moved from their big institutions ('hell') to small, independent housing ('havens' at best) – which was quite an improvement in itself – but with no integration in the community, no public home. The home making practices of gays, on the other hand, were intimately linked to social movements and collective action, resulting in public places for themselves, characterized by all elements of what the authors claim as a 'heaven', providing space for identity and visibility.

In Chapter 4 Deborah Martin revisits her earlier work on 'place framing' to consider how it functions in different contexts and in translation with public policy, especially legal discourses. Place framing draws from social movement theory to articulate how neighbourhood organizations portray activism as grounded in a particular place and scale. Martin considers in this chapter how the concept applies in Athens, Georgia in the US. Furthermore, she explores how the neighbourhood-identifying and mobilizing function of place frames encounter the legal and political discourses of public policy. The translation of place frames into political discourse reflects strategic decision-making, formal negotiation, and shifting scales of neighbourhood activism. The chapter considers those shifts and the actors influencing them, and their importance for the forms and outcomes of urban activism.

Scale, Territory and Region: Structuring Collective Interests, Identities and Resources

The second section – dealing with the related spatialities of scale, territory, and region – examines the dynamic interaction among these areal/hierarchical spatialities and social movement mobilization. Social movements develop within overlapping institutional and socio-cultural territories (e.g. neighbourhoods, cities, regions, nation-states, transnational). Territorially constituted institutions and actors typically relate to other such institutions and actors in complex and dynamic ways. These dynamic relations present particular movements with ever-changing sets of opportunities and constraints, with social movements often 'jumping' to those scales where they find the greatest opportunities and resources to advance their cause. The section stresses the dialectical relations between political territories and social movements: as geopolitical institutions establish 'rules of the game', the strategies and practices developed by movements in response to these rules may influence the ways in which states restructure and reform their territorial institutions.

The first chapter on *Scale, Territory and Region*, Chapter 5 by Martin Jones, offers both an empirical exploration and theoretical reflection on English regionalism and 'polymorphic spatial politics'. This chapter is centrally concerned with uncovering varieties of English regionalism, which have emerged through the spaces of devolution and constitutional change between 1997 and 2004. Jones discusses official and grassroots regional social movements, both from the position of those advocating territorial modes of empowerment and those resisting such spatial strategies. The chapter blends work being undertaken on the geography of regions with writings on social movements to develop the concept of 'regional social relations', i.e. connections between state forms, state actions and the social processes/practices operating through and in the shadow of the contemporary capitalist state. Key here is the dynamic between power relations, social groups and the spaces of opportunity created by state activity. Theoretically, Jones picks up from and expands upon the work he contributed to the development of the TPSN framework, reminding us that regional movements cannot be understood simply by focusing on the construction of regions, but must be understood through a polymorphic spatial framework that accounts for multiple spatialities and the complex political geometry they form.

In Chapter 6 John Agnew and Ulrich Oslender embrace recent debates in political geography that point to a key inadequacy in international relation theories, and in particular the Westphalian model of state sovereignty assumed by most, in positing the existence of new 'regimes of sovereignty' that are closely associated with ongoing processes of globalization. The idealized sovereignty of the nation-state is still, according to the authors, rigidly linked in dominant theories to the notion of a transparent territoriality or the control over a national territory clearly marked in space by established borders. In this chapter the authors propose the notion of 'overlapping territorialities' to examine sources of territorial authority other than that of the nation-state with reference to Latin America. In this context, a number of social movements (indigenous and Afrodescendent, for example) have achieved legal recognition of collective land ownership, with local communities establishing themselves as differential territorial authorities within the nation-state. Agnew and Oslender argue that empirical lessons from Latin America importantly contribute to a necessary re-thinking of the links between state sovereignty and territoriality as mediated, in this case, by the role of active social movements challenging the established spatial fabric of state-based politics.

Johan Moyersoen and Erik Swyngedouw argue in Chapter 7 that cities, social movements, activists and community organizations have often found ways to avoid conventional ways of political protest and to challenge norms and laws of the urban elite in a radical (democratic) way – for example squatter movements, critical arts movements, direct actions (such as reclaim the streets etc.). Such activism is often instigated by processes of uneven development (for example gentrification, biased tax policies, social exclusion) that create a polarized political landscape in the social as well as physical space in the city. The current dynamics of polarization and unevenness in the built environment and in the social spaces of people's everyday life have generated innovative strategies and alternatives at the local level. In addition, these social actions have instigated groups to 'jump scales' and to challenge the power constellations that (re-)produce these inequalities at the overall urban level. Although cracks in the city are expressions of uneven development and social disparity, they open up new possibilities for social, political, economic or cultural experimentation outside of the conventional modes of political participation. They often generate spaces for political empowerment, radical democracy and social innovation. However, the antagonistic nature of cracks do not set the dissenter completely free in his or her actions. This chapter unravels the social and institutional dynamics of how a small group of pioneering urban activists engaged with some of the key actors in a deprived neighbourhood of Brussels in a process of urban renewal.

Margit Mayer, in Chapter 8, looks at local activities of movements critical of neoliberal globalization and their interactions and coalitions with other, locally based, movements. Drawing on observations of recent urban mobilizations and community-based activism particularly in the US and Germany, she focuses on the ways in which these movements address the shifting scalar organization of statehood. The chapter explores the added value the 'politics of scale' debate might bring to the problems emancipatory movements are facing in dealing with the reconfiguring scalar architecture of governance. Two types of protest movements serve as empirical objects of study: on the one hand the mobilizations against the neoliberalization of social and labour market policies, against the dismantling of the welfare state, and for social and environmental justice, which have come to the forefront of urban activism over the last decade; on the other hand, a transnationally active so-called anti-globalization movement which increasingly sees localities as the scale where global neoliberalism 'touches down', where global issues become localized. Networks that are part of this transnational movement have been importing repertoires and goals from the global protest, often in collaboration with the social justice alliances characteristic of the first-mentioned type of (local) protest movements. The chapter evaluates both dilemmas and opportunities these 'glocal' movements experience, thereby contributing to the explanatory power of the 'politics of scale'.

Networks: Connecting Actors and Resources Across Space

While the concepts of scale, territory and region have gained great prominence in geography, a number of prominent scholars have argued that these concepts underemphasize the diverse connections made among political actors across space. They often advocate replacing a scalar or territorial approach to politics with one based in networks, whereby the pluralism of local actors and their intimate connections with distant others are emphasized. Without abandoning the insights of scalar/territorial approaches, the last section of the book highlights recent efforts to analyse the network dimensions of social movements through the use of a variety of forms of network theory.

The first chapter in the *Networks* section, Chapter 9 by Dingxin Zhao, addresses the relations between the built environment and organization in anti-US protest mobilization in Beijing. Zhao treats built-environment-based and organization-based mobilization as two factors of participant mobilization and tries to understand the relationships between them. By examining different styles of student mobilization during the 1999 anti-US protests in three Beijing universities, each with a similar built environment and spatial routines of the students, the chapter shows that the built environment played a crucial role in mobilization in cases where there was less organizational involvement. In short, less-organized protests may be compelled to mobilize relying upon built-environment-based mobilization tactics.

In Chapter 10 Ted Rutland examines how the movement responsible for the Portland, Oregon energy policy was formed and structured. Recent attempts to explore the relationship between networks and social movements have encountered significant problems, including: (1) difficulties in distinguishing between pre-existing social networks and the movements generated as a result; (2) inattention to what it is that forms and binds social ties over time and space; and (3) unexamined assumptions about the (inherent and timeless) character of human agency in forging and joining social networks and movements. Rutland's chapter suggests new directions for social movement studies by drawing on the 'material-semiotic' approach of science studies theorists such as Donna Haraway, Andrew Pickering and Bruno Latour. After a brief review of the approach, which stresses the co-constitution of societies and natures through networks of associations, the chapter demonstrates its analytic potential through an examination of the emergence of a remarkably effective environmental movement in Portland, Oregon.

In Chapter 11 Andrew Davies and David Featherstone argue that the prominence and visibility of transnational forms of organizing is a defining feature of contemporary contentious politics. The authors claim that there are significant histories of transnational forms of contention and organizing, and they are by no means new, despite being frequently depicted as such. The growing prominence of and interest in such transnational forms of organizing has unsettled some of the key ways of understanding the geographies of contentious politics. This has opened up a challenge to the ways that both social movement theory and political geography have been structured by an implicit assumption that the national arena is the most obvious container for political activity.

Paul Routledge, Andrew Cumbers, and Corine Nativel, in Chapter 12, pull together several of the themes of spatially attuned social movements research, arguing that movements have increased their spatial reach over the past 20 years by constructing multi-scalar networks of support and solidarity for their particular struggles, and also by participating with other movements in broader campaign networks (e.g. to resist neoliberal globalization). Rather than a monolithic and coherent 'global justice movement', the authors' findings support a conception of a series of overlapping, interacting, competing, and differentially placed and resourced networks that they term Global Justice Networks (GJNs). Routledge,

Cumbers, and Nativel argue that through such networks, different place-based movements are becoming linked up to much more spatially extensive coalitions of interest. In order to analyse how such operational logics become entangled within the workings of GJNs, the chapter analyses two particular networks, People's Global Action, an international network of grassroots peasant movements, and the International Federation for Chemical Energy Mine and General Workers' Unions (ICEM), a global union federation (GUF) which brings together around 400 affiliate trade unions.

Conclusion

In the conclusion, Byron Miller takes stock of the diverse contributions to this volume, developing a framework for analysing the ways in which multiple spatialities are co-implicated in the mobilization and suppression of social movements. He argues that social movement research in geography, like much of human geography generally, is fragmented into different approaches to sociospatial analysis based in different spatial ontologies and theories of sociospatial struggle. While all commonly employed conceptions of spatiality are relational, there remain significant debates and differences among scholars over how best, spatially speaking, to approach the study of social movements. Given that many of these differences stem from different ontological foundations – most commonly critical realist versus poststructuralist – this is not a trivial issue. Yet, the diversity of the spatially oriented social movement research has yielded a wealth of complementary insights. Is it possible to integrate or reconcile different sociospatial approaches in social movement research? Without claiming to definitively resolve this dilemma, one way forward may be to treat the production of spatialities as themselves the product of social and political struggle rather than ontologically given, with spatialities to be regarded as spatial technologies of power that are strategically and contextually employed as a central component of the 'game' of contentious politics. Attempts to transform power relations are simultaneously attempts to transform spatial relations: social and political struggle is simultaneously a struggle to transform, shift, and/or fix spatialities. Understanding the production of spatialities as both a product and a technology of struggle allows us to understand spatialities, and their co-implications, as contextual and dynamic.

References

Agnew, J. (1987) *Place and Politics: The Geographical Mediation of State and Society.* Boston: Allen and Unwin.
—— (1994) 'The Territorial Trap: The Geographical Assumptions of International Relations Theory', *Review of International Political Economy*, 1: 53-80.

—— (2002) *Place and Politics in Modern Italy.* Chicago: University of Chicago Press.

—— (2009a) *Globalization and Sovereignty.* Lanham, MD: Rowman and Littlefield.

—— (2009b) 'Territory', in D. Gregory, R. Johnston, G. Pratt, M. Watts, and S. Whatmore (eds) *The Dictionary of Human Geography 5th Edition*, Malden, MA and Oxford, 746-747.

Amin, A. (2004) 'Regions Unbound: Towards a New Politics of Place', *Geographiska Annaler*, 36B (1): 33-44.

Amin, A. and Thrift, N. (2002) *Cities: Reimagining the Urban.* Cambridge: Polity Press.

Archer, M., R. Bhaskar, A. Collier, T. Lawson, and A. Norrie (eds) (1998) *Critical Realism: Essential Readings.* Oxon and New York: Routledge.

Beaumont, J. R. (2003) 'Governance and Popular Involvement in Local Antipoverty Strategies in the UK and The Netherlands', *Journal of Comparative Policy Analysis: research and practice*, 5 (2-3): 189-207.

Beaumont, J. and W. Nicholls (2008) 'Introduction to the Symposium: Plural Governance, Participation and Democracy in Cities', *International Journal of Urban and Regional Research*, 32 (1): 87-94.

Bosco, F. (2006) 'The Madres de Plaza de Mayo and Three Decades of Human Rights' Activism: Embeddedness, Emotions, and Social Movements', *Annals of the Association of American Geographers*, 96 (2): 342-357.

Boudreau, J.A. and R. Keil (2001) 'Seceding from Responsibility? Secession Movements in Los Angeles', *Urban Studies*, 38 (10): 1701-1731.

Brenner, N. (2004) *New State Spaces: Urban Governance and the Rescaling of Statehood.* Oxford: Oxford University Press.

Castells, M. (1983) *The City and the Grass-roots: A Cross-cultural Theory of Urban Social Movements.* London: Edward Arnold.

—— (1997) *Power of Identity.* Oxford: Blackwell Publishers.

Castree, N. (2002) 'False Antitheses? Marxism, Nature and Actor-networks', *Antipode*, 34 (1): 111-146.

Coleman, J. (1988) 'Social Capital in the Creation of Human Capital', *The American Journal of Sociology*, 94: 95-120.

Collins, R. (2004) *Interaction Ritual Chains.* Princeton University Press.

Cumbers, A., Routledge, P., and C. Nativel (2008) 'The Entangled Geographies of Global Justice Networks', *Progress in Human Geography*, 32 (2): 183-201.

Davies, A. D. (2012) 'Assemblage and Social Movements: Tibet Support Groups and the Spatialities of Political Organization', *Transactions of the Institute of British Geographers*, 37 (2): 273-286.

Diani, M. (1997) 'Social Movements and Social Capital: A Network Perspective on Movement Outcomes', *Mobilizations*, 2 (2): 129-147.

—— (2005) 'Cities in the World: Local Civil Society and Global Issues in Britain', in Donatella Della Porta, D. and S. Tarrow (eds) *Transnational Protest and Global Activism.* Lanham, MD: Rowman and Littlefield Publishers.

Dikec, M. (2007) *Badlands of the Republic. Space, Politics and Urban Policy.* Oxford: Blackwell Press.

Duyvendak, J. W. (2011) *The Politics of Home: Belonging and Nostalgia in Western Europe and the United States.* London: Palgrave.

Elden, S. (2010) 'Land, Terrain, Territory', *Progress in Human Geography*, 34: 799-817.

Featherstone, D. (2003) 'Spatialities of Transnational Resistance to Globalization: The Maps of Grievance of the Inter-Continental Caravant', *Transaction of the Institute of British Geographers*, 28 (4): 404-421.

—— (2005) 'Towards the Relational Construction of Militant Particularisms: Or Why the Geographies of Past Struggles Matter for Resistance to Neoliberal Globalisation', *Antipode*, 37 (2): 250-271.

—— (2008) *Resistance, Space and Political Identities: The Making of Counter-Global Networks.* Oxford: Wiley-Blackwell.

Fung, A. (2006) *Empowered Participation: Reinventing Urban Democracy*, Princeton, NJ: Princeton University Press.

Giddens, A. (1984) *The Constitution of Society. Outline of the Theory of Structuration.* Cambridge: Polity.

Gould, R. (1993) 'Trade Cohesion, Class Unity, and Urban Insurrection: Artisanal Activism in the Paris Commune,' *The American Journal of Sociology*, 98 (4): 721-754.

—— (1995) *Insurgent Identities: Class, Community, and Protest in Paris from 1848 to the Commune.* Chicago: University of Chicago Press.

Granovetter, M. (1983) 'The Strength of Weak Ties: A Network Theory Revisited', *Sociological Theory*, 1: 201-233.

Hagerstrand, T. (1970) 'What about People in Regional Science?', *Papers in Regional Science*, 24 (1): 6-21.

Harvey, D. (2001) 'Militant Particularism and Global Ambition: The Conceptual Politics of Place, Space, and Environment in the Work of Raymond Williams', in D. Harvey, *Spaces of Capital: Towards a Critical Geography.* Edinburgh: Edinburgh University Press, 158-187.

—— (2003) *Paris, Capital of Modernity.* New York: Routledge.

—— (2006) *Spaces of Global Capitalism*, New York: Verso.

Herod, A. (2001) *Labour Geographies*, New York: Guilford Press.

Huxley, M. (2008) 'Space and Government: Governmentality and Geography', *Geography Compass*, 2: 1635-1658.

Jessop, B., N. Brenner, M. Jones (2008) 'Theorizing Sociospatial Relations', *Environment and Planning D: Society and Space*, 26 (3): 389-401.

Jonas, A. (2006) 'Pro Scale: Further Reflections on the Scale Debate in Geography', *Transactions of the Institute of British Geographers*, 31, 399-406.

—— (2012) 'Region and Place: Regionalism in Question', *Progress in Human Geography*, 36 (2): 263-272.

Jones, M. and B. Jessop (2010) 'Thinking State / Space Incompossibly', *Antipode*, 42 (5): 1119-1149.

Katznelson, I. (1981) *City Trenches: Urban Politics and the Patterning of Class in the United States*. Chicago: University of Chicago Press.

Keck, M. and K. Sikkink (1998) *Activists beyond Borders*. Ithaca: Cornell University Press.

Larner, W. (2000) 'Neo-liberalism: Policy, Ideology, Governmentality', *Studies in Political Economy*, 62: 5-25.

Lefebvre, H. (1991/1974) *The Production of Space*. Translated by Donald Nicholson-Smith. Oxford: Basil Blackwell (originally published 1974).

Legg, S. (2009) 'Of Scales, Networks and Assemblages: The League of Nations Apparatus and the Scalar Sovereignty of the Government of India', *Transactions of the Institute of British Geographers*, 34: 243-254.

Leitner, H. and B. Miller (2007) 'Scale and the Limitations of Ontological Debate: a Commentary on Marston, Jones, and Woodward,' *Transactions of the Institute of British Geographers*, 32: 116-125.

Leitner, H. and E. Sheppard (2009) 'The Spatiality of Contentious Politics: More than a Politics of Scale', in R. Keil and R. Mahon (eds) *Leviathan Undone? Towards a Political Economy of Scale*. Vancouver: University of British Columbia Press, 229-300.

Leitner, H., Sheppard, E., and K. Sziarto (2008) 'The Spatialities of Contentious Politics', *Transactions of the Institute of British Geographers*, 33: 157-172.

McCann, E. and K. Ward (2010) 'Relationality/territoriality: Towards a Conceptualization of Cities in the World', *Geoforum*, 41 (2): 175-184.

McCann, E. and K. Ward (eds) (2011) *Mobile Urbanism: Cities and Policymaking in the Global Age*. Minneapolis: University of Minnesota Press.

McFarlane, C. (2009) 'Translocal Assemblages: Space, Power and Social Movements', *Geoforum*, 40: 561-567.

MacLeod, G. and M. Jones (2007) 'Territorial, Scalar, Networked, Connected: In What Sense a "Regional World"?', *Regional Studies*, 41 (9): 1177-1191.

Mann, M. (1993) *Sources of Social Power, Volume II: The Rise of Classes and Nation-states, 1760-1914*. Cambridge: Cambridge University Press.

Marston, S., Jones, J.P., and K. Woodward (2005) 'Human Geography without Scale', *Transactions of the Institute of British Geographers*, 30 (4): 416-430.

Martin, D. (2003) '"Place-framing" as Place-making: Constituting a Neighborhood for Organizing and Activism', *Annals of the Association of American Geographers*, 93 (3): 730-750.

Massey, D. (1994) *Space, Place and Gender*. Minneapolis: University of Minnesota Press.

—— (2004) 'Geographies of Responsibility,' *Geografiska Annaler*, 86 B 1: 5-18.

—— (2005) *For Space*. London: Sage.

—— (2007) *World City*. Cambridge: Polity.

Miller, B. A. (1994) 'Political Empowerment, Local-Central State Relations and Geographically Shifting Political Opportunity Structures,' *Political Geography*, 13 (5): 393-406.

——— (2000) *Geography and Social Movements: Comparing Antinuclear Activism in the Boston Area.* Minneapolis/ London: University of Minnesota Press.

——— (2004) 'Spaces of Mobilization: Transnational Social Movements,' in C. Barnett and M. Low (eds) *Spaces of Democracy: Geographical Perspectives on Citizenship, Participation and Representation.* London: Sage Publications, 223-246

——— (2007) 'Modes of Governance, Modes of Resistance,' in H. Leitner, J. Peck and E. Sheppard (eds) *Contesting Neoliberalism.* New York: Guilford Press, 223-249.

——— (2009) 'Is Scale a Chaotic Concept? Notes on Processes of Scale Production,' in R. Keil and R. Mahon (eds) *Leviathan Undone? Towards a Political Economy of Scale.* Vancouver: UBC Press, 51-66.

Nicholls, W. (2006) 'Associationalism from Above: Explaining Failure in the Case of France's *La Politique de la Ville*', *Urban Studies*, 43 (10): 1779-1802.

——— (2008) 'The Urban Question Revisited: The Importance of Cities for Social Movements', *International Journal of Urban and Regional Research*, 32(4): 1468-2427.

——— (2009) 'Place, Relations, Networks: The Geographical Foundations of Social Movements', *Transactions of the Institute of British Geographers*, 34 (1): 78-93.

——— (2011) 'Cities and the Unevenness of Social Movement Space: The Case of France's Immigrant Rights Movement', *Environment and Planning A*, 43 (7): 1655-1673.

——— (forthcoming) 'Undocumented and Unafraid: Undocumented Youths and the Dreamers' Movement', Immanuel Ness (ed.) *Encyclopedia of Global Human Migration.* New York: Wiley-Blackwell.

Ong. A. (2006) *Neoliberalism as Exception: Mutations in Citizenship and Sovereignty.* Durham, N.C.: Duke University Press.

Paasi, A. (2009) 'The Resurgence of the "Region" and "Regional Identity"', *Review of International Studies*, 35 (1): 121-146.

Peck, J. (2010) *Constructions of Neoliberal Reason.* Oxford: Oxford University Press.

Peck, J. and A. Tickell (1994) 'Searching for a New Institutional Fix: the After-Fordist Crisis and the Global-Local Disorder', in A. Amin (ed) *Post-Fordism.* Oxford and Cambridge, MA: Blackwell, 280-315.

Purcell, M. (2006) 'Urban Democracy and the Local Trap', *Urban Studies*, 43 (11): 1921-1941.

Raco. M. and R. Imrie (2000) 'Governmentality and Rights and Responsibilities in Urban Policy', *Environment and Planning A*, 32: 2187: 204.

Rose, N. (1999) *Powers of Freedom: Reframing Political Thought.* Cambridge: Cambridge University Press.

Rosenberg, T. (2011) 'Revolution U', *Foreign Policy*, February 16: 1-13.

Routledge, P. (1997) 'Putting Politics in its Place: Baliapal, India, as a Terrain of Resistance', in J. Agnew (ed) *Political Geography: A Reader.* London: Arnold.

—— (2003) 'Convergence Space: Process Geographies of Grassroots Globalisation Networks', *Transactions of the Institute of British Geographers*, 28 (3): 333-349.

Sayer, A. (2010) *Method in Social Science*, 2nd ed. New York: Routledge.

Sheppard, E. S. (2002) 'The Spaces and Times of Globalization: Place, Scale, Networks, and Positionality,' *Economic Geography*, 76: 307-330.

Sikkink, K. (2005) 'Patterns of Dynamic Multilevel Governance and the Insider-Outsider Coalition', in D. Della Porta and S. Tarrow (eds) *Transnational Protest and Global Activism*. Lanham, MD: Rowman and Littlefield Publishers.

Smith, N. (1992) 'Geography, Difference, and the Politics of Scale', in J. Doherty, E. Graham and M. Malek (eds) *Postmodernism and the Social Sciences*. London: Macmillan, 57-79.

Soja E. (1980) 'The Socio-Spatial Dialectic', *Annals of the Association of American Geographers*, 70 (2): 207-225.

—— (1989) *Postmodern Geographies*. New York: Verso.

—— (1996) *Thirdspace*. Malden, MA and Oxford: Wiley-Blackwell.

Sparke, M. (2006) 'Political Geographies of Globalization: (2) Governance', *Progress in Human Geography*, 30 (2): 1-16.

Tarrow, S. and D.McAdam (2005) 'Scale Shift in Transnational Contention', in Donatella Della Porta, D. and S. Tarrow (eds) *Transnational Protest and Global Activism*. Lanham, MD: Rowman and Littlefield Publishers.

Tilly, C. (1964) *The Vendée*. Cambridge: Harvard University Press.

—— (1986) *The Contentious French: Four Centuries of Popular Struggle*. Cambridge: Harvard University Press.

—— (2005) *Trust and Rule*. Cambridge: Cambridge University Press.

Uitermark, J. (2010) *Dynamics of Power in Dutch Integration Politics*, Ph.D., Amsterdam: University of Amsterdam

Wolf, E. (1982) *Europe and the People without History*. Berkeley: University of California Press.

PART I
Place and Space:
Sites of Mobilization

Chapter 1

Putting Protest in Place: Contested and Liberated Spaces in Three Campaigns

Donatella della Porta, Maria Fabbri and Gianni Piazza

Putting protest in place: an introduction

Protest is defined in the sociology of social movements as a 'resource of the powerless … they depend for success not upon direct utilization of power, but upon activating other groups to enter the political arena' (Lipsky 1965: 1). A protest thereby induces indirect channels of communication in the mass-media and alliances with more influential actors. If the protest is a resource which some groups utilize during the decision-making process, it should not be viewed purely in instrumental terms. In fact, protest actions are 'sites of contestation in which bodies, symbols, identities, practices and discourse are used to pursue or prevent changes in institutionalized power relations' (Taylor and van Dyke 2004: 268). During the course of a protest both time and money is invested in risky activities, yet often resources of solidarity are also created or re-created. The protest in fact creates a sense of collective identity which is a condition for a collective action (Pizzorno, 1993). Many forms of protest 'have profound effects on the group spirit of their participants', since 'in the end there is nothing as productive of solidarity as the experience of merging group purposes with the activities of everyday life' (Rochon 1998: 115). For workers, strikes and occupations of factories have represented not only instruments for collective pressure but also arenas in which a sense of community is formed (Fantasia 1988) and the same has occurred with the occupation of schools and universities by students (Ortoleva 1988). Furthermore, in social movements the means used are very closely tied to the desired ends: 'tactics represent important routines, emotionally and morally salient in these people's lives' (Jasper 1997: 237). If protest is shaped by the social, cultural and political structures where it develops, it also in turn affects those structures.

Space is part of the social structures that protest is influenced by and also shapes. If the analysis of space has long been a deep 'silence' in social movement studies (Sewell 2001), interest in the spatial dimension of protest has recently increased. In fact, it has been observed that:

> Like time, space is not merely a variable or container of activism: it constitutes and structures relationships and networks (including the processes that produce gender, race and class identities); situates social and cultural life including

repertoires of contention; is integral to the attribution of threats and opportunities;
is implicit in many types of category formation; and is central to scale-jumping
strategies that aim to alter discrepancies in power among political contestants
(Martin and Miller 2003: 145).

For quite some time critical geographers have considered space as socially
constructed. Following Lefebvre's (1991) influential distinction, the relevance of
three types of socially produced space for social movements has been pointed at:
a) the *perceived* space (or special practices) refers to 'the material spaces of
everyday life where production and reproduction occurs' (Martin and Miller
2003: 146); b) the *conceived* space refers to the representation of space as socially
constructed through (dominant and alternative) discourses, meanings, signs; and
c) the *lived* space, or representational space, where a) and b) interact.

In particular, reflection on social movements and protest has addressed places
as 'sites where people live, work and move, and where they form attachment,
practise their relations with each other, and relate to the rest of the world' (Leitner,
Sheppard, and Sziarto 2008: 161). Places have *material* aspects as 'by shaping
social interaction and mobility, the *materiality* of space also shapes the nature and
possibility of contention' (ibid., emphasis added), but they are also imbued with
meaning and power, as they are symbolically constructed, with symbolic cues that
signal appropriate and inappropriate behaviours, ownership, etc. (what is in-place
and what is out-of-place). In fact, a place emerges from 'the coming together of
the previously unrelated' and is 'open and internally multiple' (Massey 2005: 141).

Social movements are certainly structured by the space in which they develop:
'whether as a terrain to be occupied, an obstacle to be overcome, or as an enabler
to have in mind, [space] matters in the production of collective action. Space is
sometimes the site, other times the object and usually both the site and the object
of contentious politics' (Auyero 2006: 567). Protest happens in physical spaces:
'Activists take advantage of or put up with special constraints' (ibid.: 572). Spatial
distance between potential participants thwarts mobilization, while co-presence
facilitates it (Tilly 2003). The spatial imaginary (e.g. open versus closed space;
public versus private space) of the activists as well as the spatial routines of daily
life influence the availability to join protest as well as its forms (Sewell 2001;
Wolford 2004). Following Agnew (1987), Oslender has looked at these spaces
as constituted by a) *location* as 'the physical geographical area and the way in
which it is affected by economic and political processes' (Oslender 2004: 961);
b) *locale* as the 'formal and informal settings in which everyday social interactions
and relations are constituted,' being 'actively and routinely drawn upon by social
actors in their everyday interactions and communications' (ibid: 962); and c) *sense
of place* as 'the way in which human experiences and imagination appropriates the
physical characteristics and qualities of geographical location' (ibid.).

Social movements are, however, also space producers: they manipulate
places, producing new ones. Social movement activities grow in fact in what Paul
Routledge has called *terrains of resistance*, that is, 'sites of contestation and the

multiplicity of relations between hegemonic and counter-hegemonic powers and discourses, between forces and relations of domination, subjection, exploitation and resistance' (Routledge 1996: 516). Protest itself uses and produces space. In fact:

> As a site of contestation, a terrain of resistance is not just a physical place but also a physical expression (e.g. the construction of barricades and trenches), which not only reflects a movement's tactical ingenuity, but also endows space with an amalgam of meanings – be they symbolic, spiritual, ideological, cultural or political. A terrain of resistance is thus both metaphoric and literal. It constitutes the geographical ground upon which conflict takes place and its representational space with which to understand and interpret collective action (ibid: 517).

Protest itself constitutes a sense of space, as protest is typically embedded at the local level, but also more and more often it produces global frames. At the local level, attachment to place and culture, ecological and economic practices work as sources of alternative visions and practices (Escobar 2001), but global visions might develop in action as '[a] sense of place is a political construction, created from concrete, contingent practices, in particular circumstances' (Drainville 2004: 40). As emotionally intense and cognitively innovative events, protests contribute to construct this 'sense of place' and also change the symbolic meaning of a place. Occupying spaces, assigning new meaning to them, but also creating new spaces, the dynamics of protest evolve around a territorial contestation of space-specific cultural codes: 'Because different social groups endow space with an amalgam of different meanings and values, particular places frequently become sites of conflict where the social structures and relations of power, domination and resistance intersect' (Routledge 1996: 519). Direct actions constitute performative terrains, transforming places in stages, where solidarity is generated by shared intense emotions (Juris 2008). Profanation, territorial offence and repairing ceremony take place during the interaction of protesters, counter protesters and the police (Mathieu 2008). And an effect of these battles is to change the symbolic meaning of places.

In what follows, we offer a brief introduction to the three campaigns and then show how these protests around the construction of large infrastructures not only demonstrate contestation of the use of a specific territory, with the elaboration of an alternative conception of that space, but also how new spaces were created as terrains of resistance. The symbolic contestation of the conception of space interacted with the physical occupation of some sites, that not only acquired high symbolic meaning but also had a strong effect on the protest itself, allowing for the development of intense relations up to the formation of shared (territorially based) identities. If the sense of the place (the local culture) influenced the protest, the protest produced itself a definition of the place as collective identities developed in 'liberated' as well as 'contested' spaces.

A tale of three campaigns: No TAV, No Bridge, No Dal Molin

The protest campaign against the construction of a 57 km tunnel as a part of a new High Speed Rail Line (*TAV – Treno Alta Velocità*) in Val di Susa (on the border with France) is a long lasting mobilization. Promoted in the 1990s by ecological associations and the mayors of the valley worried about the negative impact the excavation works would cause on the environment and the health of the residents, the mobilization has grown in the valley from 2000 onwards supported by No TAV citizens' committees – able to mobilize the local community – and other actors (squatted social centres, rank-and-file unions, farmers' associations, social forums, etc.). The protests became cross-issue and the protest actors networked with thousands of citizens participating in several demonstrations, on camp sites and on pickets (*presidi*) on the contested territories. In November 2005, the violence of the police, intervening to evict the occupants on the picket at the checking site, gave a national dimension to the protests, with large media coverage of the event followed by a series of rail and road blockades by No TAV in the valley and by solidarity demonstrations throughout Italy. The acute phase of the conflict ended with the partial success of the TAV opponents, the temporary suspension of works, the removal of building sites in June 2006, and the starting of negotiations between national government representatives and local politicians. Notwithstanding the ambiguity of the Prodi centre-left government (2006-2008) – the TAV remains in the political agenda – the building sites were not reopened, while the Berlusconi centre-right government reaffirms the willingness to restart the TAV implementation policy process by the end of 2009. The mobilization effort continues.

The opposition to the project of a bridge on the Messina Straits, between Sicily and Calabria, is also a long-lasting mobilization. Originating in the mid-1990s as a counter-informative campaign of a committee formed by intellectuals, environmental associations and green-radical left parties, the protest campaign started in the early 2000s, with the mobilization of citizens' committees, social forums, social centres, ecologists and local parties opposed to the construction of the bridge, decided on by the Berlusconi government (2001-2006). The protests against the bridge involved cross-issue actors that go beyond the theme of environmental protection, also claiming to modernize existing infrastructures according to the principles of eco-compatibility. In November 2002, No Bridge activists and organizations participated at the European Social Forum in Florence: the discourse of protesters extended itself to a transnational dimension and the battle against the bridge became inserted into the framework of a more general struggle against neo-liberal globalization. During the summers of 2002, 2003 and 2004, the No Bridge activists organized national and international 'Camps of struggle' on the two coasts of the Straits, with the participation of hundreds of activists from Italy and abroad. From the end of 2004, the mobilization intensified with several mass demonstrations, reaching its peak on January 22, 2006 when 20,000 people participated in the National Procession against the Bridge in

Messina, including a considerable delegation of protesters from Val di Susa, as a result of a twinning between No TAV and No Bridge. In the spring of 2006, the centre-left government suspended the implementation of the project, but in 2008 the new Berlusconi government resumed the policy process and announced the beginning of works on December 23, 2009. Consequently, the No Bridge campaign also intensified with mass demonstrations on August 8, in Messina and on December 19, on the Calabrian coast.

The No Dal Molin campaign has protested the expansion, doubling or new settlement of a US military base in the area named Dal Molin, located between the city of Vicenza and a neighbouring municipality, where the military zone and civilian airport coexist. The United States foresaw the reuniting of the 173rd Airborne Brigade there, but the plan involved the occupation of the civilian zone. The possibility of urban and environmental impact on a non-military area heightened the pacifist and anti-military frame of the protest, given that the territory is already highly militarized. The campaign stands out for its different areas of mobilization that acted together as a single unit at certain moments. The first phase of the initiative took the form of an informational campaign, but on October 26, 2006, in front of a guarded town council meeting to ratify the yes vote for the base, the '*pignatte*' ('cooking pots') movement was born. The mobilization grew from January 2007 onwards, after the centre-left government declared that it would not oppose the plan: citizens marched, some activists occupied the tracks of the railway station, and a large picket tent for the occupation was put up near the Dal Molin area. Calls increased for unified national initiatives, which turned into vast pacifist gatherings. In the meantime, the local effort consisted of different initiatives: torchlight processions and the collection of signatures but also assemblies, festivals and summertime camp-outs in the picket area. During the local elections of 2008 the Dal Molin issue stood out as a key theme in the election campaign. Even in 'a community split down the middle' the centre-left candidate's victory marked an end to non-cooperation with the local council, by calling for and managing a consultation with the public, after a series of legal developments. However, the resulting no vote against the plan was disregarded by national institutions, and today only news about clashes between the demonstrators and police seems able to reverse the dwindling media and public opinion attention. In this context, while the mayor declares to have 'raised the white flag' and pronounces the need to think about trade-offs, the mobilization holds fast to its united goal of saying no to the base.

Contested spaces

The deputy police commissioner glimpsed local politicians: my councillor and I were wearing the national flag, it was not the first time that the state turned on itself, but this time it was strange because it was the deputy commissioner that gave me orders to evacuate the streets in 5 minutes. They would have passed through anyway! At this point I began to call other politicians, various other people, asking them to come and join us. We are fortunate that we know the mountains well, the paths and mule tracks. After 10 minutes the deputy commissioner returned to ask us what we had decided, and I replied that we would not move and would defend our territory. At this point the police advanced with their shields above their heads. We conducted an entirely pacific resistance, with our hands in the air; we were retreating because we could not stand such a conflict, until behind us came reinforcements from everywhere, which helped us to resist the advance. So many people arrived that the police had to stop, and despite pushing us from the side were unable to move us. This went on until late in the evening, it was a very tough confrontation, from 7 in the morning till 8 in the evening, until the deputy commissioner, in agreement with the president of the Mountain Community, suspended their activities. At that point we decided to leave, as the police themselves were doing. That same night the police occupied the area (IVS7).

This account chronicles one of the transforming events of the protest in Val di Susa, the battle of Seghino on 31 October 2006. Often in line with an increase in the number of protesters there is a propensity for more disruptive actions: more moderate actions are attempted but are considered insufficient against the perceived 'brick wall' of the authorities (Mosca 2004). The phase of mass demonstration is often accompanied by direct actions such as the blocking of roads or railway lines, which although excluding violence, are nevertheless a challenge to the state in terms of public order. A radicalization of conflict was particularly evident in Val di Susa, around a classic mechanism of interaction with the forces of order that also attains a symbolic value. There was an escalation in the conflict with the police, centred around the building site which both sides were seeking to control and which became, in Routledge's term, a terrain of resistance.

Other sites also acquired this (not only symbolic) meaning. The Battle of Seghino; charges in Venaus on 29 November; the dispersal of the site occupation on 6 December; the re-occupation of the site on 8 December: these were the most acute moments of conflict with the police, phases of escalation which are often accompanied by waves of mobilization. The effects of the protest are in fact tied to the ability to interrupt the routine, often through non-conventional acts. Public order is often a central frame for those who oppose the protest, and our cases do not provide an exception to this rule. Above all from the second half of 2005, the No TAV protesters were accused by TAV supporters of not only being selfish but also violent. The objectives for clearing out the site of Seghino on the night of 5-6 December were, according to the then Minister of the Interior, 'to revert to minimal conditions of legality and to re-launch dialogue' by 'blocking the extremist fringe before it affected Turin and the Winter Olympics' (*R* 16/12/05). Similar views were held among the Centre-Left: the secretary of the Democratic Left (DS) in Piedmont asserted that 'there is an ever-increasing risk of a degenerating situation while the threat of a possible terrorist act grows daily' (*R* 6/11/05).

Despite the risk of stigmatization, direct action around the occupied site was perceived by those who opposed this project as an instrument that raised the visibility of a protest ignored by the mass media. As the president of the Mountain Community recalls, 'thanks to Minister Pisanu our visibility increased. I was hoping that this would happen, because from an electoral viewpoint 100,000 people count for nothing (given that they all vote differently), but if the protest spreads throughout Italy then their fear grows and their insults escalate along with the accusations of "localism". For this reason we are talking of a "large back-yard"' (IVS8). Even beyond the valley 'the attacks by the police earned the sympathy of those who knew nothing of the TAV ... for them it was counter-productive because it gave us added visibility and prompted a democratic spirit that went beyond the TAV conflict, because in a democratic country certain things should not be done' (IVS3).

However, beyond the instrumental dimension linked to increased visibility, activists underlined the positive effects of direct action in and around the contested space as a moment of growth in solidarity with the local population. It is precisely the effect of this injustice frame (Gamson 1990) that is often mentioned by protesters as a source of consensus within the larger local community and a mechanism that reinforces identification. The intervention of the police in the site occupation became the symbol of an unfair attitude towards those who were protesting peacefully, the military occupation of the area 'being seen as an arrogance that nobody could justify' (IVS4). As a local inhabitant observes, 'the explosion of the movement (and nobody expected a participation of this strength) occurred from the 31st of October 2005 onwards, the days in which the violence of the government sent the troops into the valley' (IVS2). However, the charges in Seghino represent a moment of mobilization that goes beyond mere external solidarity. 'At the site occupation there were always 100-200 people during the day. When it looked likely to be cleared out then 2,000-3,000 people arrived, staying throughout the night to defend our position' (IVS11). Participation becomes more intense when faced with a perceived external aggression, described by activists as an act of war against a peaceful community. In the words of one activist, this perceived aggression forces the community to 'join the front-line':

> People appeared in very large numbers on a week-day, they didn't go to work but went to the site occupation instead, believing that there was no use just in talking but that they should join the front-line. They all appeared with banners and flags. In Bruzolo, when the police were confronting the crowd, we joined in with our household utensils to defend ourselves. We are not afraid of anybody, we want to defend our territory in a peaceful way. Maybe you will laugh at us, but the battle is long (IVS5).

Besides, the mobilization led to a re-definition of the identity of the community, which occurs, above all, during the course of protests; the sense of belonging is perceived as being built in action, through the participation to the mobilization, rather than ascribed criteria. A No TAV activist indeed says: 'We needed an identity and maybe we found it during the struggle, on the idea that the territory is ours: this struggle is strong because it comes from a choice, not because the valley was of our fathers and grandfathers' (IVS9). The identification with community is then not exclusive; on the contrary an open and inclusive conception of community emerges through the protest, that is able to integrate different cultures and values: 'This idea of a territory that has taken people from outside, has led to different cultures, not a pre-structured culture, but a various set … this valley has allowed people who came to live here feel it like its own' (IVS9).

Injustice, arrogance and dislike are the principal narrative frames that emerge with regard to the presence of the police on the territory of the valley, perceived as a 'militarization of the valley'. Its consequences are often recalled as an act of violence on the territory and its inhabitants:

> On the night of the 29th my mobile rang. It was a friend of mine who said that they had come out and militarized the area. I set out with my heart in my mouth, the news made me feel downtrodden and tricked, with respect to a struggle that we had always conducted in the open, without subterfuge, while they carried out their actions in the middle of the night … at Venaus they would not let anyone enter, we were stopped by the riot police. The scenes I witnessed were truly shocking, lots of people that normally work at dawn could not enter. An old woman arrived saying that she had to look after her grand-daughter and the police told her to have the baby taken out of town, as she would not be able to get in (IVS4).

In the choral narrative of the protest the militarization of the valley is the 'final drop that makes the glass spill over', while the successive mobilization is the 'reaction against arrogance: The moment in which they made false moves with arrogance and even trickery there was a popular reaction, from everybody not just militants' (IVS4). In the perception of the activists it was at this point that 'the people started to get angry, there was no way of stopping them, they occupied roads and highways (the people, not the associations), they would have stayed day and night until the government gave a signal … from the 1st of November till the 6th of December it went ahead like this, then on the 6th they used force, beating old people. Two days later people shouted "Let's take back the land" and 100,000 people descended and took it back' (IVS5).

The same indignation emerged in the narratives on the dispersal of the site occupation by the police forces. In the recollection of one activist, 'they destroyed the books of the university students who were studying (after all this was time taken away from daily activities) throwing them on a bonfire. And when people were forced to leave the fields, the police went round with the No TAV banners as if they were a symbol of conquest … and they also had the cheek to destroy the food supplies that were needed to live in the camp … old people were beaten and they stopped the ambulances from coming. An old man was slumped on the floor for an hour because they never even let the stretchers in' (IVS10). There were frequent recollections of 'a police chief on a Caterpillar truck shouting on a megaphone "Crush them all!" and encouraging the driver to push ahead until it was right in front, a nasty and dangerous experience which I will remember all my life because it was the first time I was scared that something awful could happen' (IVS4).

Indignation against perceived arrogance is what emerges from these interviews and lies at the root of the growth of the movement, affecting its capacity to react. If repression increases the costs of collective action it can often have the effect of discouraging it. However, it may also reinforce the processes of identification and solidarity (della Porta and Reiter 2006). On 31 October, the re-conquest of Seghino (the place where the works were due to begin) was narrated as an epic return. In the words of one interviewee, the police charges marked the start of 'the time of fighting: The morning after in the valley there was a massive strike. The workers left their factories, the teachers never entered the schools while the parents never took their children there, and everyone went to occupy the valley, which remained so for three days. It was the time of revolt, which culminated on the 8th of December with the re-conquest of the field. It was wonderful.' The memories of the police blockade at Mompatero are added to the observation that 'even the meek in front of injustice are capable of rebellion and will not turn back, because they understand it is a question of pride and dignity. This was the most important thing' (IVS10).

The building sites promised to become places of contestation also for the other two campaigns. Even if the building works for the bridge on the Messina Straits have not yet begun, No Bridge activists have predicted that there will still be 'problems because non-violent direct action will be savagely repressed, especially if they start the building works' (IME5). At the time of writing, there is great expectation for the 'laying of the foundation stone', announced by Berlusconi for December 23, 2009 in Villa S. Giovanni on the Calabrian coast. Mass demonstrations and direct actions are planned to prevent the construction works and, probably, the building sites will become the contested spaces of the Straits.

For the No Dal Molin mobilization there is also a transforming event around a contested space that stands out as a narrative moment that determines a turning-point in the campaign through a reactive, demonstrative action. While visiting Romania on January 15, 2007, the Prime Minister of the centre-left government (Romano Prodi) communicated to journalists, in what activists call the 'edict of Bucharest', the government's choice not to oppose the base, contrary to election campaign promises. When the television news broadcast the story, the protest spontaneously went into action and activists occupied the area near the base project: a large picket tent was put up and the No Dal Molin occupation was inaugurated. In the following days, local initiatives intensified and the first national mobilization was launched. It became a large pacifist demonstration, and prompted government crisis. The citizens who protested by banging on pots beneath the windows of the town council and came together for a migrating assembly gained media visibility and political attention through direct action. In this case too, it broke the routine: it was an act of disobedience that creates a 'space for movement'.

Here it was not a reaction to police repression (which did not occur) that heightened the framework of injustice as much as indignation at the lack of transparency and consistency on the part of institutions at every level: 'outrage increases, anger increases, because besides the initial decision which was unacceptable no matter what, there is a range of unacceptable, non-democratic behaviours that causes us to react ever more strongly!' (IVI7). The birth of the No Dal Molin occupation, although within a context of competing for the few opportunities for action, did not eliminate each individual area of mobilization's identity, but instead gave physical shape to a need to participate: 'above all we are the expression of the people's need to reclaim their power and to influence things that concern their futures and the futures of their children' (IVI6).

Liberated spaces

> Our identity began to strengthen itself from June, when the government tried to initiate the works. That summer people began to stay at the site from morning till night, people from the same town became friends that were only acquaintances before. The pensioners said that we should 'do it in this way because the battle is not over', because they understood the difference between watching it on TV and organizing activities. *The people became a community* ... the site occupation became a social event and this cemented an identification between territory and citizen which is quite exceptional. Then the events of Venaus obviously emphasized the solidarity in these difficult situations. People ended up in hospital from police beatings, and a sense of community had been created (IVS8, our emphasis).

As it emerges from this interview with the President of the Mountain Community of Lower Val de Susa, the militarization of the area and the police charges were seen by activists as something that legitimized their protest through a feeling of indignation. The 'people' became a 'community' through long and intense actions, such as the site occupation or the camps, which affected the daily lives of the participants and formed arenas of communication and discussion (often tough) between its various political and territorial components. In this sense, terrains of resistance also operated as liberated spaces, where relational, cognitive and affective mechanisms changed the very nature of the protest campaigns,

In the case of Val de Susa, the struggles around the No TAV site occupation of 2005 were seen as a moment of growth for the protest not only in numerical terms but also in terms of identification with the protest. In the words of the activists, the site occupation had 'great emotional force', 'a shared intimacy', was 'wonderful as well as striking for the behaviour of the people; the diversity of those present; and the sense of serenity' (Sasso 2005: 61). In the memory of the movement there were the 'unforgettable nights of Venaus, when we had a bonfire in the fields and the snow fell, and we felt truly united' (Velleità Alternative 2006: 20). In the narrative of the activists the site occupation is remembered as a serene but intense experience which reinforced feelings of mutual trust: 'When on the night of 5-6 December the police forces went to occupy the land at Venaus ... there was a wonderful encampment under the falling snow, fires burning, children and dogs playing. There were pots full of food, young people from all over Italy – because at that point we became the focus and hope for a series of struggles. All this they stopped with batons, beatings and by destroying our tents' (IVS10).

These site occupations are in fact seen as places of strong socialization: 'real homes built on this territory, which became focal points – a wonderful thing. In the summer there were scores of people that came to talk and socialize, allowing feelings of solidarity to grow with the awareness that this struggle was for everyone' (IVS11). Participation in the protest was seen as gratifying in itself, affecting daily life: 'Throughout the whole summer there were 50-100 people that occupied three places in the valley (Borgone, Bruzolo, Venaus). In the morning, you went to get a coffee at the site occupation and not at the bar. If you wanted an alternative dinner you went to the site occupation, where you might also see a concert' (IVS5).

Allowing frequent and emotionally intense interactions, the site occupations were perceived as an opportunity for reciprocal identification, based on mutual recognition as members of a community. According to a report: 'This is the story of an unwitting revolution, says a young man, in these days we also changed, lost our prejudices and struck up friendships. People met each other that previously would have had little occasion to ... we met, listened and found that we shared a common destiny' (Sasso 2005: 62-63). In the site occupations 'you got to know people *through the struggle*, you recognized each other' (IVS10, our emphasis). In this sense the action itself constitutes a resource of mutual solidarity and reciprocal trust, which allowed capacity to withstand later moments of intense conflict.

These site occupations therefore represent arenas of discussion and deliberation, places to experience a different form of democracy, participative because they allow for the creativity of individuals. In the words of one activist: 'Everything began from these site occupations, a wonderful form of participatory democracy where people from below could have their say: They could coin a slogan, a new banner, invent a new march, a new message' (IVS5). The site occupations thus became 'political laboratories' that produced interaction and communication:

> Unity is so strong in the No TAV movement, we are so compact that we always overcome the many obstacles we have to confront ... as militants, this struggle was a political laboratory, a moment of incredible growth, because very often it is difficult to act concretely, beautiful words we utter on the world we want, the contradictions we want to eliminate. Here we threw ourselves into the game, we experiment on the things we said and we learnt a lot from these people, from their motivation, their capacities, and we had to confront the realities of our

own words which were far from the realities of political action. We concretized ourselves in a struggle of this type, and it was a moment of growth (both human and political) for all of us (IVS1).

The experience of the site occupation therefore transcended the opposition to high-speed trains, becoming a place in which 'all the small problems which must be confronted daily are resolved through discussion, with spontaneous assemblies, with mutual trust and a complicity which reinforces the sense of solidarity' (ibid). In our daily lives we experience new values, for example 'the absence of money' (Velleità Alternative 2006: 134). In the words of one activist, 'the site occupations were places inhabited by a different life, where you could eat for free because money no longer had any value, and this not only attracted people like myself, who have a vision that is not only an ideal. I believe that this reality can be applied; I believe in the possibility of radically changing this world and not only reforming it ... it was a collective hope, and when they responded with militarization the people rebelled ...' (IVS10).

A similar but distinct role is played by the construction of arenas of communication and discussion/confrontation in the 'camps', which were evident in both mobilizations. Above all in the No Bridge campaign, the camps on both sides of the Straits performed an important role in the construction and development of the protest campaign, but also in the dynamics of co-operation and conflict between these various political and territorial actors, particularly in the phase preceding the mass demonstrations. These camps facilitated contact between activists and organizations, furthering the process of cross-issue networking, particularly with regard to the subject of large-scale public works. This extension in the aims of mobilization are highlighted by participants at the camps who note the

> consolidation and amplification of networks and the co-ordination of movements which challenge those who wish to degrade our social, environmental and territorial quality of life, through large-scale public works and other operations which seek to use the places and spaces of our lives to serve the interests of monopolistic capital and speculators. It is particularly useful to extend this protest to the committees who more generally oppose large-scale public works and related issues, while connecting this with the environmental and cultural organizations present, also improving the exchange of information, news and resources between groups active in the local realities (DME23).

If the camps on both sides of the Straits are located within a broader global social movement (the first international camp was on the agenda proposed at the European Social Forum in Florence 2002), this spill-over cross-issues effect is later amplified during the protest campaign.

While representing arenas of discussion and communication that favour processes of networking, dialogue and collaboration between different political groups (both moderate and radical) and territorial ones (Messina and Calabria), the camps remain distinct from the site occupations described above in that they only represent *part* of a broader movement. In the words of one journalist-activist, not only is collaboration evident but so is territorial competition between the two sides of the Straits:

> The camp of 2002 was the first moment in which the two souls of the movement, those from Messina and those from Calabria, which have always had difficulty working and elaborating together, had a moment of synthesis and mutual recognition (hitherto they only knew each other from internet blogs or emails). They dialogued and worked together. Bear in mind that other camps (2003-2004) witnessed conflict along more geographical than political lines. This led to a following separation into two different camps located in two different places with different agendas (IME5).

The continual tension between unity of action and ideological diversity is confirmed by a Calabrian activist from the 'antagonistic' component, which consisted of social centres and autonomous organizations. On the one hand, the ability to co-ordinate protests on both sides of the Straits was seen as an example of 'dynamic inclusiveness':

> In 2002 the proposals of some comrades to develop links with the Calabrian side in the direction of dynamic inclusiveness was fully received and some of us Calabrians participated in the camp on the Sicilian coast, while others organized joint initiatives on the Calabrian coast; connecting the two sides of the Straits in a joint struggle was a winning tactic (IME7).

On the other hand, already during the first camp there were tensions, linked to the prevalence of existing organizational identities and competition between them. The perception by the activists of the squatted social centres that the Party of Communist Refoundation wanted to dominate the campsite led to a conflict between the more radical 'antagonistic' area (social centres and rank-and-file unions) of both sides of the Straits and the more moderate organizations (social forum, left parties and environmental associations) of Messina and Calabria. For this reason, the 2003 and 2004 campsites were organized separately, by the antagonistic component on the Calabrian coast and by the moderate groups on the Sicilian side. The camps are therefore also occasions for meeting between geographically distant groups, but more homogeneous in ideological terms (della Porta and Piazza 2008: 145).

Beyond these tensions and internal divisions, the camps represent important moments of deepening debate on various themes for the activists and organizations involved, and not only those closely tied to the question of the bridge. They also allow the development of alternative proposals and new strategies of protest activity, such as holding seminars, workshops and plenary assemblies'.[1] As can be read in one of the concluding documents:

> the plenary assemblies, which contributed towards deepening our scientific, technical and political knowledge on the given themes, also propose scenarios of sustainable development of the territory, which presented an alternative to the failed model of development in Southern Italy. In this sense we pledged to maintain and develop a wider movement of ideas, proposals, knowledge, conflict, rebellion, capable of confronting the devastation and misery caused by the prevailing capitalist model of development (DME23).

Even the interviewees noted their own increased political awareness and their capability for theoretical elaboration. Above all, daily face-to-face interactions created 'human and political' relationships:

> The camps have been moments of internal political growth for the activists, because they foment themselves by analysis, elaboration, human and political relations ... from the point of view of 'production' the 2nd camp produced a document ('Stop the development of capital') which was of great political value, not only regarding the bridge but also the complex contradictions of the Italian south (IME5).

In addition, the interviewees noted that the camps were also instruments to raise awareness, involve and potentially mobilize the local population, through numerous initiatives on the territory, demonstrations and continual interaction with the surrounding population. Among these positive elements was the experience of the camp in summer 2003, where the activists underlined

> the positive communication between the participants of the camp and the local population at Villa, Cannitello and Messina; the solidarity shown by the people from Gazzi in Messina during the march around the prison; the impressive adhesion to the march in Messina on the 1st August and at Villa San Giovanni on the 2nd. These elements demonstrate that it is possible to wake the local

---

1   During the 2003 camp there were four seminars, respectively on 'Environmental protection and the territorial management of fundamental resources', 'Bridges between people – the Mediterranean, Immigration and Co-operation'; 'Local Projects and Alternative Production'; 'The Politics of Large-Scale Public Works and Imperialism – what space for the Movement?' (www.noponte.org). In the course of the third meeting in July 2004 against the Bridge on the Messina Straits organized by the Messina Social Forum at Torre Faro (ME) a series of seminars were held on different themes. These included international voluntary work; fair trade and critical consumption; demilitarization of Sicily; new municipalism and participative democracy; mafia and large-scale public works; abolition of the CPT (www. terrelibere.org).

population from their slumber despite mafia threats and a powerfully distorted media campaign (DME23).

The creation of a physical space helps to develop a sense of community in the Dal Molin case as well: the site of occupation facilitates reciprocal identification, and all of its core concrete activities become a source of mutual solidarity and reciprocal trust. This modifies the participation experience itself. As an activist put it 'The permanent picket tent brings many people inside the protest, and compels them to no longer reason through the earlier lens of mediation between organizations, but instead with an eye towards sharing' (IVI8).

The picket area gained in fact visibility for the mobilization and facilitated the connection and comparison with analogous entities and similar Italian and European movements. Slogans and symbols spread and political awareness increased against the NIMBY stigma and instead promoted a 'glocal' definition of the conflict:

> Inside us, we sense the path of local conflicts that are fundamental today in reinterpreting the relationship between citizen and territory, between local and global. When you talk about Val di Susa, you're not speaking of Venaus, you're talking about the European transport network; when you talk about Vicenza, you're not talking about Via Sant Antonino, you're talking about global war. Therefore it's plain to see that we're not just confining ourselves to our own backyard (IVI9).

These mechanisms went into action mainly during the festivals and summertime camp-outs, which brought back the tradition of anti-nuclear pacifist camp-outs of the 1980s and increased the 'fun of participating' and the spill-over issues effect, especially on the anti-war theme. The picket site creates the circumstances for establishing initiative ties and a sense of 'beyond our own backyard'. However, it also positions itself as a place open to the community for debates and activates inclusive dynamics, especially between the protesters and Catholic factions. In this shared space, mutual distrust and differences in traditions and life paths are overcome, as this passage from an interview with two members of the occupation from opposite political-cultural backgrounds attests:

> IVI5: There is a great capacity for listening, I truly have never spoken so much, because I used to work with people who agreed with me, they shared my perspective, while here it's a never ending debate'...

> IVI8: We've understood that each of us must take a step backwards. This was a fundamental and beautiful thing, because when each of us enters the site we must leave something behind and open ourselves up to another point of view. At other times it never would have happened that a Catholic would be listened to

by a protester, because of his origins or background, but instead each of us has brought our experiences to the table and put them to use (IVI5, 8).

'Cross-fertilization' in action was produced on the picket site, both because the shared action led to a negotiation of meaning (Melucci 1996), and because the mutual understanding increased trust between actors and individuals who had little occasion to spend time together and limited themselves to rejecting one another. Respect grew, as does the exchange of knowledge and skills. In this way the 'life in the picket' came to influence daily life through powerful reflections on militant identities and individual lives, and connects the public and private spheres of the protest. As one of the promoters of the picket site recalls:

> Until last year I used to wear stiletto heels, a fur coat and an evening dress to New Year's, now I'm here in mountain boots and fleece pants! But I think that in life it is good to change, to realize what your priorities are! I've lost my taste for shopping, all these things, this tinsel doesn't interest me anymore, I'm interested in setting up my life in another way. Because of this I've been badly burned, because I've even lost many people since obviously many friends don't recognize me anymore, but it doesn't matter … It comes to mind that 6 years ago I used to have furious quarrels with my son, who was 14 and sometimes went to the social centre: it came to me keeping him shut at home. Now my son arrives with all of the kids (from the social centres) to have meetings at my house (IVI7).

In the protest narratives, the occupation site becomes a point of departure for awakening the city from its stupor. This change is deemed profound and long lasting, primarily in relation to the city: 'This city disgusted me for a long time but now I say that I love Vicenza, I recognize that I love it, I love my city, because it knew how to rebel against this' (IVI5).

Conclusions

Protest campaigns develop around contested spaces, but also create their own liberated spaces. In both types of terrains, resistance take place, with its forms and intensity influenced by the social, political and geographical characteristics of the site in which it is located – but it also changes them, liberating them and assigning them different uses and meanings.

The protest is therefore an arena, that is, a place where different actors meet, communicate and confront, sometimes with tensions. As with the occupation of factories in the labour movement, the site occupations (above all in Val de Susa), the protest camps (above all on the Messina Straits) and the picket site (in the No Dal Molin case) constitute terrains of resistance, where the structures and cultures of the protest changed, relations were created, affective ties intensified and identities were built. For the No TAV and the No Dal Molin activist, long-lasting and territorially rooted forms of occupation of space helped creating an inclusive identity of the local community through reciprocal identification and the perception that they are experiencing new values and practices of participative democracy. Even though limited in time, the camps (especially those on the two sides of the Straits) played an important role in the construction of the campaign: They favour the intensification and expansion of multi-thematic networking processes between activists and different groups; display a growing rate of participation; represent important moments of debate that allow the formulation of proposals and strategies for action; become instruments to raise awareness and to mobilize the local population, interacting with the surrounding territorial reality; and finally, allow the different souls of the protest to meet and work together, although through moments of tension.

The forms of protest in our three cases are varied as well as being both instrumental and symbolic, procedural and counter-cultural. The various nodes in protest nets bring their own repertoire of actions (legal actions by environmental organizations; strikes by workers; direct action by social centres; institutional pressures exerted by mayors). In this process of diffusion, various forms of action are imported from contemporary movements (e.g. boycotting certain goods, which is typical in movements for globalization from below). This same variety of forms is itself claimed as a sign of support and involvement. An analysis of the repertoires of protest confirm the innovative features of the campaigns that are not limited to known forms of action – the repertoire which Tilly (1978) observes – but become intertwined and innovative, in a process of contamination in action. In certain moments of this process the resources invested in the action multiply. As Alessandro Pizzorno (1997) observes with regard to the movements of the 1960s and 1970s, the resources of militancy are produced and re-produced in the phase of expansion of the protest, when the opportunity for action reduces internal conflict.

References

Agnew, J. (1987) *Space and Politics. The Geographical Mediation of State and Society.* Boston: Allen and Unwin.

Auyero, J. (2006) 'Spaces and Places as Sites and Object of Politics', in R. Goodin and C. Tilly (eds) *The Oxford Handbook of Contextual Political Analysis.* Oxford: Oxford University Press, 564-578.

della Porta, D. and Piazza, G. (2008) *Le ragioni del no. Le campagne contro la Tav in Val di Susa e il Ponte sullo Stretto.* Milano: Feltrinelli.

della Porta, D. and Reiter, H. (2006) 'Conclusion', in D. della Porta, A. Petersen and H. Reiter (eds.), *Policing Transnational Protest.* Aldershot: Ashgate.

Drainville, A. C. (2004) *Contesting Globalization. Space and Place in World Economy.* London: Routledge.

Escobar, A. (2001) 'Culture Sits in Places: Reflections on Globalism and Subaltern Strategies of Localization', *Political Geography*, 20: 139-174.

Fantasia, R. (1988) *Cultures of Solidarity. Consciousness, Action, and Contemporary American Workers*. Berkeley: University of California Press.

Gamson, W. (1990) *The Strategy of Social Protest*. 2nd ed. Belmont, CA: Wadsworth (original edition 1975).

Jasper, J. (1997) *The Art of Moral Protest: Culture, Creativity and Biography in Social Movements*. Chicago: Chicago University Press.

Juris, J. S. (2008) *Networking Futures: The Movements Against Corporate Globalization*. Durham, NC: Duke University Press.

Lefebvre, H. (1991) *The Construction of Space*. Oxford, Blackwell.

Leitner, H., Sheppard, E. and Sziarto K. M. (2008) 'The Spatiality of Contentious Politics,' in *Transactions of the Institute of British Geographers*, 33: 157-172.

Lipsky, M. (1965) *Protest in City Politics*. Chicago: Rand McNally and Co.

Martin, D. G. and Miller, B. (2003) 'Space and Contentious Politics', *Mobilization, an International Journal*, 8: 143-156.

Massey, D. (2005) *For Space*. London: Sage.

Mathieu, L. (2008) 'The Spatial Dynamics of the May 1968 French Demonstration,' *Mobilization: An International Quarterly*, 13: 83-97.

Melucci, A. (1996) *Challenging Codes*. Cambridge/New York: Cambridge University Press.

Mosca, L. (2004) 'Cooperazione e conflitto fra opportunità politiche e temi della mobilitazione', in D. della Porta (ed.) *Comitati cittadini e democrazia urbana*. Soveria Mannelli: Rubbettino, 138-198.

Ortoleva, P. (1988) *Saggio sui Movimenti del 68 in Europa e in America*. Roma: Editori Riuniti.

Oslender, U. (2004) 'Fleshing Out the Geographies of Social Movements: Colombia's Pacific Coast Black Communities and the 'Aquatic Space', *Political Geography*, 23: 957-985.

Pizzorno, A. (1993) *Le radici della politica assoluta*. Milano: Feltrinelli.

—— (1997) 'Le trasformazioni del sistema politico italiano 1976-1992', in F. Barbagallo (ed) *Storia dell'Italia repubblicana*. Torino: Einaudi, 303-344.

Rochon, T. R. (1998) *Culture Moves. Ideas, Activism, and Changing Values*. Princeton: Princeton University Press.

Routledge, P. (1996) 'Critical Geopolitics and Terrains of Resistance', *Political Geography*, 15 (6-7): 509-521.

Sasso, C. (2005) *No Tav. Cronache dalla Val di Susa*. Napoli: Intramoenia.

Sewell, W. H. Jr. (2001) 'Space in Contentious Politics', in R. R. Aminzadem, J. A. Goldstone, D. McAdam, E. J. Perry, W. H. Sewell Jr., S. Tarrow, and C. Tilly (eds) *Silence and Voice in the Study of Contentious Politics*. Cambridge: Cambridge University Press, 51-88.

Taylor, V. and van Dyke, N. (2004) '"Get up, Stand up"'. Tactical Repertoires of Social Movements', in D. Snow, S.A. Soule and H. Kriesi (eds) *The Blackwell Companion to Social Movements*. Oxford: Blackwell.

Tilly, C. (1978) *From Mobilization to Revolution*. Reading, MA: Addison-Wesley.
—— (2003) 'Contention over Space and Place', *Mobilization*, 8: 221-226.
Velleità Alternative (2006) 'No Tav. La valle che resiste', Supplement of *Autonomia*, n. 15.
Wolford, W. (2004) 'This Land is Ours Now: Spatial Imaginaries and the Struggle for Land in Brazil', *Annals of the Association of American Geographers*, 94: 409-424.

Interviews

Messina Straits

IME1. A. Giordano, WWF, responsible for Southern Italy. Messina, 09/05/2005.

IME3. S. Visicaro, comitato 'La Nostra Città' – 'Messinasenzaponte', Messina, 14/05/2005.

IME4. M. Camarata, 'Laboratorio contro il ponte' – Coordinamento 'Rete No Ponte', Messina, 12/12/2005.

IME5. A. Mazzeo, journalist of *Terre Libere*, Messina, 20/04/2006.

IME6. S. Bonfiglio, Provincial Secretary of Rifondazione Comunista, Messina, 08/05/2006.

IME7. M. Marzolla, Coordinamento Calabrese/Meridionale contro il Ponte – ex Social Center 'A. Cartella', Reggio Calabria, 18/05/2006.

Vicenza

IVI 5. F. Pavin, picket No Dal Molin (Disobedient)

IVI 6. M. Soccio, House of Peace, (Non-violent Movement)

IVI 7. C. Bottene, picket No Dal Molin ('independent' citizen)

IVI 8. S. Osti, picket No Dal Molin (Caritas)

IVI 9. O. Jackson, picket No Dal Molin (Greens)

Val di Susa

IVS1. Chiara, Centro Sociale Askatasuna, Val de Susa, 16/2/2006.

IVS2. C. Scarinzi, secretary of Comitati Unitari di Base, Val de Susa, 16/2/2006.

IVS3. De Masi, Provincial Councillor, Greens party, Val de Susa, 16/2/2006.

IVS4. M. Piccione, Comitato Spinta dal Bass di Avigliana, Val de Susa, 15/2/2006.

IVS5. P. Coterchio, Legambiente Piemonte, and Richetto, university professor, Val de Susa, 15/2/2006.

IVS7. M. Russo, Mayor of Chianocco, Val de Susa.

IVS8. A. Ferrentino, President of the Mountain Community of Lower Val de Susa, Val de Susa.

IVS9. M. Clerico, university professor of Environmental Security at the 'Politecnico' of Turin, Val de Susa.

IVS10. Nicoletta, Secretary of the PRC section of Bussoleno-Val de Susa, Val de Susa.

IVS11. G. Vighetti, Comitato di Lotta Popolare contro l'alta velocità di Bussoleno, Val de Susa.

Documents

DME2 – *Petizione Popolare contro la costruzione del Ponte sullo Stretto di Messina per un reale sviluppo della Sicilia.* Comitato La Nostra Città – Messina 3/5/02. www.messinasenzaponte.it.

DME17 – *Per il gemellaggio NO TAV – NO PONTE.* Coordinamento No Ponte – Rete Meridionale del Nuovo Municipio. Reggio Calabria, 29 december 2005. www.reteNo Ponte.org.

DME19 – *'10 buone ragioni per dire NO'.* Synthesis of the critical observations presented by the Greens to the SIA, in D'Agostino A. (ed.), *Lo Stretto di Messina. Il ponte insostenibile e le sue alternative*, in 'Quaderni del sud – quaderni calabresi', special number, n. 98-99, December 2005-March 2006.

DME23 – *Fermiamo lo sviluppo del capitale. Documento finale dell'Assemblea plenaria del 1° Campeggio internazionale contro il Ponte sullo Stretto* (Cannitello-Punta Faro, 28 luglio – 2 agosto 2003), in D'Agostino A. (ed.), *Lo Stretto di Messina. Il ponte insostenibile e le sue alternative*, in 'Quaderni del sud – quaderni calabresi', special number, n. 98-99, December 2005-March 2006.

Newspapers

R – la Repubblica

Websites

www.messinasenzaponte.it
www.noponte.org
www.terrelibere.org

Chapter 2

The Liberalization of Free Speech: Or, How Protest in Public Space is Silenced

Don Mitchell

Introduction

This chapter examines how the freedom to speak is regulated. A particular paradox of liberal states is that they guarantee freedoms but these freedoms require the erection of countless legal controls and regulations to ensure the social order. This apparatus of legal and repressive controls demarcates the boundaries of acceptable freedoms in liberal society, policing what, when, how and where things are said. In this sense, there are two issues that are crucial to the construction of a *liberal* theory of free speech. The first is that what makes a difference in the nature of speech is *where* that speech occurs. There would be nothing wrong, presumably, with falsely shouting fire, even in a time of war, in the middle of the wilderness, or even on a busy street, if there is adequate room to move. The trick for free speech regulation, therefore, becomes one of spatial regulation. Regulation of location, or place, becomes the surrogate for the regulation of content. The second crucial point is to distinguish between the content of speech and the possibility that such speech might have an effect.

My purpose in this chapter is to examine the intersection of these two issues to show how contemporary speech laws and policing effectively silence dissident speech in the name of its promotion and regulation. As courts in the United States have over the twentieth century moved away from a regime that penalizes what is said – in essence liberalizing free speech – it has simultaneously created a means to severely regulate where things may be said, and it has done so in a way that more effectively silences speech than did the older regime of censorship and repression. It could be argued, to put all this another way, a liberal approach to silencing opposition – to *keeping it from being heard* – is to let geography, more than censorship, do the silencing. And this is the direction American law is tending towards.

In what follows I will make my argument clear first through a brief historical geography of First Amendment law and the evolution of what is called public forum doctrine, and then by looking at three case studies that show how regulating the *where* of speech effectively silences protest. The implication of my argument is that under the free speech regime currently in force in the United States, dissent can only be effective when it is illegal.

The geography of free speech

The First Amendment to the US Constitution is straightforward: 'Congress shall make no law respecting the establishment of religion, or prohibiting the free exercise thereof; or abridging the freedom of speech, of the press; or of the right of the people to peaceably assemble, and to petition the Government for a redress of grievances.' This has rarely prevented Congress from doing exactly that, and the history of US Constitutional law is a history (at least in good part) of 'abridgements' to the right to speak and to assemble. Even more than Congress, however, state and local jurisdictions have rarely felt much compunction about limiting the speech and assembly rights of agitators, labour union members, socialists, and nearly anyone else they didn't like.

The opinions of Justice Oliver Wendell Holmes served as an important reference in developing the parameters of free speech. In his dissent in *Abrams*, Holmes wrote this:

> [W]hen men have realized that time has upset many fighting faiths, they may come to believe even more than they believe the very foundations of their own conduct that the ultimate good desired is better reached by the free trade in ideas – that the best test of truth is the power of the thought to get itself accepted in the market, and that truth is the only ground upon which their wishes can safely be carried out. That at any rate is the theory of our Constitution. It is an experiment, as all life is an experiment. Every year if not every day we have to wager our salvation upon some prophesy based upon imperfect knowledge. While that experiment is part of our system I think therefore we should be eternally vigilant against attempts to check the expression of opinions that we loathe and believe fraught with death, unless they so imminently threaten immediate interference with the lawful and pressing purposes of the law that an immediate check is required to save the country...[1]

As remarkable and stirring as that passage is, it is also deeply problematic (and as I have explored in other work: Mitchell 1996, 2003) because it puts into place – by implication in Holmes's own words, but later made explicit in a whole series of cases (see the review in Kaufman 1991) – a distinction between speech and conduct. Even 'First Amendment absolutists' like Justice Hugo Black saw nothing with the regulation of peaceful rallies if their *conduct* interfered with some other legitimate interest. This conduct could be widely interpreted.[2] For most of the

---

1 *Abrams* at 630.

2 See, for example, *Gregory v. City of Chicago* 394 US 111, 124 (1969) (Black, J. concurring). Black makes the capaciousness of regulatable conduct clear: 'I think that our Federal Constitution does not render the States powerless to regulate the conduct of demonstrators and picketers, conduct which is more than "speech", more than "press", more than "assembly", and more than "petition", as those terms are used in the First Amendment.'

first half of the twentieth century, conduct that could be prohibited included the mere act of picketing. Courts upheld numerous injunctions against picketing on the basis that the conduct it entailed was necessarily either violent or harassing (see Avery 1988/1989). Indeed, in one famous case in the 1920s, Chief Justice William Taft wrote of picketing that its very 'persistence, importunity, following and dogging', offended public morals and created a dangerous nuisance.[3] The problem with picketing, Taft thought, was twofold. First, through its combination of action and speech, it tried to convince people not to enter some establishment; second, it tended to draw a crowd.[4] To the degree it did both – that is, to the degree it successfully communicated its message – it interrupted business and, in Taft's eyes, undermined the business's property rights, and therefore could be legitimately enjoined.[5] Speech was worth protecting to the degree that it was *not* effective. Not until the 1940s did the Court begin to recognize that there might be an important speech right worth protecting in addition to the unprotected conduct.[6]

There is an additional result of Holmes's declaration about the value of speech in *Abrams*. Whereas the First Amendment is silent on *why* speech is to be protected from congressional interference, Holmes makes it clear that the protection of speech serves a particular purpose: *improving* the state.[7] Indeed, he quickly admits that speech likely to harm the state can be outlawed.[8] And neither he nor later Courts ever moved away from the 'clear and present danger' test. Speech, Holmes argued, is a good insofar as it helps promote and protect the 'truth' of the state.[9] There might be a lot of room allowed under this construction for criticism of the state, but it can still be quieted by anything that can reasonably be construed as a 'legitimate state interest' (like protecting the property rights of company subject to a strike). According to the Court's decision in the case of *Gitlow v. New York* (1925), any speech that 'engender[s] the foundations of organized government and threaten[s] its overthrow by unlawful means' can be banned.[10] Note here that speech does not have to *advocate* the overthrow of government; rather it can be banned if through its persuasiveness *others* might seek to overthrow the

---

3 *American Steel Foundries v Tri-City Central Trades Council et al.* 257 US 184, 204 (1921). This language has since, perhaps inadvertently, been transferred to, and absolutely central in, anti-panhandling laws as they have developed around the United States (Mitchell 1998, 2005).

4 *American Steel Foundries* at 205, 210.

5 *American Steel Foundries* at 204.

6 *Thornhill v. Alabama* 310 US 88 (1940).

7 Amar (1998) recognizes the silence in the text of the First Amendment but finds in the debates surrounding its framing that its authors were concerned that *the people* possess a right and the means not to *alter* or to *abolish* the state, a sense of the Amendment that is now, in part through the efforts of Holmes now completely lacking.

8 *Abrams* at 626, 630.

9 *Abrams* at 630.

10 *Gitlow* at 667.

government.[11] On such grounds all manner of manifestos, and many types of street speaking, may be banned. And more broadly, as evidenced in picketing cases like the one discussed above, a similar prohibition may be placed on speech that, again through its persuasiveness (e.g. as to the unjustness of some practice of event) rather than through direct exhortation, might incite people to violence.

Of course, speech (and its sister right, assembly) must take place *somewhere* and it must implicate some set of spatial relations, some regime of control over access to places to speak and places to listen.[12] Consequently, the limits to speech, or more accurately the means of limiting speech, become increasingly geographic beginning in 1939 in the case of *Hague v. CIO*, when the Supreme Court finally recognized that public *spaces*, like streets and parks were necessary not only to speech itself but to political organizing.[13] At issue in *Hague* was whether the rights to speech and assembly extend to the use of the streets and other public places for political purposes, and in what ways that use could be regulated. The Court based its decision in the language of common law, arguing that 'wherever the title of the streets and parks may rest, they have immemorially been held in trust for use of the public and, time out of mind, have been used for the purposes of assembly, communicating thoughts between citizens, and discussing public questions'.[14] But whatever the roots for such a claim in common law, it hardly stands scrutiny in the United States, where the violent repression of politics has been as much a feature of urban life as its promotion (Preston 1963). That makes *Hague* a landmark decision: it states clearly for the first time that 'the use of the streets and parks for the communication of views on national questions may be regulated in the interest of all ... [but] it may not, in the guise of regulation, be abridged or denied'.[15] At the same time, the Court made it clear that protected speech in public spaces was always to be 'exercised *in subordination to* the general comfort and convenience, and in consonance with peace and good order ...'.[16] The question, then, became one of finding ways to regulate speech (and associated conduct) such that order – and even 'general comfort' was always maintained.

---

11   In this sense there is a continuity in Holmes's jurisprudence with an earlier common law doctrine known as 'bad tendency' that held that speech that might tend towards bad behaviour or influence others in that direction could be quashed (Cole 1986).

12   This is no less true of electronic communication: there is a geography of inclusion and exclusion at both the speech and listening end, a geography over which some actors – whether public or private – will always have greater or lesser degrees of control. We too often forget this simple and indisputable fact and in doing so too often adopt theories of speech that assume it takes place in a vacuum, that it can possible occur *no place*. Such a stance is, literally, utopian. To allay such utopianism, Kurt Iveson (2007) argues that we need to pay attention to the conditions of possibility for *public address*.

13   *Hague v. Committee for Industrial Organization* 207 US 496 (1939).

14   *Hague* at 515.

15   *Hague* at 515.

16   *Hague* at 515, emphasis added.

The answers to that question were spatial. They were based on a regulation of urban *geography* in the name of both 'good order' and 'general comfort'. Speech rights needed to be balanced against other interests and desires. But order and comfort, it ought to go without saying, suggest a much lower threshold than does 'clear and present danger'. While recognizing in a new way a fundamental right to speech and assembly, that is, the *Hague* court in fact found a language to severely *limit* that right, and perhaps even to limit it more effectively than had heretofore been possible. To put this another way (and as I will argue more fully in a moment), the new spatial order of speech and assembly that the Court began constructing with this decision *allowed* for the full flowering of a truly liberal speech regime: a regime for which we are all, in fact, the poorer.

The rudiments of this regime are familiar enough. Starting in the 1950s, the Court began crafting what have been called the 'Roberts Rules of Order' for public space (Klaven 1965), but which are formally known as the public forum doctrine (for a review see Graber 1991). The development of this doctrine has entailed the development of a new metaphor for understanding speech. Where Holmes in the 1920s spoke of a 'free trade in ideas', by the 1960s, Justice William Brennan found himself concerned with a 'market*place* of ideas'.[17] Since all ideas need a place in which they can be expressed, the Court had to pay attention to the nature – the structure as well as the rules governing – that place. '[F]reedom of speech does not exist in the abstract', Brennan argued. 'On the contrary, the right to speak can only flourish if it is allowed to operate in an effective forum – whether it be a public park, a schoolroom, a town meeting hall, a soapbox, or a radio or television frequency.'[18] The issue Brennan is raising here is what constitutes an effective forum and how that forum can be 'regulated in the interests of all' while maintaining 'general comfort' and 'good order'. This is the issue the Court has been grappling with ever since *Hague*. Its solution has been to establish a set of rules, often the subject of litigation to this day – and of vigorous disagreement among the Supreme Court justices – that allow for the regulation of the time, the place, and the manner of speech, but not the direct regulation of the content of public and political speech. Public forum doctrine has evolved in hopes of assuring that order can be preserved while speech itself, at least formally, is protected.

The Court, therefore, has developed a typology of places. First are 'traditional public forums' – like many streets and parks – where political speech has 'traditionally' taken place. In these areas, any regulation of speech is subject to 'strict scrutiny' – that is, regulation must be shown to *protect* some vital state interest, or to *prevent* some clearly identifiable harm. Speech and assembly may be regulated as to time, place, and manner, just so long as that regulation remains 'content neutral'. Similarly, any permit system developed to assure that

---

17 *Lamont v. Postmaster General* 381 US 301 (196**) (Brennan, J. concurring), emphasis added.

18 *Columbia Broadcasting System v. Democratic National Committee* 412 US 1994 (1973) at 193 (Brennan J, dissenting).

appropriate time, place, and manner restrictions are followed must also be content neutral. The second type of places are 'dedicated public forums', which are those spaces specifically designated by a governing agency as available for First Amendment purposes. Examples include public university free speech areas – and perhaps classrooms – or the plazas in front of some government buildings. Use of these spaces for speech and assembly activities may be revoked for everyone (the dedication may be removed), but such a revocation cannot be done on either a case-by-case basis or on the basis of content. The third category is public property that is not a public forum: military bases, the insides of most government buildings, and, as we will see, airport grounds. Of course, there is also a fourth kind of place – or more accurately property. On private property, except in some very limited circumstances, speech rights simply do not pertain. Such property is of growing importance as more and more public activities occur on publicly accessible private property, like shopping malls, or redeveloped waterfronts whose titles have been ceded to the developers. The implications for speech and assembly of the growth of publicly accessible private property, as we will see below, are profound (see also Blomley 2004; Kohn 2004; Staeheli and Mitchell 2008).

But the point I want to make is that with the development of public forum doctrine, we can begin to see that the *silencing* of speech – and consequently of protest – is now affected most readily not by its outright ban, but rather through the regulation of the conduct that accompanies it and the place where it occurs. To make this point, I undertake three case studies. These studies, I think, are emblematic of the ways that, in the name of its promotion, speech is silenced in public space. They are not exhaustive, and as will be clear, they indicate that struggle – contention – over not only the right to speak, but also to be heard, continues unabated (see Mitchell 2005). It is this on-going struggle that has forced the Court into its promotion of a liberal theory of speech, and its concomitant development of public forum doctrine.

Changing geographies of protest

Case 1: The privatization of public space

The first case study concerns the privatization of public space. Note in the 1939 *Hague* decision quoted above, the phrase, 'wherever the title ... may rest'. As the geography of the public forum has shifted, that title – that is, the status of space as property – has taken on added significance. It is hardly news to point out that privately owned but publicly accessible spaces, like malls, shopping centres, and festival market places have become primary gathering places in North American cities (Garreau 1991; Goss 1993, 1996, 1999; Kohn 2004; Kowlinski 1985; Miller 2007; Sorkin 1992; Zukin 1991). Yet if public space is not only a space of politics, but also a space of sociable gathering (and, indeed, if each historically has been essential to the other: cf Calhoun 1992; Hartley 1992; Keller 2009; Iveson 2007;

Staeheli and Mitchell 2008). While the privatization of public space may not be news, it is nonetheless of incredible importance because these are the spaces where people gather. This is so, in part, because the Supreme Court has declared that the First Amendment simply does not extend to the space of the mall.[19] The property rights of owners trump the political rights of citizens. How the Court got to this determination is interesting and, interestingly, geographic.

Consider, in this regard, the case of Horton Plaza Shopping Center in Downtown San Diego. Horton Place Shopping Center is the centrepiece of San Diego's redevelopment (Staeheli and Mitchell 2008). The shopping centre – a festival marketplace-type mall – was built next to the traditional centre of downtown San Diego, Horton Plaza Park. The intent of the mall was to create a new centre for downtown, one perceived to be safe and free from the urban problems that were seen by many to plague Horton Plaza Park itself: homelessness, loitering, public drunkenness, and so forth. City planners and the Hahn Corporation – the developer of the mall – hoped that the mall would draw suburbanites and tourists into the urban core and help cement the gentrification of the nearby Gaslamp Quarter. In turn, this would lead to the removal or relocation of the soup kitchens and other social services, the X-rated bookstores, theatres, and cheap bars that served the sailor and transient populations that congregated downtown. It would also remove the boarding houses and single room occupancy hotels that housed the indigent (and largely elderly) people who were the main residents of the downtown core. The removal and relocation of these sorts of land uses, city planners hoped, would again make downtown inviting for inward investment, middle and upper class residents, and convention delegates.

When Horton Plaza Shopping Center opened in 1985, it had the effect of internalizing much of the life of the city, an effect that wasn't really overcome until after the city emerged from the deep recession of the early 1990s. In the meantime, Horton Plaza Park became a contested zone, with the mall and other nearby property owners continually working, often through the city's redevelopment agency, to find ways to remove the homeless, the poor, and others who would gather there. After all, the park served as the entrance to the mall for most of those who came by foot (which is how most workers in the neighbouring skyscraper district would arrive). By 1991, the city had put in place plans, largely related to landscaping, that simply made it impossible to hang out in the park. The benches were removed, and the lawns were replaced by prickly plants (Granberry 1991; Horstman 1990; Johnson 1990; Perry 1989). The park became a place to pass through, rather than to gather, to the satisfaction of the Horton Plaza Shopping Center management which was intent on making their mall the central gathering point downtown (Cooley 1985; Staeheli and Mitchell 2008: 52-53). According to the former president of the Gaslamp Quarter Association (and an original tenant of the mall), even with the successful gentrification of the neighbouring streets,

---

19   *Lloyd Corp v. Tanner* 407 US 551 (1972); see Friedelbaum (1999); Kohn (2004).

Horton Plaza Shopping Center remains one of – if not the – primary public spaces in the city (Staeheli and Mitchell 2008: 52).

In the meantime, however, the developer of the mall, the Hahn Corporation, was determined to clearly regulate what *kind* of gathering place (what kind of 'public' space) it was going to be. In (reluctant) deference to the *PruneYard* decision made in 1980,[20] Hahn had established a permit system that allowed groups to use the space of the mall for petitioning and leafleting. Permitted organizations were restricted to one of four small areas where they were given access to a 'community cart' (similar to those used by vendors) and two chairs. From these, shoppers could be leafleted or approached and engaged in conversation. Permits had to be requested 72 hours in advance, and not more than two people from the same group could occupy the designated leafleting space at the same time. Finally, no political activity of any type was permitted between Thanksgiving and New Year's Day.[21]

On March 12, 1986, William Phipps applied for a permit from Horton Plaza Shopping Center to perform on mall property a political skit dramatizing the effects of US-sponsored bombings in El Salvador. Phipps applied on behalf of an organization known as Playing for Real Theater and said that there would be eight people performing for about ten minutes, followed by leafleting. Despite the fact that the mall management had frequently encouraged street performers at the mall, and that it had played hosts to concerts and other large gatherings, the manager denied the request to perform a skit (citing congestion problems), but approved the request to hand out leaflets.[22]

Playing for Real and Phipps did not appeal the denial; nor did they take advantage of the opportunity to hand out leaflets. Neither did they ever attempt to circumvent the denial of permission and attempt to perform their skit anyway.[23] Nonetheless, on April 4, 1985, Horton Plaza Associates, the legal entity that owned the mall on behalf of the Hahn Corporation, went to court and sued Phipps, Playing for Real, and the San Diego chapter of the Committee in Solidarity with the People of El Salvador (CISPES), seeking injunctive relief against trespassing, interference with business, and annoyance. The mall sought a temporary restraining order against the defendants to prevent them from performing any skits at the mall. The rationale for seeking relief was that an unnamed police informant had told a detective that the group was planning a protest for April 5, to start outside the mall in Horton Plaza Park and then move inside the shopping centre.[24] The mall was granted a restraining order. Through it (in the words of the California Appellate Court decision on the case), Phipps and his associates were specifically enjoined from:

---

20   See *Horton Plaza Associates v. Playing for Real Theater* 228 Cal. Rptr. 817 (1986) at 820.

21   *Horton* at 819-820.

22   *Horton* at 820, 828.

23   *Horton* at 828.

24   *Horton* at 819, 828.

performing a great variety of expressive acts within the Center, including distributing pamphlets or other literature, soliciting signatures except as permitted by the Center, soliciting money, using furniture or other materials or displays without permission, approaching any patrons, causing a disturbance on the premises, hindering business, 'performing any dramatization or any acts or events', or 'in any other way impeding, disturbing, or interfering with the commercial activity of the Horton Plaza Shopping Center ...[25]

Two weeks later the mall won a preliminary injunction against the defendants, which, in essence, made the temporary restraining order permanent.

In support of the injunction, the mall management made its position clear as to why it was imperative to prohibit the kinds of politically expressive conduct Playing for Real engaged in. Karen Lesley Binder, the director of public relations for the mall, argued in an affidavit:

> In my opinion, Horton Plaza is particularly subject to being detrimentally affected by unregulated political activity occurring within the center. In this regard, Horton Plaza is, of course, located in downtown San Diego; in order for us to attract customers, we must overcome the public's perception that a downtown environment may not be their first choice for shopping activities. In my opinion, a visitor to this downtown shopping center is more sensitive to being subjected to political demonstrations and solicitations than at a competitive suburban shopping mall. In view of same, my objective at Horton Plaza is to create a relaxed, nonthreatening environment, and to have an atmosphere distinctive from the adjacent downtown areas. All promotional activity at the center is designed and regulated to not only entertain our customers, but to encourage shopping activity and specifically avoid any interference with customer traffic and customer shopping. I am particularly concerned that the unregulated activity such as that desired by Defendants would be contrary to those objectives, and substantially interfere with shopping activity at the center.[26]

In addition, since the mall had a programme of street performance in place, it was concerned that the public would take a skit performed by a political group to be part of the mall's own promotional activities and as such would assume that the mall endorsed the political views expressed.[27]

The Court granted the injunction and further held that Horton Plaza Shopping Center's time, place and manner restrictions were reasonable. In essence, the Court approved 'prior restraint' on the defendants' speech acts in the mall, finding in advance that they were unprotected. Both the injunction and this finding were upheld on appeal. Since the California Supreme Court ultimately declined to

---

25   *Horton* at 821, quoting the temporary restraining order.
26   *Horton* at 821-822.
27   *Horton* at 821.

review the Court of Appeal's decision, the Horton case stands, in California at least, as the final word on what sort of speech is possible in a California mall.

That speech looks like this: the appeals court found that while the state Supreme Court's decision in *PruneYard* protected leafleting and petitioning on publicly open private property, it did not protect 'expressive conduct' like performing a play.[28] The Court cited a whole series of cases to show that private property could not be considered anything like a traditional public forum. Moreover, according to the Court, other public forums existed largely because Horton Plaza was located in the city and not in the suburbs:

> Horton Plaza differs from the shopping centers involved in the [other cases where more expansive free speech rights have been upheld] because it is a downtown rather than a suburban shopping center. Decisions such as Pruneyard discuss how a suburban shopping center tends to become the only place in a large community where people can be found gathering together on foot; accordingly, such a center is indeed the most conveniently available forum for the dissemination of ideas and solicitation of signatures. Horton, however, lies in the heart of a downtown metropolitan business district where extensive pedestrian traffic occurs on the streets surrounding the center as well as within it. Under these circumstances, the Center owner cannot be said to have a monopoly of pedestrian traffic in the area such as would justify extensive interference with his right to manage his property.[29]

But the court's argument here chooses to ignore the fact that the Horton Plaza Shopping Center's whole goal, even to the point of paying for the removal of benches in Horton Plaza Park, was to move public life inside, to capture it really, for its own commercial interests.

Such *geographic* strategies on the part of private property owners to control the movements and gathering of people had little impact on the court. Through such strategies the area where political speech *can* be continues to shrink (Alexander 1999; Kohn 2004; Mitchell 2003). Protest is easily silenced when the significant gathering places in a city are on private property – unless of course we concede that the only valid form of political speech is that specifically approved of in *PruneYard*: petitioning and leafleting.

Yet this is not to say that protesters will fare any better on public property, as we will see in the next case study.

Case 2: Picketing at an airport

This second case study concerns a strike. Faced with a contract they were not happy with in late September 2000, about 85 workers on the automated baggage handling

---

28   *Horton* at 824.
29   *Horton* at 827-828.

system at Denver International Airport (DIA) voted to go on strike against Phelps Program Management, the subcontractor that operated the baggage machinery on the United concourse. The union representing the baggage handlers was the same as the one representing United's mechanics, customer service agents, and ramp workers, the International Association of Machinists (IAM). While a number of rank-and-file machinists and other non-baggage handlers said they might cross a picket line if one was established by the baggage handlers, the IAM in Washington said it might call on all its members at DIA to honour the picket line (Draper 2000h).

Meanwhile, United got the City of Denver to agree to restrict picketing to a rarely used, empty parking lot some three miles from the terminal and threatened any non-striking workers with disciplinary action if they failed to show up to work as scheduled (Griffin and Leib 2000a). The reason this location was chosen was bluntly stated by the airport's spokesman: 'We issued a permit for the union to picket in the Mt. Elbert parking lot. United's workers won't have to cross the picket line when they get to work' (Griffin and Leib 2000a). In other words, because picketing might have been effective at a more central location (see Draper 2000h), it was banished, an action that threw the unions at the airport into disarray.

Standard practice among flight attendants' and pilots' unions is to honour picket lines, but since they would not have to cross the picket line to get to their place of employment, it was doubtful that many would not show up for work. 'Our pilots will probably go to work if no one is picketing at the entrance of the terminal', noted a spokesman for the Air Line Pilots Association (Griffin and Leib 2000a). Put in even more of a bind were the brother and sister members of the IAM who worked in other jobs at the airport. The IAM recommended that non-striking workers go to work since they would not have to cross a picket line to get to it (Draper 2000g; Griffin and Leib 2000a).

When the Phelps workers walked off the job at 5 am on Tuesday, September 26, 2000, they acquiesced to the City's restriction and set up their picket line in the distant Mt. Elbert parking lot (Draper 2000i). And out at that parking lot, the silence was deafening, just as it was meant to be. Even though picketers marched and yelled at Phelps management who (as we will see later) were required to park in the lot, the *Denver Post* reported that 'a cold wind blowing across the plains south of Denver International Airport muffled their shouts, and only a handful of groggy reporters witnessed their protests'. 'Had the airport officials allowed Phelps' 85 union workers to picket at the terminal or employee parking lots', the *Post* continued, 'they might have brought United to its knees' (Griffin and Leib 2000b). The question of whether or not union brothers and sisters should honour the picket line was rendered moot by the removal of the pickets to the Mt. Elbert lot. As the vice president of the Denver council of the Association of Flight Attendants remarked, '[t]here is no physical picket line for us to cross, and so we have not crossed a lawful picket line' (Griffin and Leib 2000b).

The case law that made the banishment of the pickets, and hence assured their ineffectuality, is interesting. Labour law allows general contractors at construction sites to picket against a subcontractor to a designated location, like a secondary gate, so that the business of the general contractor can continue unimpeded (Griffin and Leib 2000a). This location becomes the point at which non-striking workers and managers from the struck company must 'enter the workplace', so that anyone going to work for the struck subcontractor must pass the picket line. No one else has to. Phelps argued that its position as a subcontractor to United allowed it to do the same thing. Phelps management (and its few non-striking, non-union employees), therefore, had to park in the struck parking lot, but they were the only ones. And since this was now the designated entry point for Phelps workers, picket lines established elsewhere at DIA, if they were effective, could be construed as encouraging an illegal secondary boycott by United employees, since these other lines would not be set in the way of Phelps workers, according to a Phelps attorney (cited in Griffin and Leib 2000a).

The IAM threatened to file a court injunction to get the picket location changed, but on the first day of the strike they did not do so, preferring instead to try to get the airport authority to designate a better site for picketing (Draper 2000b, 2000f; Griffin 2000). In the meantime the City of Denver issued a permit allowing striking Phelps workers to hand out informational leaflets, but not to picket, at employee parking lots. Only five days into the strike, a reporter for the *Rocky Mountain News* was able to report that 'while a handful of striking baggage workers maintain their picket lines nearly three miles from their place of employment, the public's interest in their strike wanes and Denver International Airport's operations continue unscathed' (Draper 2000c). The enforced silence of the picketers was having its desired effect, which was, according to the Mayor of Denver, 'to keep the airport operating' (Draper 2000c).

In hopes of reviving the public's waning interest, airport workers and the Colorado AFL-CIO staged a solidarity protest at the picket site on the seventh day of the strike (Colden 2000a; Draper 2000d). The next day the IAM finally filed suit against the city in federal court. The suit was filed on First Amendment grounds, and explicitly made the case that picketing was a form of speech that any passer-by was free to ignore: it was not a form of compulsion and hence the city could not claim the denial of a speech right on the grounds of protecting the operation of the airport. This was a difficult argument because both the city and the union agreed that the airport was a 'nonpublic forum' and that the city had the right to restrict speech on airport property (Abbott 2000; Colden 2000b; see Berg 1993; Schutte 1993).

After a hearing, the federal judge found for the city and ordered picketers to remain where they were. The judge's argument is important: 'If this case is viewed through the lens of the First Amendment, I see no injury here'. The strikers 'were given a reasonable time, manner, and place for pickets'. Moreover, they had been provided with an opportunity to distribute informational leaflets elsewhere at the airport. Finally, the judge ruled that the probable effect of allowing pickets at the airport would be to 'create chaos at DIA' by encouraging a secondary boycott. Indeed, the judge argued that strong circumstantial evidence suggested that encouraging an airport-wide strike was the union's goal in the first place and that the First Amendment argument it was making was a smokescreen (Draper 2000e; McPhee 2000).

The argument, then, is that it is both reasonable and good to move strikers to a place where they could not be effective, where their speech would have no possibility of being heard, if that means protecting the financial and other interests of the city. Or to put that another way, people can speak and protest and picket all they want, just so long as that speaking, protesting, and picketing has no chance of being effective. The silencing at work here, though, is not only the silence of banishment through time, place, and manner regulations, and a permit system that benefits the collective employers (Phelps, United, DIA, other airlines, etc.) over the collective workers (IAM and other unions). It is also a silencing accomplished by making the transmission of *intended* messages impossible. This is made clear in one of the last articles published on the strike. A reporter for the *Rocky Mountain News* noted that '[a]lthough the strike hasn't affected airport operations, it garnered a lot of attention because of its potential to shutdown airport operations, if thousands of other airport employees ... honored the baggage workers' picket line' (Draper 2000a).[30] If the point of a strike is to gain attention, which in part it is, then the strikers were effective. But if the point of a strike, and the picketing that goes with it, is to apply pressure, and to win demands, then they were not very effective. They were not effective because public forum doctrine, together with loaded labour law, allows for the banishment of strikers when it fits the desires of employers. If workers had been able to picket at their normal place of work – where employees entered the terminal, or in the normally used employees' parking lots – then their message would have had meaning. The content of their speech – that there is a strike in progress and it is the obligation of all those who stand for the rights of workers to honour it – would have had a chance of being heard, of making a difference. First Amendment law, in this regard, does not so much promote dissident speech as effectively silence it. It does so by assuring that the appropriate *place* for such speech is only a place that renders speech meaningless. Geography creates *de facto* content restrictions.

But airports, like malls, might in some sense be considered exceptional cases. The airport, after all, is a nonpublic forum, and the mall is private property. What of those traditional public forums that still remain: the streets and parks of the city?

Case 3: Zoning protest

The third case therefore returns to the paradigm of public forums, the streets, where protesters find the city as a whole more and more segregated into a series of protest and no protest zones – that is, they find a city looking more and more like the

---

30   The strike was settled on October 23, 2000. Part of the settlement included the union agreeing to drop an appeal of the federal judge's decision restricting picketing to the Mt. Elbert lot (Griffin and Kirskey 2000).

forlorn parking lots of Denver International Airport. In the streets, public forum doctrine encourages police and other city, state, and national officials to construct designated 'protest zones' and 'no protest zones' outside major international meetings, events like political conventions, and, perhaps most uncomfortably for progressives, outside abortion clinics (Mitchell 2005; Staeheli and Mitchell 2008: Chapter 1). Recent commentators have suggested that the zoning of protest is 'a post-9/11 device' of social control (Keller 2009: 4). A basic knowledge of the origins of the 1964 Berkeley Free Speech Movement (FSM) (cf. Mitchell 1992) gives life to this suggestion, as does the analysis that follows.

'Free speech zones' were created on a number of public college campuses in the wake of the FSM and other campus uprisings of the 1960s and 1970s. By 1984, if not earlier, free speech and *no*-free speech zones were standard parts of the landscape at major political conventions, such as the Democratic National Convention in San Francisco that year. But perhaps the most intriguing place to examine this dynamic of protest zoning is in Seattle at the height of the demonstrations against the World Trade Organization in November and December, 1999. The creation of a no-protest zone during these demonstrations was less an intentional policy of the city, than an outcome of the city's failure to prevent protest in other ways. As with many cities, Seattle has long had a permit process for parades and protests. During the WTO meeting, 'official' protests had been permitted for a number of areas around town. Labour unions, for example, rallied near the old King Dome; other protesters gathered at the Space Needle. But, on November 30, 1999, some 20,000 un-permitted protesters gathered downtown, many of whom were deployed in a well-organized plan to block key intersections and, it was hoped, to disrupt the ability of the WTO delegates to get to the meeting site. With a small number of protesters smashing windows and engaging in other acts of violence against property, and with the labour unionists beginning their march to join the protesters downtown, Mayor Paul Schell declared a state of emergency, stopped the labour parade before it reached downtown, and asked the governor to mobilize the National Guard to clear the streets. The next day, Mayor Schell issued an order that barred any person from 'enter[ing] or remain[ing] in a public place' within a 50 block zone downtown. Exceptions were granted for WTO delegates, business owners and employees, emergency personnel, and, interestingly, shoppers (Perrine 2001).[31] Protest was quite simply banned from downtown. While order was restored to the streets quickly after the order was issued, it was not rescinded until after the meeting was over.

The city staked the legitimacy of its emergency order on the precedent established in the case *Madsen v. Women's Health Center*.[32] The *Madsen* case,

---

31   Herbert (2007) provides a compelling analysis of the multiple state motives in its actions in implementing and defending the creation of the no-protest zone; here I focus on the more narrow question of what such zoning *does*: how it works to quiet dissent, whatever the motives may be.

32   *Madsen v. Women's Health Center* 512 US 753 (1994).

and one that followed it (*Hill v. Colorado*[33]), established the legitimacy of creating 'bubbles' around abortion clinics within which protest, picketing, leafleting, and 'sidewalk counseling' was either forbidden or altogether or severely restricted. The goal, of course, was to protect the rights and safety of women entering the clinics, as well as those of the clinic's employees.

Following *Madsen*, the City of Seattle argued not only that creating a fifty block no-protest zone around the WTO was legitimate as an emergency measure, but also that it was legitimate on its face: it was simply a reasonable time, place, and manner restriction on speech and assembly that was tailored to advance a compelling state interest: protecting the delegates to the WTO from the disruptive protests around them (Perrine 2001: 640). The nature of this compelling state interest is unclear. While women have a constitutionally protected, if increasingly fragile, right to abortion that the *Madsen* decision seeks to balance against the rights of protesters, it is impossible to guess what right WTO ministers have to traverse the streets and use the public buildings of Seattle without having to encounter even peaceful protesters.[34] And so, in fact, in its arguments in support of the continuance of the order, the City argued instead that the order was a critical means not only for restoring order, but for maintaining it.

The National Guard and other police forces, armed with the emergency order and their batons, successfully cleared the streets and kept all protesters out of downtown on December 1. Here is how one news article reported the scene:

> A crackdown that put National Guard troops, state troopers, and police officers in head-to-toe black on every corner of downtown yesterday all but ended impromptu protests against the World Trade Organization while allowing the group's meetings and permitted processions to go on as scheduled. Standing shoulder to shoulder or marching in unison, Seattle police officers wearing gas masks and carrying batons charged the demonstrators and pushed them out of what was dubbed the 'no protest zone' near the convention center where President Clinton addressed the WTO delegates (Gorov 1999, A1).

The interesting word in that passage, of course, is 'impromptu'. Seattle officials averred over and over that they had no interest in stopping parades and protests that had been sanctioned through its permit process. The no protest zone was a response to those – some violent, the vast majority not – who sought to protest without the city's permission: who sought, that is, to exercise their right to assembly without clearing it with the government first.

In the end, a federal judge upheld the city's position, seeing no illegitimate abridgement of protesters' rights in the city's establishment of a no protest zone.

33　*Hill v. Colorado* 120 S. Ct 2480 (2000), for an analysis, see Mitchell (2005), for a fuller examination of its relevance to Seattle, see Perrine (2001).

34　Recall that the emergency order remained in effect after order was restored and lasted the length of the meetings.

The judge stated, plainly enough, that 'free speech must bend to public safety' (quoted in Ith 2001: A1). In this case it had to bend for 50 blocks, and right out of downtown – even though in *Madsen*, the court had found a 36-foot exclusion zone to be reasonable, but both a 300-foot zone in which approaching patrons and workers of clinics, and a 300-foot no protest zone around residences of clinic workers to be *too great* a burden on free speech, ordering a much smaller no-protest bubble to be drawn.[35] Given this sort of specificity in the Supreme Court's *Madsen* decision, it seems unlikely that such a large protest exclusion zone as Seattle's could withstand scrutiny.

But there is another issue at work too. The judge in Seattle supported the city's contention that *sanctioned* protest was acceptable. The no protest zone was necessary because of *impromptu* protests. But, of course, the very effectiveness of the protests was their (apparent) spontaneity.[36] This is what caught the media's – and the public's – imagination; and that is what allowed for the massive upsurge of political debate, in the US and around the world, that followed.

Perhaps, tactically, Seattle's 'mistake' was to not establish designated protest and no-protest zones in advance of the meetings. Such a move had been effective in the 1996 Democratic and Republican Conventions (and earlier ones too). And in subsequent years and events it has become standard practice, as with the 2000 National Conventions, the annual meetings of the World Bank and the IMF in Washington, and the World Economic Forum in New York in February 2002, where protesters have been kept out of certain areas by fences, barricades, and a heavy police presence (D'Arcus 2006; Staeheli and Mitchell 2008). In the case of the 2000 Democratic National Convention in Los Angeles, it was the protesters who were fenced off, with the city establishing an official 'protest zone' in a fenced parking lot a considerable distance from the convention site (Rabin 2000a, 2000b; Rabin and Shuster 2000). The rationale was, of course, 'security', a rationale backed by appeals to the authority of the Secret Service. The ACLU, among others, sued the city, eventually winning a decision that invalidated the city's plans. The city was forced to establish a protest zone closer to the convention centre, with the judge chiding the City of Los Angeles for failing to consider the First Amendment when it established the rules for protest and security around the event. 'You can't shut down the 1st Amendment about what might happen', the judge said. 'You can always theorize some awful scenario' (quoted in Rabin and Shuster 2000: 1).

This victory should not be considered very large. Its effect, and the effect of other cases like it, has largely reduced the ACLU and other advocates of speech rights to arguing the fine points of geography, poring over maps to determine just *where* protest might occur (cf. Staeheli and Mitchell 2008: 10). Protesters are put entirely on the defensive, always seeking to justify why their voices should be heard

---

35   *Madsen* at 773-774.

36   Of course, much planning went into unsanctioned protest; its effectiveness, however, was predicated precisely by catching the city, the WTO delegates, and the world by surprise.

and their actions seen, always having to make a claim that it is not unreasonable to assert that protest should be allowed in a place where those being protested against can actually hear or see it, and always having to 'bend' their tactics – and their rights – to fit a legal regime that in every case sees protest as subordinate to 'the general order' (which, of course, really means the 'established order').

Conclusion: the liberalization of free speech

The *Oxford English Dictionary* indicates that at the time the Bill of Rights was written, to 'abridge' meant both to 'curtail' and to 'cut short; to reduce to a small size'. The dual meaning of the term probably did not escape the authors of the Bill. They could have written 'curtail' if that was all they meant. But they meant 'abridge'. As the Supreme Court finally came to recognize this fact in the mid-twentieth century, and sought to limit the ability of federal, state, and local governments to ignore the rights embodied in the First Amendment, it found ways to liberalize free speech. Even so, it has nonetheless allowed for the continued abridgement of political speech by other means. It is hard to argue that the rights of picketers at Denver International Airport were not very much 'reduced to a small size'. And as the importance of publicly accessible private property to public life in the United States has grown, the room to effectively exercise rights of speech and assembly has concomitantly shrunk; it too has been 'reduced to a small size'.

Such a result was implicit in the very language Justice Holmes used when he launched the project of liberalizing free speech. Since he was concerned with the effects of what people said – how mere talk (or later, other forms of expressive activity) could constitute a 'clear and present danger' – he had to focus not so much on speech itself but on context, both social and geographical. 'We admit that in many places and in ordinary times', he wrote, 'the defendants ... would have been within their constitutional rights. But the character of the act depends on the circumstances in which it is done ...'.[37] And he meant it. This is the real foundation of public forum doctrine, and so therefore of the liberalization of free speech. Or, to say the same thing, it is the precise legal justification for silencing speech – to make it so it cannot have any effect – through geography.

The growing liberalization of First Amendment law, a liberalization that has suggested that even the most seditious speech might be protected (just so long as it was ineffectual [cf. Klarman 1996]), has required regulation to develop in new ways. Instead of focusing on exactly what is said (as was the case in the World War I cases), court rulings and police practices since the 1930s have begun exploring the ways that the meaning of what is said – its effectiveness – is a result of where it is said – its geography. And with this, courts have sought, under the rubric of balancing competing needs and rights to determine a set of rules for regulating space in order to regulate speech. Simultaneously, the privatization of

---

37 *Schenck* at 52.

public space, the movement of sociability into the private property of the mall, as demonstrated with the case of Horton Plaza, shows that the right to speech and assembly can be undermined through something as mundane as changing property regimes (Staeheli and Mitchell 2008).

To the degree that the state has learned – or been forced – to restrain itself from regulating just what is said and thought, some significant (if still very tenuous) victories have been won. But these victories have been steadily eroded through a new legal regime in which courts, rather than Congress, have taken the lead in abridging the rights to speech and assembly by assiduously segregating speech and protest, by zoning it, so that its very effectiveness can be minimized and perhaps eliminated. It is clear that in the US, the right to speak does not imply a right to be even potentially effective, to have a chance to make a difference.[38] Dissident speakers have to remain outside the mall that has become the new public space of the city (Kohn 2004); they must remain at a distance from the politicians and the delegates they seek to influence; they must picket only where they will have no chance of creating a meaningful picket line. But no matter what the courts say and no matter how carefully police and courts together draw lines of protest, creating a geography of rights that can be frankly oppressive, Seattle and later Quebec, and perhaps especially events in Genoa and the massive marches and rallies in opposition to the war against Iraq, have shown that there are ways to transcend geography, to speak out loudly and incessantly against those who would effectively silence protest in public space. These ways are, more and more, necessarily illegal.[39]

References

Abbott, K. (2000) 'Baggage Workers Sue City: Strikers Fight Decision to Sideline Pickets', *Rocky Mountain News*, 4 October, 4A.

Alexander, M. (1999) 'Attention Shoppers: The First Amendment in the Modern Shopping Mall', *Arizona Law Review*, 41: 1-47.

Amar, A. (1998) *The Bill of Rights: Creation and Reconstruction*. New Haven: Yale University Press.

Avery, D. (1988/1989) 'Images of Violence in Labor Jurisprudence: The Regulation of Picketing and Boycotts, 1894-1921', *Buffalo Law Review*, 1: 1-117.

Berg, B. (1993) 'Diminishing the Freedom to Speak on Public Property: International Society for Krishna Consciousness, Inc. v. Lee', *Creighton Law Review*, 26: 1265-1300.

---

38 Assuming the speakers do not have the wherewithal to control the means of media production.

39 The costs of breaking the laws that create this geography of silence, however, are steeply on the rise. The USA PATRIOT Act, its revisions, and its state-level correlates, have made it now possible to charge protesters with terrorism for what formerly might have drawn a charge of disorderly conduct. See Chang (2002); Mathiesen (2002).

Blomley, N. (2004) *Unsettling the City: Urban Land and the Politics of Property.* New York: Routledge.

Calhoun, C. (ed.) (1992) *Habermas and the Public Sphere.* Cambridge, MA: MIT Press.

Chang, N. (2002) *Silencing Political Dissent: How Post-September 11 Anti-Terrorism Measures Threaten Our Civil Liberties.* New York: Seven Stories.

Colden, A (2000a) 'Unions Rally at DIA as Strikers Picket in the "Wilderness": Distant Site Annoys Baggage Handlers', *Denver Post*, 3 October, C1.

—— (2000b), 'Pickets Request More Visible Spot: Judge Hears DIA Baggage Handlers' Plea Today', *Denver Post*, 4 October, C1.

Cole, D. (1986) 'Agon at Agora: Creative Misreadings in the First Amendment Tradition', *Yale Law Journal*, 95: 857-905.

Cooley, K. (1985) 'Bums Out, Shoppers In: Downtown's Horton Plaza Park Gets a Shiny New Face', *Los Angeles Times* (San Diego Edition), 6 September, 2.1.

D'Arcus, B. (2006) *Boundaries of Dissent: Protest and State Power in the Media Age.* New York: Routledge.

Draper (2000a) 'Baggage Handlers Open Talks: Negotiations Resume with Union on Strike for 2 Weeks', *Rocky Mountain News*, 11 October 4B.

—— (2000b) 'Bag Handlers Threaten Court Action: Union Wants Right to Picket at Terminal', *Rocky Mountain News*, 30 September, 3B.

—— (2000c) 'Bag Workers Strike Raises Legal Issues', *Rocky Mountain News*, 30 September, 3B.

—— (2000d) 'City Scolded Over Pickets: Pilots, Labor Leaders Hold "Solidarity" Rally to Back Bag Workers', *Rocky Mountain News*, 3 October, 1B.

—— (2000e) 'DIA Picket Line Will Stay Put: Moving Baggage Workers' Strike Would Likely "Create Chaos" at Airport, Judge Rules', *Rocky Mountain News*, 5 October.

—— (2000f) 'Strike at DIA Stays Put', *Rocky Mountain News*, 28 September, 2B.

—— (2000g) 'United Baggage Handlers to Strike', *Rocky Mountain News*, 26 September, 4A.

—— (2000h) 'United Dreads Sympathy Strike: If 100 Baggage Workers Walk, 4,500 May Honor Pickets', *Rocky Mountain News*, 23 September, 1B.

—— (2000i) 'United Says Strike Isn't Affecting DIA', *Rocky Mountain News*, 27 September, 1B.

Friedelbaum, S. (1999) 'Private Property, Public Property: Shopping Centers and Expressive Freedom in the States', *Albany Law Review*, 62: 1229-1263.

Garreau, J. (1991) *Edge City: Life on the New Frontier.* New York: Double Day.

Gorov, L. (1999) 'A Crackdown Calms Seattle: Action Taken to Prevent Confrontation', *Boston Globe*, 2 December, A1.

Goss, J. (1993) 'The "Magic of the Mall": An Analysis of Form, Function, and Meaning in the Contemporary Built Environment', *Annals of the Association of American Geographers*, 83: 18-47.

—— (1996) 'Disquiet on the Waterfront: Reflections on Nostalgia and Utopia in the Urban Archetypes of Festival Marketplaces', *Urban Geography*, 17: 221-247.

—— (1999) 'Once-Upon-a-Time in the Commodity World: An Unofficial Guide to the Mall of America', *Annals of the Association of American Geographers*, 89: 45-75.

Graber, M. (1991) *Transforming Free Speech: The Ambiguous Legacy of Civil Libertarianism*. Berkeley: University of California Press.

Granberry, M. (1991) 'Face-Lift Crew Deposes Park Vagrants', *Los Angeles Times* (San Diego Edition), 21 June, B1.

Griffin, G. (2000) 'Strikers Remain Far Away: DIA Workers Haven't Filed for Closer Area', *Denver Post*, 27 September, C1.

Griffin, G. and Kirskey, J. (2000), 'Baggage Contract Approved "Overwhelmingly": OK Ends Strike at DIA', *Denver Post*, 24 October C1.

Griffin, G. and Leib, J. (2000a) 'DIA Baggage Union to Strike: Walkout Begins at 5 a.m. Today', *Denver Post*, 26 September, A1.

—— (2000b) 'Strikers Exiled at DIA: Airport's Decision Weakens Pickets', *Denver Post*, 27 September, A1.

Hartley, J (1992) *The Politics of Pictures: The Creation of the Public in the Age of Popular Media*. London: Routledge.

Herbert, S (2007) 'The "Battle of Seattle" Revisited: Or, Seven Views of a Protest Zoning State', *Political Geography*, 26: 601-619.

Horstman, B. (1990) 'City OKs Removal of Grass, Benches at Horton Plaza', *Los Angeles Times* (San Diego Edition), 9 October, B3.

Ith, I. (2001) 'Court Vindicates City's WTO Riot Measures, Judge Says Schell Acted Appropriately', *Seattle Times*, 31 October, A1.

Iveson, K (2007) *Publics and the City*. Oxford: Blackwell.

Johnson, G. (1990) 'City Yanks Benches from Under Undesirables Downtown: Council Adopts Plan to Replace Grass, Remove Benches in an Effort to Rout Criminals and Drunks at Horton Plaza', *Los Angeles Times* (San Diego Edition), 3 July, B2.

Kaufman, S. (1991) 'Note and Comment: The Speech/Conduct Distinction and First Amendment Protection of Begging in Subways', *Georgetown Law Journal*, 79: 1803-1830.

Keller, L. (2009) *Triumph of Order: Democracy and Public Space in New York and London*. New York: Columbia University Press.

Klarman, M. (1996) 'Rethinking the Civil Rights and Civil Liberties Revolutions', *Virginia Law Review*, 82: 1-67

Klaven, H. (1965) 'The Concept of the Public Forum: *Cox v. Louisiana*', *1965 Supreme Court Review* 1.

Kohn, M. (2004) *Brave New Neighborhoods: The Privatization of Public Space*. New York: Taylor and Francis.

Kowlinski, W. (1985) *The Malling of America: An Inside Look at the Great Consumer Paradise*. New York: William Morrow.

McChesney, R. (1999) *Rich Media Poor Democracy: Communication Politics in Dubious Times*. Urbana: University of Illinois Press.

McPhee, M. (2000) 'No Pickets at DIA Terminal, Judge Tells Baggage Workers', *Denver Post*, 5 October.

Mathiesen, T. (2002) 'Expanding the Concept of Terrorism?' in P. Scranton (ed.) *Beyond September 11: An Anthology of Dissent*. London: Pluto Press, 84-93.

Miller, K. (2007) *Designs on the Public: The Private Lives of New York's Public Spaces*. Minneapolis: University of Minnesota Press.

Mitchell, D. (1992) 'Iconography and Locational Conflict from the Underside: Free Speech, People's Park, and the Politics of Homelessness in Berkeley California', *Political Geography*, 11: 152-169.

—— (1996) 'Political Violence, Order, and the Legal Construction of Public Space: Power and the Public Forum Doctrine', *Urban Geography*, 17: 158-178.

—— (1998) 'Anti-Homeless Laws and Public Space I: Begging and the First Amendment', *Urban Geography*, 19: 6-11.

—— (2003) *The Right to the City: Social Justice and the Fight for Public Space*. New York: Guilford.

—— (2005) 'The S.U.V. Model of Citizenship: Floating Bubbles, Buffer Zones, and the Rise of the "Purely Atomic" Individual,' *Political Geography*, 24: 77-100.

Perrine, A. (2001) 'Notes and Comments: The First Amendment vs. the World Trade Organization: Emergency Powers and the Battle in Seattle', *Washington Law Review*, 76: 635-688.

Perry, A (1989) 'San Diego at Large: Shop 'til You Drop, But Don't Expect a Bench', *Los Angeles Times* (San Diego Edition), 13 January, 2.1.

Preston, W. (1963) *Aliens and Dissenters: Federal Suppression of Radicals, 1903-1933*. New York: Harper and Row.

Rabin, J. (2000a) 'Officials Scramble to Draft New Security Plan', *Los Angeles Times*, 21 July, 1.

—— (2000b) 'Panel OKs 3 Protest Marches', *Los Angeles Times*, 19 July, 1.

Rabin, J. and Shuster, B. (2000) 'Judge Voids Convention Security Zone', *Los Angeles Times*, 20 July, 1.

Schutte, S. (1993) 'International Society for Krishna Consciousness, Inc, v. Lee: The Public Forum Doctrine Falls to a Government Intent Standard', *Golden Gate University Law Review*, 23: 563-598.

Sorkin, M. (ed.) (1992) *Variations on a Theme Park: The New American City and the End of Public Space*. New York: Hill and Wang.

Staeheli, L. and Mitchell, D. (2008) *The People's Property?: Power, Politics and the Public*. New York: Routledge.

Zukin, S. (1991) *Landscapes of Power: From Detroit to Disneyland*. Berkeley: University of California Press.

Chapter 3

Struggling to Belong: Social Movements and the Fight to Feel at Home

Jan Willem Duyvendak and Loes Verplanke

Introduction

In this chapter, we pay attention to social movements explicitly aiming at new 'homes'. A first example is the movement fighting against 'total institutions' for people with psychiatric and intellectual disabilities. The people involved in this movement favour *community* care: patients should come to live in so-called 'normal' neighbourhoods and find their place, their home, among 'normal' people instead of being tucked away in huge institutions 'in the woods'. Societal discrimination – resulting at the micro-level in people being outcasts, living isolated lives – should be countered by caring communities.

A parallel story can be told about gays and lesbians: since their family homes were often quite hellish for them – with family members rejecting their sexual preferences and often even pushing them out of their homes – they often had to look for new places that could become their home. In their search to create new homes, they often ended up in the big cities, where their anonymity helped to maintain invisibility if necessary. At the same time, metropolitan conditions favoured the flocking together of the like-minded, providing space and place for very public gay communities. In the development of gay neighbourhoods like the Castro in San Francisco, self-organization of gays and lesbians has played a decisive role.

Social movements made claims for both groups of people who previously experienced home as some kind of 'hell': political and social discrimination either forced their invisibility at home (gays had to remain in the closet) or made them invisible for society by locking them out and up (in separate institutions). In both cases, the groups involved wanted to move out of these nightmare scenarios to find new places which they could really and literally call home. In this chapter, we will look into the fight of these two social movements for new practices of home making.

Notwithstanding these similarities, we chose to research these two movements for their differences. As we will see, the role these social movements played in home making was not identical at all and, consequently, the outcomes in terms of 'home' diverge enormously as well. The new home making practices at the community level were far less successful for people with handicaps than for gays. In practice, the former moved from their big institutions ('hell') to small, independent housing ('havens' at best) – which is quite an improvement in itself – but with no integration in the community, no public home. What they got were new 'havenly' places that felt like 'home'; secure, safe, comfortable, and very private, even somewhat isolated places – not exactly the type of new 'home' that the caring community movement was originally aspiring for. The fact that the movement in favour of community care was not so much a movement of people with disabilities themselves but of their – often self-proclaimed – spokespersons, that it demobilized at an early stage and that the actual implementation of reform was directed by professionals within a bureaucratic state, might well explain this outcome. The home making practices of gays, on the other hand, were intimately linked to social movements and collective action, resulting in public places for themselves, characterized by all elements of what we could call a 'heaven', providing space for identity and visibility. In this sense, the gay movement was critical in creating a home that was directly linked to the construction of a socially, ideologically, and politically cohesive community.

Conceptualizing feeling at home: hell, haven and heaven

Before we present the empirical data that corroborate these claims, we first have to better understand the possible meaning of home in relation to collective action. Though the study of 'home' seems to address micro-phenomena, it should incorporate wider structural forces that influence feelings of home. One cannot separate questions of how people inscribe space with meaning from social struggles involving class, race, gender and sexuality. Contrary to many psychological and culturalist studies, our analysis of home therefore explicitly focuses on power relations.

But what do we mean by 'home'? In contemporary social theory, images abound of exile, diaspora, time-space compression, migrancy and 'nomadology'. However, the concept of home – the obverse of all this hyper-mobility – often remains un-interrogated', according to certain scholars. Though it is surely correct to say that 'home' is often under-theorized, over the past decades, many books, special issues and articles have appeared on 'home', 'feeling at home' and 'belonging' that provide excellent overviews of the research to date (Blunt and Dowling 2006: 67; Bozkurt 2009; Després 1991; Gieryn 2000; hooks 2009; Mack 1993; Masscy and Jess 2003; Moore 2000; Porteous and Smith 2001; Rybczynski 1986; Saunders 1989; Saunders and Williams 1988; Somerville 1997).

Nevertheless, not all social scientists, let alone members of the public, make use of 'home' in a very reflective way. One problem with home is its very familiarity; people speak in terms of 'belonging' and 'feeling at home' all the time. For sociological understanding, this familiarity is both an advantage and a disadvantage. On the one hand, everybody can participate in the debate on 'home'; on the other, many already claim to know what 'home' is and how it feels. Curiosity becomes rare. Paradoxically, this familiarity does not necessarily produce articulate ideas about what 'feeling at home' is, related to a peculiar aspect of 'home' and 'feeling at home': while everyone initially agrees that we know what it is to feel at home, the moment we have to describe what it means to us, we begin to stutter. Feeling at home, then, is one of those rare emotions that elude words. People may reveal, when urged to do so, that they feel 'at ease' when they feel at home, that they feel 'safe', 'secure' and 'comfortable', at 'one with their surroundings'. If one feels at home, one is at peace – a rather passive state where things are self-evident because they are so familiar.

In other words, feeling at home is not only a familiar sentiment to us all; familiarity itself is one of its key defining aspects. Environmental psychologists in particular, who have carried out much of this research, stress the importance of 'familiarity' in their definition of home. From their *phenomenological* point of view, home is perceived as a safe and familiar space, be it a haven or shelter, where people can relax, retreat and care. Following the Indo-European notion of *kei*, meaning 'something precious' – from which the German word for home (*Heim*) is derived (Hollander 1991; cited in Mallet 2004: 65) – attachment to a home place is seen as a primordial sentiment (Fried 2000) created by familiar daily routines and regular settings for activities and interactions.

Sociologist Pierre Bourdieu acknowledges the importance of familiarity in 'feeling at home' as well. Whereas the unfamiliar is 'out of place', home is the place 'to be' – a place so familiar that it feels almost like a 'natural' place. Bourdieu writes: 'The agent engaged in practice knows the world ... He knows it, in a sense, too well ... takes it for granted, precisely because he is caught up in it, bound up with it; he inhabits it like a garment or a familiar habitat' (Bourdieu 1999: 142-143). For Bourdieu, however, this 'naturalness' of feeling at home is not natural at all: it is culturally created. Bourdieu wants to understand why people experience places as natural – as 'home' – and criticizes scholars who fail to reflect on this 'naturalizing' effect of the familiar. Indeed, many environmental psychologists employ natural metaphors, in particular botanical ones: one is home where one is 'rooted'. 'Such commonplace ideas of soils, roots, and territory are built into everyday language and often also into scholarly work, but their very obviousness makes them elusive as objects of study' (Malkki 1992: 26). In a wonderful article, Malkki points out that many 'spatial' metaphors carry this air of naturalness as well: 'Metaphors of kinship (motherland, fatherland, *Vaterland, patria*, ...) and of home (homeland, *Heimat*, ...) are also territorializing in this same sense; for these metaphors are thought to denote something to which one is naturally tied' (Malkki 1992: 27-28).

'Familiarity' is a necessary but often insufficient condition for feeling at home. Other factors that may play a role resemble those aspects Rybczynski lists in *Home: A Short History of an Idea* (1986) that make a house feel like home: intimacy and privacy, domesticity, commodity and delight, ease, light and air, efficiency, style and substance, austerity, and comfort and well-being. If we leave out the material, house-bound elements, we have a list that resembles those of many other authors as well. In their intriguing *Domicide: The Global Destruction of Home (2001)*, Porteous and Smith discuss the classifications of scholars who have been struggling with the many possible meanings of home. Summarizing their findings as well as the meanings we came across in hundreds of articles and books, we come to the following basic classification of the 'elements of home':

1. *Familiarity:* 'Knowing the place'.
2. *Haven: secure, safe, comfortable, private and exclusive:* Physical/material safety; mentally safe/predictable; Place for retreat, relaxation, intimacy and domesticity.
3. *Heaven: public identity and exclusivity:* A public place where one can be, express and realize oneself; where one feels free and independent. Home here embodies shared histories; a material and/or symbolic place with my own people and activities.

This, to be sure, is a rudimentary typology (in a similar vein, Setha Low (2004) distinguishes between a 'fort' and a 'castle'), but it suffices in our quest to understand the struggle of the two social movements we discuss in the chapter. 'Haven' covers aspects of home that pertain to feelings of safety, security and privacy, which most often relate to the micro-level of the house. Those aspects of home that come under the heading of 'heaven' (Porteous and Smith 2001: 44) are more outward-oriented and/or symbolic: they help individuals to 'be', develop and express themselves, and to connect with others, often through the creation of intentional communities and as a result of collective action. In that sense, as we will show, the 'quality' of collective action determines the type of 'home' that is experienced. But as we will see, whether experienced as haven or heaven, feeling at home is a highly selective emotion: we don't feel at home everywhere, or with everybody. Sentiments of feeling at home, as produced by and within collectivities such as social movements, seem to entail including some and excluding many.

From hell to haven: home making by people with psychiatric or intellectual disabilities

In many Western countries, movements have mobilized in the past 25 years for policies of deinstitutionalization for psychiatric patients and people with learning disabilities. These people should no longer be banished to institutions in the countryside; it would be better for them to once again be a part of society, to live in ordinary neighbourhoods in towns and villages. While there would be additional support for these individuals, the idea was that they would live in their own houses (instead of institutions) as independently and autonomously as possible. Since the late 1990s, this policy has broadly been referred to as *community care* (Means and Smith, 1998).

Prior to the 1970s, psychiatric patients and people with learning disabilities were viewed as patients in need of continuous nursing and tucked away in countryside institutions. At the time, the therapeutic ideal prescribed that the best place to care for these people was in large institutions far from their former daily environment. Patients could be cared for and supervised 24 hours a day; they would find peace and quiet, ample space and a well-regulated life. In the 1970s, patient organizations as well as professionals and academics began to criticize this 'medical regime', asserting that remote institutions only served to isolate people from 'normal' communities. These institutions were not only deemed discriminatory; they failed to make people less ill or disturbed. *Asylums: Essays on the Social Situation of Mental Patients and other Inmates* (1961), the iconic work by the American sociologist Erving Goffman, was a source of inspiration for the critics of institutionalization. Goffman compared psychiatric hospitals to other 'total institutions' such as prisons, barracks, convents and even concentration camps. Their 'total' nature was embodied in barriers such as locked doors, high walls, electric fences, water and woodland that precluded contact with the outside world. The worst feature of the asylum was that the inmate's 'self is systematically, if often unintentionally, mortified' (Goffman 1961: 15). Goffman and other influential critics, including the psychiatrists Laing and Szasz, stated that it was not so much institutionalized inmates who were ill or mad, as society itself.

In the 1970s, social movements postulated a new ideal that not only tolerated deviant behaviour, but even stated it was a healthy reaction to a sick society (Duyvendak 1999). It was therefore also in the interests of society that psychiatric patients or people with learning disabilities were part of it.

The era of deinstitutionalization

The reaction of policy-makers to this criticism by social movements and social scientists was surprisingly responsive: they introduced a policy of deinstitutionalization, offering extramural support and treatment for patients who needed long-term care but who no longer lived in residential institutions (Kwekkeboom 2004). Several western countries (the USA, the UK, Italy and the Scandinavian countries) closed down many psychiatric hospitals and institutions for people with learning disabilities, replacing them with small facilities in ordinary communities providing local extramural care. Norway and Sweden introduced legislation that entitled anyone with any kind of disability to live in a house in an ordinary neighbourhood; in fact patients had no choice as these countries no longer maintained residential institutions. In the Netherlands, policy-makers interpreted the criticism of institutions mainly as one of scale and type of housing: the size and impersonal nature of the institutions became a thing of the past and 'small' became the maxim of the 1980s and 1990s. Small-scale sheltered living units were established, first in the grounds of institutions, and later, beyond the institutions' confines in residential neighbourhoods in towns and villages (Means and Smith 1998; Overkamp 2000; Welshman 2006).

Having your own home

We interviewed about 70 people with psychiatric or learning disabilities living on their own in 'normal' neighbourhoods in the Netherlands. Most of the interviewed psychiatric patients had spent considerable periods of their lives in psychiatric hospitals. Of the respondents with learning disabilities, half had previously lived in institutions run by professionals; the others had lived with their parents. Most received a house in the town where they had grown up. Respondents had no explicit expectations about how it would be to live in their own place, nor any definite expectations about the atmosphere of their new neighbourhoods, e.g. whether they would feel welcome or if their neighbours would help them settle in.

What do we know about the 'landing' of these groups? First of all, respondents unanimously appreciated having their own houses where they could do what they wanted. They mentioned advantages such as not being constantly disturbed by others, being in control of what and when they eat, their bedtimes, pets in the house, having more autonomy, etc.: 'Finally I am in control over the remote control'. No one wanted to return to their former living situation.

> Once you are free in your own house, that's really terrific. It's just positive. Even when the weather is bad, it still seems as if the sun is shining. That's my feeling here (man with learning disabilities, 30).

> I decided that it was enough with all those non-stop intakes in hospital. I really wanted to have a life in a place of my own. And here I am now: I am really calmer now that I am not continuously in and out of the institution and don't have to live in a group anymore. I have the tendency to adjust myself always to other people around me and I'm happy now that it's not necessary anymore (woman, 45, psychiatric patient).

> For many years I lived in institutions with a lot of people constantly around me. But it is no good for me to be with so many people all the time, because my head becomes too busy then. Maybe I get mad one day. That's why I have asked for a home of my own. And finally that worked out fine, because now I live here on my own and I like that very much (man with learning disabilities, 33).

Other research (e.g. Kwekkeboom 2006; Kwekkeboom and Weert 2008; Overkamp 2000) has also concluded that most individuals with psychiatric problems or learning disabilities prefer to have their own accommodation, due to the privacy and autonomy this allows. In this respect, the quality of their lives has beyond any doubt substantially improved thanks to de-institutionalization policies.

At home in the community?

In general, the interviewees have very little, if any, contact with neighbours or other locals in their new neighbourhoods. Most did not introduce themselves to their neighbours after they moved in; nor did supporting professionals suggest they do so. Contact with neighbours was usually limited to saying hello, and, at best, to brief chats on the street. There was very little contact, such as occasionally drinking a cup of coffee together or helping each other with small tasks. Some interviewees mentioned unpleasant experiences with neighbours.

Interviewees' indoor visitors are mainly relatives and personal caretakers, who are particularly crucial for people with few family contacts. Respondents looked forward to their daily or weekly visits when they could talk about what was going on in their lives and what was bothering them. In these cases the caretaker was often called 'the most important person in my life'. We asked all respondents where and to what extent they felt at home, and whether they felt a sense of belonging to their new neighbourhoods. Many immediately began to point around them, indicating they felt at home within their own houses. An important reason for this strong feeling of homeliness in one's house has to do with the fact that most rediscovered a place for themselves, free of disturbances, after having lived in groups for many years in different types of institutions.

As for the neighbourhood, most interviewees did not mention definite feelings of attachment. The neighbourhood for most of them has no meaning whatsoever. They do not know their neighbours and do not participate in the life of the neighbourhood. Only in cases where they were born and raised in this (part of the) city do respondents mention an attachment to their environment that resembles a sense of belonging.

Clearly, the people that we have interviewed have a '*haven*' – conception of home: they mainly associate this feeling with safety, security, comfort, domesticity and intimacy. Whereas the social movements and, later, policy-makers tend to privilege the '*heaven*' interpretation – the community as a warm bath, where everybody can publicly be her- or himself – many psychiatric patients or people with intellectual disabilities mostly experience a feeling of belonging when they feel safe and secure, when they are with people like themselves, and when they are in familiar surroundings. It is this last aspect they have difficulty achieving, as they do not manage to establish meaningful contacts with neighbours and other locals.

Where was the movement?

The majority of the psychiatric patients and people with intellectual disabilities we interviewed tend to live as solitary individuals in their communities (or on little islands in the case of clustered accommodation). They are happy with their autonomy. They feel at home in their houses. Where their houses are located has limited relevance for them because they tend not to have contact with other locals.

The outside world penetrates their houses almost exclusively via television, for here they can control the remote control – the outside world at a distance.

In retrospect it is rather surprising that in the planning of deinstitutionalization so little attention was given to the social context these people would end up living in. In the 1970s, the idealistic critics of total institutions perhaps naively assumed that society as a whole would benefit from the arrival in local communities of psychiatric patients and people with learning disabilities. Moreover, though the home making process was triggered by social movements, these movements disappeared in the late 1980s, early 1990s, and therefore the actual implementation of reform was directed by professionals within a rather bureaucratic context. Whereas these movements – often composed of spokespersons for people with disabilities – paid attention to the social needs of the latter, this contextual aspect got lost afterwards. People with disabilities were not really able to collectively stand up for their wishes, let alone to proudly demand special treatment (as the gay movement was, see next paragraph). While movements originally pushed for reform, professionals had the power over the realization of the reform, within a financially constrained context (one of the reasons for governments' responsive reaction to social movements' demands of deinstitutionalization was the idea that extramural housing would be cheaper…).

Moreover, whereas the social movements' idea of autonomy and self-realization was explicitly a social and contextual one – it was about living autonomously in the community – these ideas became part of neoliberal framing in the 1980s: living an independent life in the community became an indisputable principle because this ideal for people with psychiatric and learning problems fitted into the dominant ideal applicable to all citizens: living as independently and autonomously as possible. As a result, a rather individualized vision of home was used to guide strategies of how to create residential units for patients. In this instance, the emphasis of professionals was not on creating connections to the community but on helping individual patients to live their life as autonomously as possible. The strong presence of professionals and the absence of social movements meant that the construction of home reflected an individualized vision of home.

From hell to heaven: moving to the Castro, a gay neighbourhood in the making

In homophobic countries many adolescents who develop a same-sex preference want to leave their 'hometowns', in order to escape from fights with their families who refuse to accept their sexual orientation. In particular, when they develop same-sex relationships young gay people quite often are literally pushed away from the house and community of origin and drawn to urban areas (Rosenfeld and Byung-Soo 2005: 559). As they leave their domiciles of origin, the places of their past where they were born and bred, and look for a new home in their domiciles of destination, gay people experience a strong sense of up-rootedness. In the United

States, where negative opinions regarding homosexuality were predominant until the late 1990s, gay men and lesbians born in the mid-West often fled to cities considered to be safe havens, such as New York and San Francisco. Their move to these cities clearly had a non-voluntary character. Homosexuals felt forced to leave the places where they grew up in order to feel free, 'find' their true selves and meet others with the same sexual preference. 'Coming out' stories are not coincidentally often framed in terms of 'coming home'. Travelling to a new place finally provided the opportunity to act out their sexual selves.

Meanings of and attachments to gay neighbourhoods are the result of complex relations between social marginality and geographical (im)mobility. Below, after reconstructing how these places came into being, we analyse the shifting meanings gay men ascribe to the habitual space of the Castro, the extent to which they consider it to be a home place and how they *collectively* construct and visually display a public sense of home to the outside world.

Mobility and marginality: where do they come from, where do they go?

The process of gay men moving from rural communities and smaller cities and towns to inner-city neighbourhoods is linked to push and pull factors. During the past decades gay men considered only two options for living a homosexual life. They could sublimate their sexual and erotic identities while staying in their hometown or they could opt for leaving and enmeshing themselves in the anonymity of urban life.

But where to go? Recently historians reconstructed how San Francisco developed into a gay capital, 'a symbolic homeland of an identity and a city that was a haven for institutional support unknown in most American cities' (Meeker 2006: 189). During the 1950s and 1960s especially, San Francisco's reputation as a comfortable place for nonconformist sexual life extended to the image of a gay capital (Boyd 1997: 88-89). Obviously, the media played an important role in the creation of the city as a Mecca for gay men and lesbians, as they understated the notion of San Francisco as the centre of the gay world and thereby perpetuating gay life as a geographically fixed phenomenon (Meeker 2006: 190-191). Likewise, books played a similar role. Apart from assisting gays and lesbians in imagining communal identity, writers also, by describing gay places, 'pointed people toward what might imaginatively be thought of as a gay homeland' (Meeker 2006: 67).

Although we cannot provide here a detailed history of the Castro (but see Armstrong 2002; Leyland 2002; Stryker and Buskirk 1996), for our ambitions in this chapter it suffices to underline that as during the 1960s the Castro turned into a gay community, many gay men literally articulated a pronounced sense of neighbourhood pride and attachment. The neighbourhood was central to many activities, such as the yearly Gay Parade in June, the Castro Fair, Halloween and other public events that showcased how gays (and some lesbians) love to live in the Castro. Furthermore, gay men's identification with the Castro as a home place was reinforced by the naming of distinctive restaurants and cafes such as *Home* and *Welcome home*. Many gay men indeed considered the Castro as an ideal place to live – a heaven – as is expressed in the following excerpt.

> In San Francisco, we're the world, as much as anybody is. And you can carry that with you, ... to ... anywhere. There's a reference point, an actual place you can go and see. You get up in the morning and go out and live in it. Stores, papers, billboards, people on the street, everywhere you fucking look. ... Bite down: It's there. ... Hey: It's home (Tate 1991: 276).

Many gay men experienced moving and settling into the Castro as an escape from *hell*. They felt they could stop dreaming about a better place, since it was right in front of them. Though they often felt they had no other option than moving to the Castro, many gay men did not experience the Castro as a ghettoized place but rather as a place to be finally free, to expose one's true self and openly display their gay and lesbian sexual preferences. This public visibility of homosexual life marked the quintessence of the Castro as a non-conformist place to live and work. Since hiding one's sexuality was what almost all gay men had done before coming to the Castro, to 'come out' and to be able to expose themselves in public – both to each other and to the homophobic outside world – became the corner stone of a newly acquired identity. 'Being oneself' was not something to be lived individually or to be confined to the private and intimate sphere of the house: not a 'haven' but a 'heaven'. Gay men were able, most of them for the first time in their lives, to make home amidst a public of likeminded people (Elwood 2000; Johnston and Valentine 1995: 100).

The Castro

In a sense, the Castro became a new home because it embodied the exact opposite of the hometowns people had escaped from. The Castro was consciously constructed as opposite to the repressive and unwelcoming outside world. By proudly affirming their identity, gays challenged the norms and family values of the 'other' America. Whatever reminded them of mainstream America was considered as non-Castro-like. The social mobilization of gay men in the Castro thus centred around a severe wish to feel, be and act differently and to be respected for that. Gay men considered the Castro to be a *special* place, to be defended against invasions of the *generic*. For example, in October 1999 an initiative called 'Save the Castro' mobilized against the arrival of a Starbucks chain shop and the extension of the cable car F-line to the neighbourhood. Protests against these developments were cast in a distinctive line of argumentation, such as in the following statement headed 'A kind of War'.

> There is a war going on in our neighbourhood. It is a war for your dollars and businesses. Every time you shop at one of the new Chains stores, you take away business from local merchant owned shops and restaurants. *Starbucks is*

symbolic of the kind of mega-corporation that buys up other chains and squeezes out competition. *They are the McDonalds of coffee and are a part of anyplace USA.* The Castro is unique in the world (Wiggin 1991).

The consequences of the cable car extension are depicted in the same way:

> It is possible that millions of people will be coming to this district very shortly, on these 'historic cars', designed we think to lure tourists onto them. ... *The problem is that this is the only neighbourhood like it in the world.* The Castro has a kind of 'mythic regard' overseas, and we are the 'guardians' of this place for future generations (Wiggin 1991, emphasis in original).

This statement makes clear that the outside world has to remain exactly that: outside. The writer pronounces a strong sense of fear of losing a particular, exclusive home environment at the very moment outsiders 'invade' the neighbourhood. Interestingly though, the author appeals to an *ethos of solidarity* of gay men in San Francisco, not by emphasizing the local and rooted character of the Castro but by putting their struggle in a universal perspective. Only in case the Castro remains theirs, gay men in the rest of the world are able to consider the Castro as their home. Ron Wiggin, organizer of the 'Save the Castro' project, warns in a letter responding to his critics that this homely sentiment runs the risk of disappearing:

> And if the Castro fades under the throngs of tourists from the not needed historic trolley car ..., and the rents go up more and the condemning attitudes of the out-of-towners coming walking down our streets, and the stores change to sell to them, and we don't want to go there anymore because we don't feel comfortable (Wiggin 1991).

The Castro had to remain special, not just for the inhabitants but for gays all over the world who looked at the Castro as their 'own' place (Chabot and Duyvendak 2002). The Castro example shows how home-making practices at a very *particular* place facilitate the *world-wide* diffusion of an imaginary gay home (Adam et al. 1999). The Castro becomes a symbolic place, a symbolic home to gays and lesbians all over the world. As sub-cultural codes start to travel around the world, the local sense of familiarity becomes detached from its geographical vicinity. Processes of cultural diffusion accommodate feeling at home with others far away instead of with those randomly available next door.

The political context and the presence of the movement

Local policy makers in San Francisco did not do much to hinder the development of a gay neighbourhood, not only out of respect for the housing market (in which the government is not expected to intervene) but also because the idea of a place-bound community fits in the self-understanding of the San Francisco government as 'A city of many neighbourhoods' (Godfroy 1988; Pamuk 2004).

Although the city government for a long time did not protect homosexuals against harassment and persecution (Beemyn 1997: 87), since the 1960s, gays and lesbians in San Francisco have become one among many 'ethnic minorities' (Gamson 1995) and thereby fit into the ideology of the city as a mosaic of geographically separated social worlds. During election periods territoriality has played an important role, as city council members who are elected by districts and neighbourhoods favour those communities that live relatively concentrated. As during the 1970s the Castro neighbourhood also gained its first homosexual representative, the famous late Harvey Milk, gays and lesbians were able to fully participate in regular political organization in order to further pursue a process of social integration and emancipation. The gay community seized an existing opening in the political system – voting per district – to promote its interests and to protect the fragile position of its members.

The gay movement has been successful in creating a place that was accommodating to their 'own' people, a place characterized by amenities, contact points, and networking opportunities that made group socialization a central aspect of this new home. The centrality of the social movement helped create a particular identity for those in the place, reinforcing the willingness of individuals to defend their collective home against all real and perceived threats. In this sense, collective action was critical in creating a cohesive community. Such high levels of connectedness have in turn contributed to giving home-dwellers much more power – when compared to the individualized psychiatric patients and people with mental disabilities – in defining their 'extended' homes and reproducing their 'public' homes over time.

Conclusion: contested communities

In both cases, social movement activists had high hopes of the role the community could play in the 'coming out' of their respective constituencies: people with a handicap had to come out of their institutions and to live 'normal' lives among 'normal' people in welcoming communities; gays and lesbians had to come out and live according to their sexual preference – as everybody else does – at places of their own choice. While these movements addressed different issues, they were both nevertheless struggles to make new homes for marginalized people.

The outcomes of these movements were quite different. Though people with mental and intellectual disabilities indeed left their 'total institutions' (and sometimes at an alarming rate!), for many the place of arrival was not a warm, welcoming community. Depending on national and local policies, some ended on the street – this particularly occurred in many cities in the US – whereas elsewhere people got their own places but did not integrate into the community or society at large (this was often the case in welfare states like the Netherlands). In all instances it shows that 'normal' communities are not necessarily very receptive to newcomers in the neighbourhood. Since social movements campaigning for community care seriously weakened during the 1990s, there was no pressure anymore to open up communities to be more welcoming to various groups of people with handicaps. Moreover, professionals implementing de-institutionalization policies framed the former movement ideals of autonomy more and more in neoliberal terms of self-reliance. In spatial terms this implied that the ambition to construct public 'homes' for 'deviant' people got lost, or better; it was transformed in an individualized conception of private homes. The lack of social movement pressure to fundamentally change the public space in order to facilitate newcomers to 'mix' with longer-term residents, caused this narrowing down of ideals from community care to individual independent living. Since people with psychiatric problems and those with mental disabilities are not exactly easy 'mixers' themselves, the absence of a welcoming context implied their social isolation and their experience of home as a safe 'haven' at best – but certainly far from a 'heaven'.

The Castro in San Francisco shows, on the other hand, that ongoing collective action can materialize in a public 'home', a 'heaven', in this case for gay men. Gays (and some lesbians) were successful in establishing their 'own' home in the 1970s and 1980s. Moreover, as our analysis has shown, real and perceived threats were countered by permanent mobilization: the success of the Castro – its transformation from a rather run-down neighbourhood into a very fashionable and popular one – attracted goods and people (Starbucks, the cable car and heterosexuals) that were perceived as 'alien' to their 'home', as undermining their home feelings. In the perception of those involved, collective action was the only guarantee to preserve a home for those who actually live in the Castro, but also for gays and lesbians worldwide for whom this particular San Francisco neighbourhood has become a universal symbol for the right to have your own public place; the right to belong to a 'heavenly' home.

References

Adam, B. D., Duyvendak, J. W., and Krouwel, A. (eds) (1999) *The Global Emergence of Gay and Lesbian Politics: National Imprints of a Worldwide Movement*. Philadelphia: Temple University Press.

Armstrong, E. A. (2002) *Forging Gay Identities: Organizing Sexuality in San Francisco, 1950-1994*. Chicago and London: The University of Chicago Press.

Beemyn, B. (1997) *Creating a Place for Ourselves: Lesbian, Gay, and Bisexual Community Histories*. New York and London: Routledge.

Blunt, A., and Dowling, R. (2006) *Home*. London: Routledge.

Bourdieu, P. (1999) *Pascalian Meditations*. Cambridge: Polity.

Boyd, N. A. (1997) 'Homos Invade S.F.! San Francisco's History as a Wide-Open Town', in B. Beemyn (ed.) *Creating a Place for Ourselves*. New York and London: Routledge.

Bozkurt, E. (2009) *Conceptualising 'Home': The Question of Belonging Among Turkish Families in Germany*. Frankfurt and New York: Campus Verlag.

Chabot, S., and J. W. Duyvendak (2002) 'Globalization and Transnational Diffusion Between Social Movements: Essentialist Diffusionism and Beyond', *Theory and Society*, 31 (6): 697-740.

Després, C. (1991) 'The Meaning of Home: Literature Review and Directions for Future Research and Theoretical Development', *The Journal of Architectural and Planning Research*, 8 (2): 96-115.

Duyvendak, J. W. (1999) *De planning van ontplooiing*. 's Gravenhage: Sdu.

Elwood, S. A. (2000) 'Lesbian Living Spaces: Multiple Meanings of Home', *Journal of Lesbian Studies*, 4 (10): 11-27.

Fried, M. (2000) 'Continuities and Discontinuities of Place', *Journal of Environmental Psychology*, 20: 193-205.

Gamson, J. (1995) 'Must Identity Movements Self-destruct? A Queer Dilemma', *Social Problems*, 42 (3): 390-407.

Gieryn, T. (2000) 'A Space for Place in Sociology', *Annual Reviews Sociology*, 26: 463-496.

Godfroy, B. J. (1988) *Neighborhoods in Transition: The Making of San Francisco's Ethnic and Nonconformist Communities*. Berkeley: University of California Press.

Goffman, E. (1961) *Asylums. Essays on the social situation of mental patients and other inmates*. Harmondsworth: Penguin.

Hollander, J. (1991) *Yearning: Race, Gender, and Cultural Politics*. London: Turnaround.

hooks, b. (2009) *Belonging: A Culture of Place*. New York and London: Routledge.

Johnston, L., and G. Valentine (1995) 'Wherever I Lay My Girlfriend, That's my Home: The Performance and Surveillance of Lesbian Identities in Domestic Environments', in D. Bell and G. Valentine (eds) *Mapping Desire: Geographies of Sexualities*. London: Routledge, 99-113.

Kwekkeboom, M. H. (2004) 'De waarde van vermaatschappelijking', *Maandblad Geestelijke volksgezondheid*, 59 (6): 500-510.

—— (ed.) (2006) *Een eigen huis... Ervaringen van mensen met verstandelijke beperkingen of psychiatrische problemen met zelfstandig wonen en deelname aan de samenleving*. Den Haag: SCP.

Kwekkeboom, M. H., and C. Weert (2008) *Meedoen en gelukkig zijn. Een verkennend onderzoek naar de participatie van mensen met een verstandelijke beperking of chronische psychiatrische problemen*. Den Haag: SCP.

Leyland, W. (ed.) (2002) *Out in the Castro: Desire, Promise, Activism*. San Francisco: Leyland Publications.

Low, S. (2004) *Behind the Gates: Life, Security, and the Pursuit of Happiness in Fortress America*. New York: Routledge.

Mack, A. (ed.) (1993) *Home: A Place in the World*. New York: New York University Press.

Malkki, L. (1992) 'National Geographic: The Rooting of Peoples and the Territorialization of National Identity among Scholars and Refugees', *Cultural Anthropology*, 7 (1): 24-44.

Mallet, S. (2004) 'Understanding Home: a Critical Review of the Literature', *The Sociological Review*, 52(1): 62-89.

Massey, D., and P. Jess (eds) (2003) *A Place in the World?* Oxford: Oxford University Press.

Means, R., and R. Smith (1998) *Community Care. Policy and Practice*. Basingstoke and London: Macmillan Press.

Meeker, M. (2006) *Contacts Desired: Gay and Lesbian Communications and Community, 1940s-1970s*. Chicago and London: The University of Chicago Press.

Moore, J. (2000) 'Placing Home in Context', *Journal of Environmental Psychology*, 230: 207-217.

Overkamp, E. (2000). *Instellingen nemen de wijk*. Assen: Van Gorcum and Comp.

Pamuk, A. (2004) 'Geography of Immigrant Clusters in Global Cities: A Case Study of San Francisco, 2000', *International Journal of Urban and Regional Affairs*, 28(2): 287-307.

Porteous, J. D., and S. E. Smith (2001) *Domicide: The Global Destruction of Home*. Montreal and Kingston: McGill-Queen's University Press.

Rosenfeld, M. J. and K. Byung-Soo (2005) 'The Independence of Young Adults and the Rise of Interracial and Same-Sex Unions', *American Sociological Review*, 70: 541-562.

Rybczynski, W. (1986) *Home: A Short History of an Idea*. New York: Penguin Books.

Saunders, P. (1989) 'The Meaning of "home" in Contemporary English Culture', *Housing Studies*, 4 (3): 177-192.

Saunders, P., and P. Williams (1988) 'The Constitution of the Home: Towards a Research Agenda', *Housing Studies*, 3 (2): 81-93.

Somerville, P. (1997) 'The Social Construction of Home', *Journal of Architectural and Planning Research*, 14 (3): 226-245.

Stryker, S., and J. v. Buskirk (1996) *Gay by the Bay: A History of Queer Culture in the San Francisco Bay Area*. San Francisco: Chronicle Books.

Tate, L. (1991) 'San Francisco, California', in J. Preston (ed.) *Hometowns*. New York: Penguin, 273-276.

Welshman, J. (2006) *Community Care in Perspective. Care, Control and Citizenship*. Basingstoke: Palgrave Macmillan.

Wiggin, R. (1991) *Save the Castro Collection*. San Francisco: Can be found at: GLBT Historical Society, 657 Mission St., Suite 300, San Francisco, CA 94105.

Chapter 4

Place Frames: Analysing Practice and Production of Place in Contentious Politics

Deborah G. Martin

Introduction

Some time ago I argued that social movements generally, and territorially defined movements and organizations in particular, articulate their goals, agenda, and shared identities in terms of place frames (Martin 2003a). This chapter revisits the concept of 'place framing', suggesting it is a useful heuristic for analysing contentious politics by focusing on perspectives within a movement. I do so not to suggest that place is the most important geography or spatiality that activists encounter or express, but to suggest that it offers an approach for accessing the multiple geographical processes that inform and motivate activisms among a group of people.

Considerable scholarly attention by geographers and others sensitive to spatial thinking continue to refine our understandings of the interconnections between geography and social movements. But this scholarship, by and large, seeks to spatially analyse contention, rather than to understand the geographies – or geographical awareness – *within* contention (Martin and Miller 2003; Leitner, Sheppard, and Sziarto 2008; Nicholls 2009). Following Moore (2008), I make a distinction between analysis and practice: While much of geographical theorizing has focused on socio-spatialities of contention as analytical categories, this chapter suggests that socio-spatial dynamics should also be considered as part of the practice of contentious politics. Revisiting place frames offers a means to focus on practice and discourses of activists, emphasizing the self-understandings of those engaged in contentious politics. Such an approach attends to the meanings produced in and through activism, highlighting the saliency of geography for those engaging in activism as well as those examining it.

In the sections that follow, I briefly identify key directions in scholarship on the convergences of geographical analysis and the study of contentious politics, focusing on discourses of spatiality. I use the term contentious politics, drawing on McAdam, Tarrow, and Tilly (2001), but accepting the broader, less state-centric definition of Leitner et al. (2008: 157), which emphasizes contentious politics as 'social and political action … to challenge dominant systems of authority'. I reintroduce the social movement concept of framing in order to consider how frame analyses offer a means to access multiple spatialities of contentious politics as they are debated, deployed, and contested by specific, situated actors and through powerful semiotic tropes (such as the imaginary of 'neighbourhood' (Purcell 2001; Martin 2003b). I advocate focusing on place frames, which can include a variety of socio-spatial relations, from multifaceted notions of place to scale, networked relations, and socio-spatial positionality (the latter term is from Leitner et al. (2008), denoting identity difference and inequality within and across activisms). Drawing on previous research (Martin 2003a, 2003b, 2004), I examine the spatialities that place-frame analysis can access, pointing especially to the interplay of place identity with political structure and the flexible socio-spatial positionalities and scales that result. I conclude by arguing that attention to framing offers an important avenue to analysis of activists' own multivalent conceptualizations of their socio-spatialities.

Space, place, contention: an overview of the literature

Spatialities and contention

A plethora of work by geographers in recent years challenges traditional geographical delineations of concepts such as space, place, and scale, by emphasizing the relationality of each concept, questioning conventional distinctions among them (Amin 2004; Massey 2004, 2005). In particular, geographers have emphasized the multivalent, overlapping, mutually constitutive and relational aspects of space/place/scale, and have more broadly pointed to the importance of a variety of spatialities in shaping political economies, contention, and social relations (Leitner et al. 2008; Jessop, Brenner and Jones 2008; Nicholls 2009). A major theme of this literature is to question the *a priori* privileging of any particular geographical concept as foundational to social relations (or some subset thereof, such as politics). Much of this work seeks ways to analytically categorize activisms according to specific geographical concepts –such as place, territory, or scale. This ordering of spatialities in activism is analytically powerful, and enhances our understandings of geographical concepts, and their explanatory power in social movement analyses. But it does not necessarily do a very good job of accessing the kinds of spatial thinking that actors engage in as part of contentious politics. Such analyses are of course immensely important to geographical thinking but not necessarily for unpacking normative geographies expressed in and through activism, by activists – or those engaging with (against) them. Spatial thinking may obviously be at play in some contention, such as conflicts over neighbourhood change or land use, as I will illustrate here. Knowing and recognizing that geographies constitute meaningful conflicts, however, does not explain how spatial positions are taken, and to what effect. Place frame analysis offers a means to do so.

Leitner et al. (2008) provide a specific, concrete example of multivalent and intersecting socio-spatialities of contentious politics through their discussion of the Immigrant Workers' Freedom Ride. In describing the Freedom Ride from its origination among a Los Angeles union local, the symbolic sites visited along the trip to Washington, DC, and the connections and shared identities forged on the trip, they convincingly demonstrate the multiple spatialities of the movement. Specifically, *places* offered sites of common racialized and citizenship struggles; *scale* was evident in movement strategies of visiting national symbols, orienting workers' rights claims to local and national officials, and in the economic and political structures producing immigrant labour markets; *mobility* was a central strategy of the ride, and a condition of immigrant access to work; *networks* were forged throughout the ride among workers, fostering new socio-spatial identities (or *positionalities*) and linkages for further activism.

The case study from Leitner et al. powerfully demonstrates the multiple spatialities of activism, but it does not clearly identify how (or whether) movement members thought explicitly about these socio-spatial dimensions of activism, and how that might have shaped their grievances and strategies. One vignette in their description of the ride does highlight a quite explicit geographical strategy: The Freedom Riders had prepared for potential border patrol checks, and when one was initiated on a bus in El Paso, Texas, the riders showed only their Freedom Ride name tags, eschewing legal papers (Leitner et al. 2008: 167-168). In doing so, the Ride explicitly challenged the saliency of territory to divide workers from a shared struggle.

This example of not carrying citizenship documents is telling because it demonstrates a deliberate challenge of territorial authority and effort to reconstitute place identities away from citizenship. It reveals the ways that activists in the Immigrant Workers' Freedom Ride actively constituted themselves in relation to territory and socio-spatial positionality. At the same time, however, in reading the account we do not know how the riders planned for this territorial challenge, how they conceptualized it strategically, and how or whether their considerations of border checks affected their positionalities as citizens or their ideas (individual and collective) about nation. Place-frame analysis offers analytical and methodological strategies to highlight these and other conceptualizations of activists in relation to the geographies that produce them in particular – in this example, politico-legal – ways. Analysing discourses of Freedom Riders in relation to their preparation for and representation of identity/document checks – within the group and to outsiders such as news media – would make clearer activists' own conceptualizations of their geographical identities. Interviews with activists would likewise provide yet another insight on these themes.

A nuanced, multidimensional approach to geographical analyses of contentious politics allows scholars to better conceptualize and understand socio-spatial relations, as argued by both Leitner et al. (2008) and Jessop et al. (2008). Seeking to provide a conceptual schema for investigating 'polymorph[ic] … sociospatial relations' Jessop et al. (2008: 392) call for attention to 'territory (T), place (P), scale (S), and (N) networks, [or …] the TPSN framework'. They argue that this schema does not privilege any single spatiality, and that their framework encourages systematic integration of multiple dimensions of socio-spatial relations into analyses of social relations and events which are always socio-spatialized, in multiple forms.

There is no question that the framework of Jessop et al. (2008) or similar frameworks (such as that of Leitner et al. 2008) are useful for nuanced theorizing of socio-spatial relations. However, this focus on socio-spatial theorizing primarily offers a means for assessment *of* contentious politics, rather than analyses of socio-spatial relations produced *by and in* contentious politics. That is, if we shift the lens *from the spatialities the analyst sees*, we can refocus our investigations *to how activists respond to, shape, and alter the socio-spatial relations* that produce the grievances, the identities, and the strategies (and outcomes) of contentious politics. My plea for such a shift is related to Adam Moore's (2008) critique of the (scholarly) conflation of scale as analytic category with scale as practice. Moore (2008: 213) suggests that ontological attention to geographical concepts such as scale (and whether it exists, as per Marston et al. (2005) misses the focus of inquiry on 'everyday scalar practices and their material effects ... reorientat[ing] research toward the consequences of [geographical] processes'. Rather than focusing our investigations of contentious politics primarily according to analytic categories of socio-spatial relations – place, scale, network, mobility, territory, positionality, etc. – we ought to also consider how (or indeed, whether) these and other socio-spatial relations are practised, or produced, through contentious political actions.

Collective action frames and place frames

As I have described elsewhere (Martin 2003a),[1] the concept of collective action frames denotes how social movements articulate issues, values, and concerns in ways that foster collective identity and activism (Snow et al. 1986; Snow and Benford 1992; Benford 1993; 1997). Beyond its application to activism, 'framing' refers to how individuals organize experiences or make sense of events (Goffman 1974). Collective-action framing makes sense of events in ways that highlight a collective set of values, beliefs, and goals for some sort of change. Frames are also discourses – 'frameworks that embrace particular combinations of narratives, concepts, ideologies and signifying practices' (Barnes and Duncan 1992: 8) – oriented to collective action. Frames are collective organizational narratives negotiated from a combination of cultural values within a movement – and derived from broader socio-cultural structures – and from some degree of deliberate framing choices by social movement leaders (Roy 1994; Benford 1997; Paige 1997). Collective action frames are not fixed and internally cohesive, but they may contain contradictions and they certainly change over time (Giddens 1976).

---

1   Parts of the following section of this chapter appeared in Martin, D. G. (2003a) '"Place-framing" as Place-Making: Constituting a Neighborhood for Organizing and Activism', *Annals of the Association of American Geographers* 93:3, 730-750. Used by permission of Blackwell Publishers.

Frame analyses are necessarily focused on discourse, practices, and especially rhetorical expressions of meaning and intention in contestation. Such a lens does not take language as spontaneous or wholly independent expressions of fully autonomous agents, however. Discourse embeds and enacts power structures and practices, in shaping the terms, words, and imaginaries of what can be spoken, and by whom (Foucault 1972; Barnes and Duncan 1992). Frame analysis attends to interactional relations among agents, but ought to also consider how relations come to be structured in certain ways.

There are three core analytic elements to collective action frames: motivation, or characterizing and defining the activist community including calls to action; diagnostic, describing problems and assigning blame; and prognostic, advocating certain types of action to solve problems (Snow and Benford 1988, 1992). As described in Martin (2003a), *motivation place frames* refer to the physical and social elements underlying the coming together of a group of people around a set of concerns; quotidian experiences which form imaginaries of place meanings among activists or potential active participants, fostering recognition of common, locationally oriented commonalities (Bourdieu 1977; Anderson 1991). *Diagnostic place frames* involve identification of problems in relation to particular territories or places, imagined or constructed and structured by habits, laws, and practices (of political economy, socio-spatial interactions, etc). Diagnostic frames should also involve a normative spatial imaginary: what a place or landscape or territory should be like. Purcell's (2001) investigation of homeowner activism in Los Angeles involved what he described as a shared 'suburban ideal' (which was derived from structurally articulated norms about nuclear families, private spaces, and home-work separation (Fishman 1987). This suburban imaginary also represented a diagnosis; by identifying landscapes that violated the ideal, activists were framing a problem requiring a remedy. Diagnostic place frames articulate the tensions between geographical ideals and grievances about failures to attain them. *Prognostic place frames* identify the actions that ought to be undertaken; the solutions activists propose to solve the problems that they have identified (Snow and Benford 1988, 1992).

Place frames specifically orient scholars to socio-spatial articulations and relations. Activism is no doubt a 'spatio-temporal event' (Massey 2005: 130) that draws upon and enacts a place: place frame analysis highlights the discourse and practices that represent place(s) –always already scaled – within collective action frames.

Place and framing: accessing socio-spatial practices

Scholars of contentious politics who are interested in the spatialities of such politics can use the concept of place framing to ask questions about the geographical structuring of activism. I use the term place framing not because I privilege place as an analytic category but because the idea of place – as a socio-spatial-temporal but also implicitly and already (negotiated) *scalar* configuration of grievances, activists, claimants, and political resistances – offers a multifaceted way of asking about socio-spatial relations and especially, of practice. By asking whether and how people talk about their grievances in geographical terms, place frame analysis attends to the spatial imaginaries that may be present in contention. In this framework, place is not local or empirical or especially 'real', but it is related to a notion of situatedness, a placing of self, conflict, and/or other. Following Haraway's (1991) argument that any view is from somewhere, place-frame analysis seeks to unpack situated actors, their grievances and conflicts in terms of their settings and relations, and norms or ideals about possible settings and situations.

I draw on Massey's (2005: 130-131) notion of place as '[a] spatio-temporal event ... woven together out of ongoing stories, as a moment within power-geometries, as a particular constellation within the wider topographies of space, and as in process'. To this idea of place, Massey also adds the non-human, citing as examples mountains or the sea, and human and natural history(ies). Massey's conceptualization of place captures its multiplicities: as a coming together of a physical environment; time-space paths of many intersecting individuals, situated within a set of common and individuated structures that converge; always temporary but with seeming longevity. Yet the convergences and settlings that 'place' implies are not simple and uncontested. As Massey states, 'the throwntogetherness of place demands negotiation' (2005: 141). Place frame analysis demands that scholars of contentious politics examine place negotiations in activism: the understandings, discourses, norms, and perspectives of activists. Attention to how place, or geographical thinking and positioning broadly, informs activism, expands understandings of how contention comes about; what sort of activists are included; whom is excluded; why activists focus their contention where they do; and why alternative narratives and sites or claimants of contention are not pursued. These place-framings will, of course, both express and enact power. Consideration of motivation, diagnostic and prognostic place framings highlights processes of exclusion, territoriality, scaling, the pervasiveness of place imaginaries, and how contestation reworks or solidifies place-based or transformative communities. Such analyses will no doubt invoke frameworks such as that of Jessop, Brenner and Jones (2008) or Leitner, Sheppard, and Sziarto (2008) in delineating particular spatialities shaping or evolving from contention. Place framing starts, however, with the spatialities expressed and produced in and through contention.

Other scholars have examined different aspects of geographical framing – Kurtz's (2003) scale frames is an important example. I privilege place frames because the process of discursively identifying a combination of situatedness, throwntogetherness, and shared constellations is inherently already a scalar defining; place has no pre-given scale, both are co-instituted through experience and discourse. Following Brubaker's caution against focusing on nations, but rather, *nationhood* (1996), I argue that place frame analysis ought to focus on how activists produce places as bases for and sites of contention; the processes in/to which they respond, use, and discourses they employ. The focus is not on place as some fixed entity, but on ideas of place as a grounding or situatedness for some sort of activism. In my work, the focus has been on understanding the many productions of neighbourhood as a discursive ideal, a basis for 'local' politics. Place-framing is not always locality-making, however; it can produce a variety of scaled – yet grounded – activisms and imaginaries. Drawing on Brubaker (1996) and Moore (2008), I suggest scholars should ask how and why place – not fixed in particular scope or scale, but imagined in particular spaces-times – is a meaningful experience, and how it is produced, or made, as an event and a locus of action. Doing so requires analytical attention to multiple, co-implicated socio-spatial relations and practices, as delineated in part through place framings.

In the following section, I review a case of activism in Athens, Georgia, to illustrate the multiple spatialities place frame analysis helps to make evident. Athens is a small city of just over 100,000 people (US Census Bureau) that in 1999 and the early 2000s was struggling with growth and land use changes, punctuated by some high profile conflicts between neighbourhood-based groups and first, a major drug-store chain, then a local hospital (Martin 2003b, 2004 describes these conflicts in more detail). In reconsidering the case of Athens here, I want to focus on the neighbourhood hospital conflict in order to highlight how place frame analysis allows scholars to focus on place as practice.

The conflict was in an urban neighbourhood, over the extent – and fact – of a hospital expansion. The dispute was explicitly over meanings of place, or rather, appropriate uses of land in a particular place. As a way to analyse practices and productions of place, then, it offers discourses that overtly invoke and define place. I further examine how, and to what effect, place frame analysis offers a means for deploying a range of socio-spatial analytical categories in assessing contentious politics. I draw on archival research from neighbourhood organizational documents and media reports, as well as 18 interviews conducted in 2000 with stakeholders in the hospital conflict. (Unless noted below, information about the events that unfolded is drawn from accounts in the interviews.) The primary goal of the initial research was to assess how neighbourhood-based activism defined the neighbourhood (see Martin 2003b) and enrolled – or sought to enrol – the local state in land use conflicts (see Martin 2004). My focus here is to investigate framings in light of practising, and claiming, place in activism.

Neighbourhood ideals, political identities, and the polity: place frames in Athens

Athens: challenging political structures, defending neighbourhood identity

Activism in Athens, Georgia from the late 1990s and early 2000s demonstrated perhaps a rather traditional model of episodic neighbourhood-based activism, in which residents banded together to respond to a perceived threat to existing neighbourhood identity and community. Over a time period of just a few years, two mostly white, middle class neighbourhoods in Athens organized against urban development: the first case was opposition to the opening of a national drugstore chain, while the other was opposition to the 20-year expansion plan of a major hospital.

The Athens Regional Medical Center (ARMC) is a mid-sized (315 bed) quasi-public hospital and health care organization in the Normaltown neighbourhood of Athens, Georgia (www.armc.org). Normaltown is an in-town neighbourhood (close to downtown, within the highway loop road circling the city) comprised of mostly single family homes, ranging from small brick and wood-frame bungalows to large, Victorian and historically significant listed houses, bordered by small businesses on a commercial street, and containing the hospital. As part of a city-wide land use planning initiative, the hospital unveiled its 20-year land-use plan in the spring of 1999. The hospital initially intended to share the plan with city commissioners (Athens has a merged city-county form of government) over a lunch meeting. Area residents heard about the meeting ahead of time from friends, neighbours, and colleagues who worked or heard from hospital officials or government officials. Residents objected to a private meeting of public officials in a state with an open records law. After hearing these complaints, hospital officials reconfigured the meeting to allow for open attendance (although lunch was limited to hospital and government officials), and held a public follow-up forum the following week, at which neighbourhood residents turned out in force (Soto 1999). Some residents came in costumes and with signs to mock the hospital (many residents had experienced contentious interactions with hospital officials years before, over previous expansions and street/traffic reconfigurations).

Neighbourhood residents were appalled at the scope of the plan, which included physical expansion of the hospital by over 20 acres, and demolition of over 60 houses. Over the course of the summer of 1999, the neighbourhood residents organized themselves as Citizens for Healthy Neighbourhoods (CHN), created visible opposition to the hospital via yard signs, and hired an attorney to challenge the hospital's legal right of eminent domain (through its quasi-public status, stemming from its origins as the public county hospital), as well as to stop the county approval of a bond issue to finance the first phase of construction. Yard signs, participation at and media coverage of public meetings, and letters to the city newspapers, represented the primary and most visible forms of contention.

The yard signs expressed the fundamental disagreement between the hospital and its residential neighbours in terms of place imaginaries. In big black letters on a yellow background, the signs exhorted, 'ARMC! Don't Destroy Neighbourhoods.' One resident interviewee described the fundamental disjuncture in place definition: 'It was actively articulated at one of the early meetings by one of the guys from the hospital board, that in-town Athens was a slum.' Residents challenged this negative imagery of their neighbourhood with the yard signs – displayed prominently in many yards throughout Normaltown – in meetings with hospital officials, and publicly in local media. Speaking in the media in regard to perceptions of her street, one resident was quoted, '[t]o lose houses to the hospital's expansion would start the arguments all over again that Oglethorpe is not really a residential street'

(McCommons 1999b). In a letter to the city manager made public in a local news magazine, a group of residents wrote:

> [t]his plan eliminates 18-20 houses ... which are at least 95% owner-occupied ... For in-town neighborhoods, density equals safety ... it is definitely not desirable for the residential community (McCommons 1999a).

These quotes illustrate the emphasis on residential land uses in Normaltown, and the rootedness – through ownership – of residents.

Although it is somewhat artificial to separate dynamic, integrated organizational discourses about the neighbourhood and activism into motivational, diagnostic, and prognostic elements, theoretically such a move permits an examination of the construction of place-based agendas for activism. Separating the components of framing discourses reveals how Normaltown activists invoked place primarily in motivational and diagnostic frames (Table 4.1).

Table 4.1   Place frames in Normaltown neighbourhood, Athens, Georgia

Motivational Frames *Exhort action, and define/ describe the community (who, what)*	**Diagnostic** Frames *Describe problems and assign cause/blame*	**Prognostic** Frames *Solutions (specific actions)*

Exhortations:
• Limit sprawl
• Plan for growth
• Protect 'in town' neighbourhoods
• Preserve community, walkability
Descriptions:
• Diversity of income, tenure, age, occupation, housing
• Families, children
• Walkability
• Connections, socializing
• Historic houses
	• Hospital plan:
	  o Auto-oriented sprawl
	  o Destroys housing
	  o Displacement, change in land use
	  o Loss of neighbourhood character
	• Hospital, city disrespect/ fail to acknowledge neighbourhood
		• Stop the hospital via legal challenges, public opposition
		• Develop a better planning process:
		  o Consultative with neighbourhood
		  o Preserve housing
		  o Architectural compatibility
		  o More urban design (versus suburban low density)

Motivation place frames in Athens as evident in the media and through interviews sought to define the community and exhort people to act against the hospital – and for the hospital to develop a different plan – in terms of the urban, residential character of the area. As indicated in Table 4.1, neighbourhood resident spokespersons referred to nuclear families, homeownership, and social diversity.

Their choice to emphasize families and homeownership demonstrates a calculated strategy to challenge the perception of the neighbourhood as, if not a slum exactly, a place where only tenants, or college-students, lived. Instead, residents linked the houses earmarked for demolition with perceptions of permanence and commitment via the statistics on homeownership quoted in the newspaper. The emphasis on ownership represented a challenge to negative – and transient – images of the neighbourhood; not by celebrating students and tenants – clearly a part of the neighbourhood and also represented in the membership of CHN – but to replace the 'temporary resident' image with that of a more permanent, and established (establishment?) resident.

Media coverage of the dispute, as well as interview accounts from residents involved in CHN, indicate a neighbourhood galvanized by concern about the loss of a large number of single-family homes, and a shared orientation towards a walkable, urban, diverse yet residential neighbourhood. Residents involved in CHN viewed the expansion of the hospital as necessarily requiring a destruction of the neighbourhood – because of the planned demolition of many houses; in the land-use plan of the hospital, the two entities could not continue to co-exist. Some of the most involved residents were more than simply concerned about housing, however; they also had social capital to draw upon in the city. Residents had connections either with local political leaders – past and present – or professional expertise relating to law or media, or status as employees of the University of Georgia, also located in Athens.

While the leaders of CHN were not necessarily the residents with the most professional expertise, such residents were nonetheless valued for their perceived ties or knowledge. One resident who was a professor at the university expressed a perception of being valued as a member of the group because of media expertise. This resident helped to develop the yard sign design and strategy. At the same time, however, resident activists worked to overcome their differences along age, tenure and class; CHN leaders came from very different sub-areas (and property valuation) within the neighbourhood, and worked to recognize and highlight their diversity discursively. In interviews, some activists acknowledged the concerns of older residents who might welcome the opportunity to sell their home to the hospital as part of a retirement strategy. Nonetheless, these activists also used the ties and commitments of these same elderly residents to the residential character of Normaltown in building a common neighbourhood-oriented identity among residents.

After a couple of months of open contention and conflict between the hospital and CHN, the two parties entered into a consultative planning process, creating a joint advisory committee to review the hospital's plans and work to compromise on various elements of design and scope. The acceptance of the neighbourhood residents into a consultative planning process reflected a great victory for CHN, in that they established neighbourhood input as an important element of planning for the hospital. The hospital board members, by agreeing to negotiate the details of the land-use plan, acknowledged the importance of the residential identity and character of the neighbourhood even as they sought hospital expansion within it.

The initial neighbourhood activist strategy, however, had attempted to legally prevent the hospital from acting as a special use public authority, by way of formal legal action and political lobbying of the merged city-county commission government. Due to their status as 'mere' residents lacking formal policy input, CHN had to try to create the structures to establish their legitimacy. When the legal challenges failed, they were left with a (very savvy) public opinion campaign, waged through the yard signs and letters to the editor of the local newspaper, which embarrassed the hospital sufficiently that the latter was willing to compromise and seek more constructive engagement. As illustrated in Table 4.1, one of the main goals for CHN was to challenge and alter the direction of development within the neighbourhood. So the diagnoses and prognoses were about challenging that opposition and reorienting the hospital's planning process and design imaginary. The consultative planning process – which persisted into the mid-2000s – offered a means for residents to continually rework their visions of the neighbourhood in tandem with the visions of the hospital board for their institution, and to develop shared place imaginaries that could be complementary, or at least, allow the neighbourhood and hospital to co-exist.

While CHN sought to foster an alternative negotiation and governance structure – which they did achieve – their success was not at the level of the formal political structure. One of the problems with their legal tactics – primarily oriented around stopping the Athens-Clarke County government from issuing revenue bonds for the hospital expansion (a failed effort) – was that the legal structures brought to bear on the problem could not acknowledge or adequately recognize the motivational frame of the neighbourhood as a political actor: No matter how much community and neighbourhood identity that was expressed or fostered through activism, there was no legal argument that neighbourhood community should be supported or preserved. The Athens-Clarke County government's relationship to ARMC, in which the hospital provided county health services in a quasi-public institutional capacity, provided no means for recognizing neighbourhood identity or residential land-uses in any sort of planning calculus. Neighbourhoods have no formal representative status in Athens government, other than along existing electoral district lines (which divides Normaltown into two).

The tactics which were actually successful for CHN were those that publicly and discursively confronted the hospital as hostile to an urban residential neighbourhood, a particular kind of place. In order to do that, CHN had to mobilize residents to appear at public meetings, write letters to the editor, and display the prominent yard signs. Characterizing a neighbourhood, via discourses about the social and physical landscapes, were essential parts of their motivational place framing. Analytically, then, place framing clearly plays a crucial role for some movements, particularly those addressed to changes of land use or characteristics of the physical landscape.

The questions for scholars, then, focus on how these motivational frames are heard within the polity, and how they are attached to particular diagnostic and prognostic frames. The effectiveness of motivation place framing highlights a key disjuncture in place-based grassroots activism: that prognoses and diagnoses based on such activism may not match the scale of motivation framing. The Athens case provides mixed evidence that the polity responds to place frame prognoses; Athens-Clarke County failed to recognize Normaltown as a political actor or as a basis for constituent political identities among voters/residents. The ARMC, however, did recognize the power of place imaginaries because they were publicly targeted as being hostile to their own local neighbourhood. Neighbourhood residents in CHN essentially by-passed a political structure insensitive to neighbourhood-based involvement and directly engaged the media-conscious hospital board (Martin 2004). Competing implications follow: (1) Place-based activism is too 'militantly particular' (Williams 1989; Harvey 1996) to effectively engage structural political change; or (2) Place articulations offer a means to reorganize the political structure and incorporate place orientations into political decision-making, albeit in an ad hoc form. Place frame analysis suggests a way to assess the successes and failures of overtly place-oriented claims as articulated within or by movements. Further socio-spatial and relational analyses of place framing is needed, incorporating multiple spatialities analyses in tandem with place frame analysis of claims from within movements.

Conclusions

In this chapter I have argued that place frames offer a conceptual heuristic to investigate how movements situate their notions of and discourses about activist communities, grievances, and potential actions to solve them, in ways that reference socio-spatial relations, especially articulations grounding activism in specific socio-temporal events. The case study from Athens, Georgia, focuses on neighbourhood-based activism, but a place frame perspective does not assume a particular type of territory or scale. Rather, the central questions for analysis are about the socio-spatial relations and framings of the movement: Why (and how) do activists reference and (re)produce these places/territories/scales, and what sort of contentious politics and places/territories/scales/networks do such framings enact? The Athens case suggests a fiercely local orientation, with some success, but does not provide a framework for how multi-scalar place framings could be equivalently powerful for motivation framings. Nationalistic movements, however, suggest that place-based conflicts provide a powerful, if always territorially constrained, imaginary.

Framings from particular movements or contentious political events are not the only source of empirical data about activism, but they offer a starting point for identifying and examining the multivalent spatialities of the politics that result. My call to focus on the framings of contention – particularly their motivational, prognostic and diagnostic spatialities – is not a call to assume a fixed place geography that activists are formed in, come out of, and respond to: place frame analysis is not a local ontology. Rather, it is a call to pay attention to what activists say about place (what form such 'places' take, how certain agents get to be the ones speaking about it); their conceptualizations of their spatialities, rootedness, flows, and connections; and how these shape their activism. Further, a serious examination of frames does not posit activists or organized contention as the only site or expression of spatiality: analyses of contention that seek to take geography seriously have to examine the place frames within a given politics in order to situate it, to examine why certain spatialities – be they localized, grounded expressions of a social group or networked, globally attentive and connection-seeking or something in between – are produced, and what comes of those productions. Doing so allows scholars to simultaneously investigate the socio-spatial relations actively engaged and produced through contentious politics by political actors, as well as to better examine and understand socio-spatial processes shaping and affecting the developments and outcomes of such politics.

References

Amin, A. (2004) 'Regions Unbound: Towards a New Politics of Place', *Geografiska Annaler B*, 86: 33-44.

Anderson, B. (1991) *Imagined Communities*, revised ed. London and New York: Verso.

ARMC.org. 'Athens Regional Medical Center' website. <https//web1.armc.org> (home page), accessed 26 April 2010.

Barnes, T. J. and Duncan, J. S. (1992) 'Introduction: Writing Worlds', in Trevor J. Barnes and James S. Duncan (eds) *Writing Worlds: Discourse, Text and Metaphor in the Representation of Landscape*. London and New York: Routledge.

Benford, R. D. (1993) 'Frame Disputes Within the Nuclear Disarmament Movement', *Social Forces*, 71: 677-701.

—— (1997) 'An Insider's Critique of the Social Movement Framing Perspective', *Sociological Inquiry*, 67: 409-430.

Bourdieu, P. (1977) *Outline of a Theory of Practice*. Cambridge: Cambridge University Press.

Brubaker, R. (1996) *Nationalism Reframed*. Cambridge: Cambridge University Press.

Fishman, R. (1987) *Bourgeois Utopias: The Rise And Fall Of Suburbia.* New York: Basic Books.

Foucault, M. (1972) *The Archaeology of Knowledge*. New York: Pantheon Books.

Giddens, A. (1976) *New Rules of Sociological Method*. London: Hutchinson.

Goffman, E. (1974) *Frame Analysis: An Essay on the Organization of Experience*. Cambridge, MA: Harvard University Press.

Haraway, D. (1991) *Simians, Cyborgs and Women: The Reinvention of Nature.* New York: Routledge.

Harvey, D. (1996), *Justice, Nature and the Geography of Difference.* Cambridge, MA: Blackwell.

Jessop, B., Brenner, N. and M. Jones (2008) 'Theorizing Sociospatial Relations', *Environment and Planning D: Society and Space*, 26: 389-401.

Klandermans, B., Kriesi, H. and S. Tarrow (eds) (1988) *International Social Movement Research, Vol. 1: From Structure to Action: Comparing Social Movement Research Across Cultures.* Greenwich, CT: JAI Press.

Kurtz, H. (2003) 'Scale Frames and Counter-Scale Frames: Constructing the Problem of Environmental Justice', *Political Geography*, 22: 887-916.

Leitner, H., Sheppard, E. and K. M. Sziarto (2008) 'The Spatialities of Contentious Politics', *Transactions of the Institute of British Geographers*, NS 33: 157-172.

McAdam, D., Tarrow, S. C. and Tilly (2001) *Dynamics of Contention.* Cambridge: Cambridge University Press.

McCommons, P., (1999a) 'Destroying to Save: The Hospital is Taking an Even Bigger Bite out of Athens', *Flagpole Magazine*, 14 April.

McCommons, P., (1999b) 'The Hospital that Ate Normaltown: Athens Regional Hospital is Growing ... Again', *Flagpole Magazine*, 7 April.

Marston, S., Jones, J. P. and K. Woodward (2005) 'Human geography without scale,' *Transactions of the Institute of British Geographers*, 30: 416-430.

Martin, D. G. (1999) *Claiming Place and Community: Place Identity and Place-Based Organizing in Inner-City Neighborhoods.* Unpublished PhD Dissertation. Minneapolis, MN: University of Minnesota.

—— (2000) 'Constructing Place: Cultural Hegemonies and Media Images of an Inner-city Neighborhood', *Urban Geography*, 21(5): 380-405.

—— (2003a) '"Place-Framing" as Place-Making: Constituting a Neighborhood for Organizing and Activism', *Annals of the Association of American Geographers*, 93: 730-750.

—— (2003b) 'Enacting Neighborhood', *Urban Geography*, 24(5): 361-385.

—— (2004), 'Reconstructing Urban Politics: Neighborhood Activism in Land Use Change', *Urban Affairs Review*, 39(5): 589-612.

Martin, D. G. and B. Miller (2003) 'Space and Contentious Politics', *Mobilization: An International Journal*, 8(2): 143-156.

Massey, D. (2004) 'Geographies of Responsibility', *Geografiska Annaler B*, 86: 5-18.

—— (2005) *For Space* . London and Thousand Oaks, CA: Sage Publications.

Moore, A. (2008) 'Rethinking Scale as a Geographical Category: From Analysis to Practice', *Progress in Human Geography*, 32(2): 203-225.

Morris, A. and C. McClurg Mueller (eds) (1992) *Frontiers in Social Movement Theory.* New Haven: Yale University Press.

Nicholls, W. (2009) 'Place, Networks, Space: Theorising the Geographies of Social Movements', *Transactions of the Institute of British Geographers*, 34: 78-93.

Paige, J. M. (1997) *Coffee and Power: Revolution and the Rise of Democracy in Central America*. Cambridge, MA: Harvard University Press.

Purcell, M. (2001) 'Neighborhood Activism Among Homeowners as a Politics of Space', *Professional Geographer*, 53: 178-194.

Roy, B. (1994) *Some Trouble with Cows: Making Sense of Social Conflict*. Berkeley and Los Angeles: University of California Press.

Snow, D., Benford, R., Rochford, E., and S. Worden (1986) 'Frame Alignment Processes, Micromobilization, and Movement Participation', *American Sociological Review*, 51(4): 464-481.

Snow, D. A. and Benford, R. D. (1988) 'Ideology, Frame Resonance, and Participant Mobilization', in Klandermans et al. (eds).

—— (1992) 'Master Frames and Cycles of Protest', in Morris and McClurg Mueller (eds).

Soto, J. (1999) 'Residents lash out against ARMC plan', *Athens Daily News and Banner Herald*, April 22.

US Census Bureau, 'American Fact Finder', *US Bureau of the Census*, <http://www.census.gov> (home page), accessed 22 December 2010.

Williams, R. (1989) *Resources of Hope*. London: Verso.

PART II
Scale, Territory and Region: Structuring Collective Interests, Identities and Resources

Chapter 5
'Polymorphic Spatial Politics':
Tales from a Grassroots Regional Movement

Martin Jones

> Without finding a geographical formula which offers the benefits of
> regionalization to all parts of the Country the Government will find it impossible
> to form a devolved framework of regional governance. It also needs to offer what
> might be described as 'democratic control' (Cornish Constitutional Convention
> 2005: 9-10).

Introduction

This chapter has three aims. I would like to empirically demonstrate, by using the
example of a region in England, how new state institutions and policies create both
new grievances and opportunities to mobilize at the regional scale. Second, I then
seek to illustrate how existing regional networks and identities are employed for these
types of struggles. I conclude the chapter by making connections between regions,
debates on socio-spatial relations, and what I would like to call 'polymorphic spatial
politics'. This concept recognizes the multidimensional intersections between space
and politics, which is mutually constitutive and relationally intertwined.

The chapter is organized into three main sections. I firstly examine the resurgence
of regions in academic discourse and political practice and suggest a need to
consider two contested processes. I draw a distinction between a state-driven and
top-down *functional regionalization* and an often pre-existing and more bottom-up
civil society regionalism. In a second section, I explore the emergence of functional
regionalization in England and use the East Midlands region to ground my concerns.
I focus on attempts made by institutions in this region over the last half century
to promote a state-centred socio-economic identity. Section three discusses how
regionalization is being contested by *Devolve!* – formerly, as Movement for Middle
England (MFME), a Midlands-based regional devolution movement; latterly, as
an English devolution movement, still active in the East Midlands. Section four
concludes the paper and discusses the notion of 'polymorphic spatial politics'.

Rise of new regionalisms: territory and socioeconomic change

Academic debates have been suggesting for some time that regions are becoming
central players in economic, social, political and cultural life (for extensive

reviews, see Harrison 2007, 2008a, 2008b, 2010). On the one hand, some authors see regions as providing the atmosphere for economic prosperity within a forever globalizing age. On the other hand, other authors connect the rise of regions to the demands of sub-national groups that are challenging the cultural and political sovereignty of the Westphalian national-state system. Keating's work on 'new regionalist movements' has been important for drawing our attention to these trends and for highlighting the connections between renewed territorial forms of political mobilization, social action through cultural expression, and state responses via the institutional frameworks of devolution and constitutional change (Keating 1998; see also Carter and Pasquier 2010). For Keating, regions are sites for generating post-national identities and instilling social cohesion in a world increasingly dominated by global and market forces. Cited examples often include Galicia, the Catalan and Basque regions, Brittany, Lombardy, Upper Silesia, and more recently the Celtic fringe of the United Kingdom.

Critical commentators, though, are suggesting an urgent need to bring spatial clarity to these parallel debates and are specifically questioning what is meant by the term 'region'. By drawing on the work of the 'new regional geography' – where regions are products of state practices, economic processes, cultural relationships, and senses of place – MacLeod (2001) argues that regions need to be seen as a multifaceted territorial phenomenon, straddling economics, culture, political and policy systems. One way of bringing clarity to these debates and getting within what I call 'actually-existing regions' (Jones 2004) is to explore a recent distinction between 'regional spaces' and 'spaces of regionalism' (Jones and MacLeod 2004). The former emphasizes the work of mostly economists and economic geographers, who posit an emerging *functional regionalization* of economic space within globalization. Functional regionalization also relates to the practices of political parties and state personnel who construct regions for delimiting particular policy and political problems – often to create the conditions for those cluster-based economic developments referred to in the above academic literatures.

By contrast, 'spaces of regionalism' uncovers those political science and political geography concerns with claims to citizenship, cultural expression, and political mobilization, often associated with the implementation of regional government and governance. Research on 'social capital' has been important here for pointing out that networks of co-operation and trust can foster territorial awareness, civic identity and, in turn, can help to secure prosperity and instil democracy. Regionalism, therefore, 'involves an act of faith – faith in the prime importance of the hidden patterns and a strong element of hope, hope for the institutional change that will unlock and reveal regional identity and give it a political voice and/or administrative form' (Garside and Hebbert 1989: 2). Building on Jones (2004), the remainder of this chapter explores these latter concerns through notions of *civil society regionalism*, to step outside the parameters of the state apparatus and allow attention to be focused on grassroots regional movements as 'secession groups' – groups which have turned their back on mainstream political parties, operate within the shadow of conventional politics, and prefer to pursue more unconventional ways of making themselves heard.

Functional regionalization in the East Midlands of England

The East Midlands – now covering the counties of Lincolnshire, Nottinghamshire, Derbyshire, Leicestershire, Northamptonshire, and Rutland – provides the context for exploring these issues. With its mixture of medium-sized industrial towns and predominately rural hinterland, Fawcett's early account on this 'Trent province' noted the difficulties of managing this region as an administrative space due to the absence of industrial clustering and the economic disunity (Fawcett 1960). Related to this concern, in stark contrast to its industrial neighbour, the West Midlands (which played a leading role in the regional planning movement of the late 1940s), the East Midlands regional voice was dormant until activated by modernist planners. Developing out of the 'North Midland' town and country planning region, an artificial 'East Midlands Standard Region' was created in 1965 to house the activities of the East Midlands Regional Economic Planning Council (EMREPC). In contrast, though, to more forward-looking regions such as the North East, which fostered an emerging regional consciousness through the limited spaces of economic planning, EMREPC did little to break out of its functional regionalization role and rather like its Yorkshire and Humberside neighbour it represented 'a blind alley along the long and slow route towards regionalism' (Pearce 1989: 128).

During the 1980s, the regional planning machinery was swept aside and replaced by local-level experiments in economic governance. After the 1992 general election, though, the Conservatives introduced a new round of administrative regionalization to bring coherence to the activities of government departments operating in the regions and also to control some of the community-based economic developments taking place within the local state. Integrated Government Offices for the Regions came into being as part of the 'new localism' in British urban policy (Jones 2004).

The East Midlands Standard Region became home to the Government Office for the East Midlands (GOEM) – bringing together (by 2001) the following functions: transport, regional development, education and skills, culture, media and sport, Home Office, and Environment, Food and Rural Affairs – and also charged with working 'in partnership with local people to maximize the competitiveness, prosperity and quality of life of the region' (GOEM 1996: 3). GOEM became a strategically significant actor in creating an East Midland's regional consciousness within which *central government* has a 'common economic and social purpose at the regional level' (GOEM, 1997: 2).

Set within an ambitious UK-wide constitutional change and devolution agenda – alongside elected Assemblies for Wales, Northern Ireland, and London, and a Parliament for Scotland – a further round of regionalization arrived with the launch of the Regional Development Agencies (RDAs) under the New Labour administration between 1997 and 2010 (see Jones 2001). The East Midland's RDA (EMDA) became active in April 1999 across the East Midlands Standard Region, enacting a series of supply-side policies – such as skills training, enterprise support, cluster-based innovation strategies, and land management – to help raise the wealth and prosperity of this region. EMDA's economic strategy presented a socioeconomic vision for this region where, 'By 2010, the East Midlands will be one of Europe's top 20 regions. It will be a place where people want to live, work and invest because of our vibrant community, our healthy, safe, diverse and inclusive communities [and] our quality environment' (EMDA 1999: 2).

With ongoing pressures being exerted on central government by 'civic regional' activists, especially in the North East of England, which had been campaigning throughout the 1990s for directly elected regional assemblies, after 2001 the Labour Party announced a 'new regional policy' to incorporate the people of England within regional structures of governance. In the words of the state:

> By devolving power and revitalising the regions we bring decision-making closer to the people and make government more efficient, more effective and more accountable ... This is a radical agenda to take us forward fully into the 21st century, where centralisation is a thing of the past. It responds to the desires many regions are already expressing and sets up a framework which can take other regions forward if they wish. Better Government, less bureaucracy and more democracy, and enhancing regional prosperity: proposals from a Government confident that it is people within our regions who know what is best for their region (DTLR 2002: foreword).

Commentators, though, noted several weaknesses in this agenda: the boundaries related to Central Government created 'standard regions'; the membership of the proposed regional assemblies was limited; their influence over government spend, restricted; and the main players within the regions were the Government Offices for the Regions (Jones and MacLeod 2004). As I discuss below, this thinking opened up lively debates, between 2001 and 2007, within civil society for thinking about regions and politics.

Alternatives to functional regionalism in the East Midlands

The Movement for Middle England

> Towards the end of 1995 I received a leaflet in the mail, from a group calling itself 'The Movement for Middle England'. Headed 'Middle England Awake' and displaying a quotation from G.K. Chesterton, 'For we are the People of England that have never spoken yet', it announced a conference to be held in Oxford, and invited people to 'work for a society you'll be proud to live in'. The

Movement for Middle England was in favour of devolution and autonomous English regions ... Was this the beginning of a genuine resistance movement, I wondered, a grass-roots resurgence in which bamboozled and browbeaten Anglo-Saxons would finally throw off the yoke of the centralised and, as the leaflet put it, Norman British state? Was it a harmless fantasy addressed to the retired colonels and morris dancers of the deep English shires? Or was it a primitive historical cult convened more in the spirit of Goose Green? (Wright 1998: 26-27).

Devolve! – known until the year 2000 as Movement for Middle England (MFME) – was founded in 1988 by a number of individuals involved with: grassroots activism; co-operative networks; þa Engliscan Gesiðas (The English Companions – an Anglo-Saxon historical and cultural society); and bottom-up models of 'direct democracy' – associated with diverse neo-distributist and anti-modernist thinkers such as G.K. Chesterton, Robert Blatchford, Peter Kropotkin, and Peter Cadogan. The winter 1988 edition of *The Regionalist* introduced MFME to the Seminar and detailed its founding aims (*The Regionalist* 1988: 4):

1. To be non-party political.
2. To work for the full autonomy of Middle England within a devolved England/ Europe.
3. To have power vested in the smallest appropriate unit (i.e. no decision to be taken at a higher level than necessary).
4. To respect the right of any regional group or area to define its relationship with the whole.
5. To establish 'shadow' institutions outside the present political system, rather than contest elections within that system.

The naming of MFME, which 'expresses the "character" of the region and also serves to reproduce it [through] powerful feelings of identification' (Bialasiewicz 2002: 119), originates from a 'third-space'-like identification with both the Greater Midlands area (East and West Midlands combined) and the smaller 'Middle Angles' territory – an Anglo-Saxon Mercian dependency that existed between AD 500 and 700, centred on Leicester (Robyns 1983), but 'rehistoricized and read anew' in the twentieth century (Bhabha 1994: 37). The Middle Angles territory of the East Midlands also possesses a unique 'historical and cultural unity' (Fawcett 1960) and for some MFME members this unity acted as a distinct reminder of a world not touched by globalization and consumer capitalism. MFME mobilized this consciousness through their literature and emblem – a deliberate subversion of the Cross of St George, with the broken-cross creating four inward pointing arrows: signifying the great regions of England co-operating without being tied in the centre by London. The oak tree in the top left quadrant symbolized the once great forests of Middle England: Sherwood, Charnwood, Needwood and Arden. This 'cultural regionalism' attracted several members from the historic county of Rutland and also provided the stimulus for heated debates on English culture within a multicultural and postcolonial world (see MFME 1992, 1996).

The initial technique of MFME was to set up stalls in market areas of towns across the Midlands, with literature, badges, car stickers, etc. This generated some income, promoted the ideas and more importantly provided direct feedback on the ideas and views of a wide cross-section of people. At this time membership was increasing, there were various workshops and social events, an open newsletter plus an ideas exchange journal for members only, then called 'Views of Middle England' (see Banks 1997). The membership structure allowed for Active members, who were expected to participate in necessary work and meetings, tithe themselves (2.5 per cent of their disposable income) and steer the movement, and Support members (whose main social interests might lie elsewhere), who helped when they wanted to, paid a modest annual subscription and were kept informed.

One particularly notable pamphlet, *Wall-to-Wall Democracy*, promoted a model of direct democracy that draws on Anglo-Saxon principles of justice and societal organization. In opposition to the rising unelected state, *Wall-to-Wall* discusses the historical and contemporary virtues of 'moot democracy', with power flowing from the citizen to the 'Regional Witan' – 'the hallmark of a liberated England' (MFME 1991: 17). According to Wright's account, 'Spurning party politics, it favoured belonging – "taking root in your region and helping to run it". It wanted to "encourage local moots of around 50 householders", and to establish them as "building blocks of future democracy"' (Wright 1998: 26). With the launch of EMDA in 1999, MFME issued a 'Charter for the Midlands', with individuals signing a petition to introduce elements of moot democracy and ensure 'regional democracy in the Greater Midlands with true subsidiarity at every level and full representation as a region of Europe' (MFME 1999: 1). Later internal debates around New Labour's proposals for directly elected regional assemblies resulted in the moot model being presented for the twenty-first century within a 'new constitution for England' (*Devolve!* 2000a).

From MFME to Devolve!

During the 1990s some members withdrew from the Movement for Middle England, wanting to define a more radical, historical and 'organic regionalism' – building up a region 'from geology and topography as a counter-modern unit' (Matless 1998: 129) – that advocated outright opposition to developments such as the Government Offices for the Regions. The resulting new Mercia Movement offered an anti-modernist manifesto 'for the future inspired by the past' and based on the justice principles of communalism, organic democracy, ecological balance, and the re-creation of Mercia as an autonomous and sustainable bioregion within an English confederation (Mercia Movement 1997). For these organic regionalists, then, 'modern administration violates natural boundaries' (Matless 1998: 130-131). The Mercia Movement participated in 1998 in the formation of the Confederation for Regional England (CfRE), a radical counterpoint to the well-funded Campaign for the English Regions (see below). Later (2001) it sought a broader dialogue in the Mercian Constitutional Convention (MCC), though without wanting to compromise its radical ideals. The MCC became the new form of the Mercia Movement, gaining local media coverage for its general historical antics and fringe political strategies: in Birmingham on 29 May 2003 it formally declared the whole of Mercia – from Cheshire to the Thames – independent of the United Kingdom (Mercia Movement 2003). A 'register of citizens of Mercia' was created to drive this forward.

Meanwhile, the Movement for Middle England maintained a more pragmatic stance, with emphasis on empowerment rather than history. It continued to work at the grassroots, in the shadow of formal electoral politics and, later, of institutions like the Government Office for the East Midlands (GOEM) and EMDA. A twin strategy of regional moderation and democratic (or social justice) radicalism emerged. And with English regionalism rising high on the political agenda at that time, MFME members called a special general meeting in June 1999 to discuss their future aims and objectives – with debate focused on either being a straightforward campaign group for a Greater Midlands region, or broadening its activities to promote cultural, democratic and regional devolution *across* England. Members present voted for the latter (a 'triple devolution of power from the centre') and subsequently renamed their organization *Devolve!* (formerly Movement for Middle England) to: signify a move towards an inter-territorial and multidimensional approach to regional politics (dropping its own territorial claim in the Midlands); provide more meaningful engagements with policy-makers, practitioners and civic regionalists involved in campaigns for elected regional government *across* England (collectively termed 'institutional regionalism'); and pay more attention to local and direct action initiatives for empowerment across Europe (cf. *Devolve!* 1999).

From 2000 *Devolve!* sharpened its intellectual thinking with four key tenets (see *Devolve!* 2001a): *regional devolution* (bringing power closer to people through regionalism to allow and encourage greater diversity); *democratic devolution* (real democracy through rotated delegates within a moot system); *cultural devolution* (nurturing English culture while respecting other cultures and identities); and *economic devolution* (co-operation instead of competition based on local economic networks). For each tenet, steering group members convene working groups to debate these different dimensions of devolution, with each tenet also covered by several pages on the developing web-site to attract further recruits (see www.devolve.org).

Devolve! *as networked regional forum*

An important aspect of *Devolve!* has been its role in debating territory and politics by providing channels and forums for other grassroots civil society regional movements to get their voices heard (see *Devolve!* 2001b). *Devolve!* has hosted three 'Whose Regions' conferences in recent years to debate the future of England.

The first was convened by MFME and involved activists from across England and generated a fruitful dialogue with policy-makers involved in regional development agencies (see MFME 1998). The second created a productive exchange between the Campaign for the English Regions (CFER), a civic regional movement lobbying for regional assemblies through cross-party constitutional conventions, and activists campaigning for a Cornish Assembly (see *Devolve!* 2000b).

A third 'Whose Regions' conference in 2003 debated the future of 'The South' within the Labour Party's programme of English devolution. The 'Spotlight on the South' conference – chaired by the godfather of Scottish devolution, Canon Keynon Wright, involving a presentation by the former minister for regions Alan Whitehead MP, and drawing representation from leading stakeholders within the southern political space – granted *Devolve!* (by a vote of the attending conference participants) a mandate to form a 'Continuing Commission on the South' to discuss: 1) regional representation and funding; 2) resource allocation, planning, infrastructure, and networking across regions; 3) citizen involvement, local government and very local democracy; and 4) identities and boundaries (*Devolve!* 2003).

Following the rejection of further regional devolution, after the North East voted by 77.9 per cent in the 2004 devolution referendum against a regional assembly proposal, *Devolve!* members have been *getting back to place* and developing 'very local democracy' at the urban scale in the city of Leicester, a form of 'regional microcosm' (Cornish Constitutional Convention 2005), encompassing: common ownership and stewardship structures; total and partial income sharing; residents groups and local forums; challenging – in the High Court – electoral destruction of local communities; holding local councillors to continuous account through new ward-based Community Alliances; and lobbying GOEM for community involvement in urban policy and budgets. With the advent of a city-regional agenda after 2006 (see Harrison 2010), the 'Continuing Commission on the South' became the 'Commission on Devolution' and *Devolve!* hosted a number of workshops and seminars on 'functional sub-regionalism' and options for devolved local government (see *Devolve!* 2006). With the advent of 'new-new localism' (Jones and Jessop 2010; Cooper 2011) under a Conservative-Liberal Democrat Coalition Government, which advocated abolishing the RDAs and replacing them with Local Enterprise Partnerships (LEPs), *Devolve!* energy is being focused on creating a Regional Autonomy Project (RAP) – a national database for 'local people' interested in 'promoting and developing regional autonomy' (*Devolve!* 2010; cf. McIvor 1998).

'Polymorphic Spatial Politics'

I have identified two processes operating through the English regions. First, a functional regionalization approach being deployed by policy-makers to meet the challenge of globalization and deliver public policy through state administered spatiality. Second, I have highlighted reactions to this by civil society regionalists, who argue for new forms of socio-economic development, alongside the reworking of territory and the politics of representation. By using the East Midlands region as a case study, I have discussed how these two processes are being contested and have explored how movements like *Devolve!* have sought to challenge the emerging political systems in England's regions – trivial though it may seem to some.

These non-trivial empirics, however, can and should be connected to conceptual debates on politics and spatiality in geography and beyond. Regions can be seen as active products of reciprocal relationships between economic behaviour, the politics of representation and identity, state power geometries, and the sedimentation of these practices in time-space. This, in turn, illustrates the need for approaches that stress region building and its politics as active and ongoing processes – rich in political strategy, territorial awareness and cultural expression – which may allow researchers and regional strategists to uncover the very formation of those untraded and interdependent conventions within particular regions (see MacLeod and Jones 2001; Paasi 2009).

There is also clearly an argument for keeping hold of territorial readings of political economy. By contrast, geographers 'thinking space relationally' have suggested the need for a relational, networked and mobile spatiality that transcends scalar and territorial social, economic and political action (for an overview of debates, compare Amin 2004, 2007; Beaumont and Nicholls, 2007; Jones 2009; MacLeod and Jones 2007). Networks should not be seen as non-spatial and without 'geographical anchors', and on the other hand, territories and scales should not be viewed as closed and static. Agnew's suggestion for moving beyond this impasse is to focus on:

> politics as organized in terms of places where most people live their lives; settings that are linked together and across geographical scales by networks of political and economic influence that have been, and still are, bounded by but decreasingly limited to the territories of national states (Agnew 2002: 2).

Agnew addresses this challenge (which he calls 'mapping politics theoretically') through research on connections between agency, difference, association, and socialization (Agnew 2002: 27-35). Low expresses similar sentiments by adopting an 'areal' approach to the politics of place, which recognizes the bounded and unbounded confrontations, struggles and compromises that occur within territory (Low 1997). Beaumont and Nicholls (2007: 2563) have a similar line of argument. They talk about the 'geo-organizational character' of social movements and the links between institutional context, spatial forms, and the maintenance of such spatial forms through 'organizing environments'.

These debates are currently being taken forward by researchers arguing for polymorphy in socio-spatial theory and associated attempts to develop a more complex and multi-dimensional analysis of these connections on the ground.

Here, an argument is made for moving away from one-dimensional thinking, i.e. seeing everything in singular spatial relations – such as territory, place, scale, or network – *in isolation* from each other, and towards frameworks capturing the organization of socio-spatial relations in multiple forms and dimensions. Jessop et al (2008: 393) capture one-dimensional thinking in Table 5.1, where they outline several forms of one dimensional spatial conceptualization. The consequences of such one-dimensionality include theoretical amnesia and exaggerated claims to conceptual innovation; the use of chaotic concepts rather than rational abstractions; overextension of concepts and their imprecise application; concept-refinement to the neglect of empirical evaluation; and an appeal to loosely defined metaphors over rigorously demarcated research strategies.

According to Jessop et al. (2008: 396), these limits can begin to be avoided by more systematic and reflexive investigations of the interconnections between the different dimensions of social relations, or the 'TPSN schema' as they put it. Table 5.2 is their attempt to sketch this out. The 16 cells capture some coordinates of analysis associated with the TPSN framework rather than concrete applications of the latter. These cells have been produced by cross-tabulating each socio-spatial dimension considered as a structuring principle with all four socio-spatial dimensions considered as fields of operation of that said principle. The table should not be seen as the product of 'taxonomic folly' (ibid.: 396); it only has a heuristic purpose. Jessop et al. then put this to work in research on 'state/space' to discuss the applicability of the TPSN framework to several realms of inquiry into socio-spatial processes under contemporary capitalism.

Table 5.1   The sites of one-dimensionalism (Jessop et al. 2008: 393)

Concrete-complex point of exit / Abstract-simple point of entry	TERRITORY	PLACE	SCALE	NETWORKS
TERRITORY	Methodological Territorialism			
PLACE		Place-Centrism		
SCALE			Scale-Centrism	
NETWORKS				Network-Centrism

Table 5.2   Beyond one-dimensionalism: conceptual orientations (Jessop et al. 2008: 395)

Fields of operation / Structuring principles	TERRITORY	PLACE	SCALE	NETWORKS
TERRITORY	Past, present, and emergent frontiers, borders, boundaries	Distinct places in a given territory	Multi-level government	Inter-state system, state alliances, multi-area government
PLACE	Core-periphery, borderlands, empires, neo-medievalism	Locales, milieux, cities, sites, regions localities, globalities	Division of labour linked to differently scaled places	Local/urban governance, partnerships
SCALE	Scalar division of political power (unitary state, federal state, etc.)	Scale as area rather than level (local through to global), spatial division of labour (Russian doll)	Vertical ontology based on nested or tangled hierarchies	Parallel power networks, non-governmental international regimes
NETWORKS	Origin-edge, ripple effects (radiation) Stretching and folding cross-border region, inter-state system	Global city networks, poly-nucleated cities, intermeshed sites	Flat ontology based on horizontality with multiple entry points	Networks of networks, spaces of flows, rhizome

The TPSN schema can be advanced as a heuristic mechanism to identify forms and modalities of resistance and in doing so can provide a backdrop for interpreting and reinterpreting empirical developments similar to those outlined in this chapter. One of the driving forces behind recent interests in territory, place, scale, and networks, of course, has related and deeply intertwined interests in progressive (and, indeed, regressive) political struggles and possibilities (see Massey 1994, 2007; Amin, 2004, 2007; Amin et al. 2003). The example of *Devolve!* has revealed some of these struggles. Table 5.3 accordingly provides a provisional TPSN mapping of forms of subaltern or counter-hegemonic politics across the 16 coordinates of analysis. Again, the populated cells are only illustrative, not exhaustive, and are designed to stimulate future research agendas. The TPSN schema is being used here to categorize socio-spatial principles, strategies, and practices in the field of 'contentious politics', seen as covering different forms of struggle and contestation from 'below' and 'within', regardless of their social bases, identities, interests, or objectives. There are two sorts of applications here: a) locating the observer's account within the grid, and b) locating the strategies and tactics of participants in contentious politics. The first application can be illustrated by the celebration of nomadism (territory-territory), or multitude (network-network), or scale-jumping (scale-scale). The second application is illustrated by militant particularism (place → scale), the freedom march on Washington etc. (place + scale + network), and so on ...

Table 5.3   'Polymorphic spatial politics': horizons of contention

Structuring principles \ Fields of operation	TERRITORY	PLACE	SCALE	NETWORKS
TERRITORY	NOMADISM	Secession Separatism Irredentism	Dual Power Anti-Imperialism	Wars of position
PLACE	Peasant wars, Migration, Asylum	RED BASES REDNECK AREAS	Council communism, Soviets Communes	Militant Particularism
SCALE	Subsidiarity	Countryside surrounds towns Siege warfare	SCALE-JUMPING	World Social Forum, international solidarity movements
NETWORKS	Mobile Tactics	Movements of Homeless and Dispossessed	Localism Factory Egoism Anarchism	MULTITUDE

Important research is beginning to take place in line with this thinking. Sheppard (2002) and Leitner et al. (2008) have proposed a multidimensional framework for studying 'contentious politics' with scale, place, positionality, networking and mobility shaping each other and in turn influencing the trajectory of politics. Nicholls (2009) has taken some of this further in ideas of 'social movement space', which seeks to capture how networks are forged in places, and vice versa. Similarly, the work of Hardt and Negri (2004, 2009) on 'altermodernity' as a politics of connection and 'left assemblage' (Tampio 2009) is seeking out the many spaces of social movements that can be joined up and their energy channelled in a more sustainable manner than has previously been the case. Woods' (2007) work on relational rural politics, explicitly the connections between place and scale, human and non-human, negotiation and configuration, illustrates both a theoretical and methodological framework for doing all this (see also Panelli 2007).

I would argue that all this represents the growing terrain of 'polymorphic spatial politics' that recognizes the multidimensional intersections between space and politics, which is mutually constitutive and relationally intertwined. As I have argued elsewhere (Jones 2009), this opens up further a field of *political geometry* whereby researchers can track multifarious spatial synchronizations. The future challenge, raised in a recent paper by Jones and Jessop (2010), is to tease out the connections, compatibilities, incompatibilities, contradictions and dilemmas involved in specific dimensions of TPSN socio-spatiality, i.e. their mutual articulation or not, and their implications for the success and failure of political struggles, goals, and gains.

The story of *Devolve!* in this chapter has provided some initial insights into this challenge. Their latest statement detailed below, titled *Fellow Regionalists (Devolve!* 2010: 1-2), highlights a movement fully aware of these multifarious socio-spatial horizons of policy and politics, but I would argue that without the resources (members and finance) to assemble points of coordination and grasp the connections, *Devolve!* will struggle to make progress beyond being a critical commentator on contemporary capitalism. *Fellow Regionalists* suggests that:

Having managed to get a number of small groupings together, (we ourselves have put in our publications/websites) we have got agreement on the collective promotion of regional autonomy. As such a simple project is in the making, operating as the Regional Autonomy Project or RAP for short. Supportive of a number of ideas: Regionalism, Decentralism and Alternative Green etc, the aim is simply to develop a working group to develop and promote regional autonomy. From this it is hoped a number of long term projects/organisations will come forth, but we shall see. Looking to build a more community based approach; we will look to develop grassroots movement building. Looking now to drop the flag waving antics of the past and concentrate on local development, be it Wessex/ Mercia or Warwickshire/ Yorkshire etc. The aim is to get this down even further to local communities and the establishment of community groups / TRA's (Tenants / Residents Associations). With regard to politics of the main parties, we are looking at taking our views to MP's / councillors and see what benefit we will get. At present this has been done before at local meetings with mixed results, so something we need to talk over more. Recently we have lost a lot of the Tory inclined English patriots who still have belief in fighting national elections/ campaigning. We view this both as a complete waste of time as small parties will get nowhere fighting national elections until we get PR [proportional representation], not AV [alternative voting] as is at present being put forward. Campaigning too we view as a waste of time, as looking at much larger campaigns, most have failed even with massive support/ finance, much better we believe to reject campaigning for change via parliament and build bottom up, from the council estates via community groups through to councillors. Looking to promote a modern regionalist answer to the crass society we live in today, we are now developing a national closed egroup/ mailing, a website and a number of promotional items that local people may use and distribute. All will have a space for local contact details.

References

Agnew, J. A. (2002) *Place and Politics in Modern Italy*. Chicago: The University of Chicago Press.

Amin, A. (2004) 'Regions Unbound: Towards a New Politics of Place,' *Geografiska Annaler: Series B*, 86: 33-44.

—— (2007) 'Re-thinking the Urban Social,' *City* 11: 100-114.

Amin, A., Massey, D., and N. Thrift (2003) *Decentering the National: A Radical Approach to Regional Inequality*. London: Catalyst.

Banks, J. C. (1997) 'Grass-roots Regional Movements in England 1974-1996', *Mimeograph*, Cheltenham, Secretary of the Wessex Regionalists.

Beaumont, J. and W. Nicholls (2007) 'Between Relationality and Territoriality: Investigating the Geographies of Justice Movements in The Netherlands and the United States', *Environment and Planning A*, 39: 2554-2574.

Bhabha, H. K. (1994) *The Location of Culture*. London: Routledge.

Bialasiewicz, L. (2002) 'Upper Silesia: Rebirth of a Regional Identity in Poland', *Regional and Federal Studies*, 12: 111-132.

Carter, C. and R. Pasquier (2010) 'Introduction: Studying Regions as 'Spaces for Politics': Re-thinking Territory and Strategic Action', *Regional and Federal Studies*, 20: 281-294.

Cooper, H. (2011) 'Made Redundant in the Regions', *Guardian*, Society, 30th March, 7.

Cornish Constitutional Convention (2005) *Devolution's Future: A New Proposal for Regional Devolution*. Truro: Cornish Constitutional Convention.

Devolve! (1999) *Annual General Meeting: Proposal on Change of Name*. Leicester: *Devolve!*

—— (2000a) *A Constitution for England*. Leicester: *Devolve!*

—— (2000b) *Whose Regions? 2. Report of a Conference held at Lanceston, Kernow on 17th June 2000*. Leicester: *Devolve!*

—— (2001a) *Information Sheet: Objectives*. Leicester: *Devolve!*

—— (2001b) *How Devolve! Works to Support Regional Movements*. Leicester: *Devolve!*

—— (2003) *Continuing Commission on the South: Proposal*. Leicester: *Devolve!*

—— (2006) *Beyond City Regions: New Options for Devolved Local Government, Conference Proceedings, Oxford*. Leicester: *Devolve!*

—— (2010) *Fellow Regionalists*. Leicester: *Devolve!*

DTLR (2002) *Your Region, Your Choice: Revitalising the English Regions. Cm 5511*. London: The Stationery Office.

EMDA (1999) *Prosperity Through People: Economic Development Strategy for the East Midlands 2000-2010*. Nottingham: EMDA.

Fawcett, C. B. (1960) *Provinces of England: A Study of Some Geographical Aspects of Devolution.* 2nd edn. London: Hutchinson.

Garside, P. and M. Hebbert (1989) 'Introduction', in P. L. Garside and M. Hebbert (eds) *British Regionalism 1900-2000.* London: Mansell, 1-19.

GOEM (1996) *Government Office for the East Midlands: Annual Report 1996/97.* Nottingham: GOEM.

—— (1997) *Government Office for the East Midlands: Summary Operational Plan, April 1997-March 1998.* Nottingham: GOEM.

Hardt, A. and M. Negri (2004) *Multitude: War and Democracy in an Age of Empire.* New York: Penguin.

—— (2009) *Commonwealth.* Harvard: Harvard University Press.

Harrison, J. (2007) 'From Competitive Regions to Competitive City-regions: New Orthodoxy, But Same Old Mistakes', *Journal of Economic Geography*, 7: 311-332.

—— (2008a) 'Stating the Production of Scales: Centrally Orchestrated Regionalism. Regionally Orchestrated Centralism', *International Journal of Urban and Regional Research*, 32: 922-941.

—— (2008b) 'The Region in Political Economy', *Geography Compass*, 3: 814-830.

—— (2010) 'Networks of Connectivity, Territorial Fragmentation, Uneven Development: The New Politics of City-regionalism', *Political Geography*, 29: 17-27.

Jessop, B., Brenner, N. and M. Jones (2008) 'Theorizing Sociospatial Relations,' *Environment and Planning D: Society and Space*, 26: 389-401.

Jones, M. (2001) 'The Rise of the Regional State in Economic Governance: 'Partnerships for Prosperity' or New Scales of State Power?', *Environment and Planning A*, 33: 1185-1211.

—— (2004) 'Social Justice and the Region: Functional Regionalization and Civil Society Regionalism in England', *Space and Polity*, 8: 157-189.

—— (2009) 'Phase Space: Geography, Relational Thinking, and Beyond', *Progress in Human Geography*, 33, 487-506.

Jones, M. and B. Jessop (2010) 'Thinking State / Space Incompossibly,' *Antipode*, 42 (5): 1119-1149.

Jones, M. and G. MacLeod (2004) 'Regional Spaces, Spaces of Regionalism: Territory, Insurgent Politics and the English Question,' *Transactions of the Institute of British Geographers*, 29: 433-452.

Keating, M. (1998) *The New Regionalism in Western Europe: Territorial Restructuring and Political Change.* Cheltenham: Edward Elgar.

Leitner, H., Sheppard, E., and K. M. Sziarto (2008) 'The Spatialities of Contentious Politics,' *Transactions of the Institute of British Geographers*, 33: 157-172.

Low, M. (1997) 'Representation Unbound: Globalization and Democracy,' in K. Cox (ed.) *Spaces of Globalization: Reasserting the Power of the Local.* New York: Guilford Press, 240-280.

McIvor, M. (1998) *Grass-Roots Pressure for Constitutional Reform in Britain.* London: Andrew Wainwright Trust.

MacLeod, G. (2001) 'New Regionalism Reconsidered: Globalization and the Remaking of Political Economic Space,' *International Journal of Urban and Regional Research*, 25: 804-829.

MacLeod, G. and M. Jones (2001) 'Renewing the Geography of Regions,' *Environment and Planning D: Society and Space*, 19: 669-695.

—— (2007) 'Territorial, Scalar, Networked, Connected: In What Sense a 'Regional World?', *Regional Studies*, 41: 1177-1191.

Massey, D. (1994) *Space, Place and Gender.* Cambridge: Polity.

—— (2007) *World City.* Cambridge: Polity.

Matless, D. (1998) *Landscape and Englishness.* London: Reaktion Books.

Mercia Movement (1997) *The Mercia Manifesto: A Blueprint for the Future Inspired by the Past.* Cotes Heath: Witan Books.

—— (2001) *A Draft Constitution for Mercia.* Cotes Heath: Witan Books.

—— (2003) *The Constitution of Mercia: The Mercian Constitutional Convention.* Cotes Heath: Witan Books.

MFME (1991) *Wall to Wall Democracy: The View from Middle England.* Leicester: Movement for Middle England.

—— (1992) *Race, Culture and Identity: The View from Middle England.* Leicester: Movement for Middle England.

—— (1996) *Replies to Middle England: Alternatives Views on Race, Culture and Identity.* Leicester: Movement for Middle England.

—— (1998) *Whose Regions? Report of a Conference held at Dr Johnson House, Birmingham on 27th June 1998.* Leicester: MFME.

—— (1999) *Charter for the Midlands.* Leicester: MFME.

Nicholls, W. (2009) 'Place, Networks, Space: Theorizing the Geographies of Social Movement', *Transactions of the Institute of British Geographers*, 34: 78-93.

Paasi, A. (2009) 'The Resurgence of the 'Region' and 'Regional Identity': Theoretical Perspectives and Empirical Observations on Regional Dynamics in Europe', *Review of International Studies*, 35: 121-146.

Panelli, R. (2007) 'Time-space Geometries of Activism and the Case of Mis/placing Gender in Australian Agriculture,' *Transactions of the Institute of British Geographers*, 32: 46-65.

Pearce, D. C. (1989) 'The Yorkshire and Humberside Economic Planning Council 1965-1979,' in P.L. Garside and M. Hebbert (eds) *British Regionalism 1900-2000.* London: Mansell.

Robyns, D. (1983) 'The Midlands: Special Supplement to the Regionalist,' *The Regionalist*, No. 2, 20-23.

Sheppard, E. (2002) 'The Spaces and Times of Globalization: Place, Scale, Networks, and Positionality', *Economic Geography*, 78: 307-330.

Tampio, N. (2009) 'Assemblages and the Multitude: Deleuze, Hardt, Negri, and the Postmodern Left', *European Journal of Political Theory*, 8: 383-400.

The Regionalist (1983) 'About Regionalism', *The Regionalist*, No. 2, 1-2.
—— (1988) 'A New Regionalist Movement', *The Regionalist*, No. 12, 4-5.
Woods, M. (2007) 'Engaging the Global Countryside: Globalization, Hybridity and the Reconstitution of Rural Place', *Progress in Human Geography*, 31: 485-507.
Wright, P. (1998) 'An Encroachment Too Far', in A. Barnett and R. Scruton (eds) *Town and Country*. London: Jonathan Cape, 18-33.

Chapter 6

Overlapping Territorialities, Sovereignty in Dispute: Empirical Lessons from Latin America

John Agnew and Ulrich Oslender

Introduction

Recent debates in political geography have pointed to a key inadequacy in international relation theories, in particular the Westphalian model of state sovereignty, by positing the existence of various 'regimes of sovereignty' (Agnew 2005). Yet, the idealized sovereignty of the nation-state is still rigidly linked in dominant theories to the notion of a transparent and singular territoriality or control over a national territory clearly marked in space by long established and stable borders. In this chapter we propose the notion of 'overlapping territorialities' to examine the intersection of sources of territorial authority, other than the nation-state, with that of states. States are rarely if ever the neatly defined entities with homogeneous powers over their territories that typical stories allege them to be. This is particularly so when much statehood is a history of contested acquisition and conquest rather than consensual union.

The notion of overlap is not entirely new, of course. Other authors have pointed to its relevance in contemporary and historical contexts. To historical sociologist Michael Mann (1986: 1), 'societies are constituted of multiple overlapping and intersecting sociospatial networks of power'. Meanwhile, Osiander (2001) and Krasner (1995) point to the failure of the Westphalian sovereignty model to effectively end the medieval organization of overlapping and competing authorities. And commenting on some of Colombia's authority alternatives – a case which we will examine in detail below – Mason (2005: 40) notes how contemporary global order is composed of 'multiple and overlapping jurisdictions'.

What we do here is to flesh out the notion of overlapping territoriality by examining specific cases, in which the contestation of space by non-state actors has found expression not only in alternative authority regimes but in concrete re-territorialization processes that imply the drawing of boundaries *within* nation-state territory. In other words, we propose to map out some of the new territorial authority regimes that have emerged over the last decades as a result of the political contestation of space and that challenge the nation-state's supposed container-like exclusive territorial sovereignty. This, we argue, will do two main things. First, it will further problematize the continued projection of the Westphalian model, which, if not obsolete as a category, is increasingly less able to account for the dynamic nature of contemporary territorialization processes and sovereignty. Second, by mapping out these 'overlapping territorialities', we wish to draw attention to the ways in which local and national struggles manage to redefine the very meaning of the contemporary nation-state.

These processes are particularly evident in Latin America, where indigenous and Afro-Latin American social movements have powerfully carved out political, cultural and economic spaces over the last two decades. Their achievements have been reflected above all in the legal recognition of collective land ownership. In many parts of Latin America, local indigenous and black communities have established themselves *de jure* as differential territorial authorities within the nation-state (Assies et al. 2000; Sieder 2002; Van Cott 2000). The contestation of space by these movements has resulted in concrete territorial gains. At the same time, however, their achievements are *de facto* under threat, as these alternative territorialities are often perceived by other actors, such as parastatal organizations and transnational capital, as challenging the dominant occidental territorial model. Thus, complex processes of de- and re-territorialization are set in motion that often take violent forms, including massacres, selective killings, and forced displacements.

We argue that these empirical lessons from Latin America importantly contribute to a necessary re-thinking of the links between state sovereignty and territoriality as mediated by social movements challenging the established spatial fabric of state-based politics. The chapter moves from a general discussion of ideas about sovereignty and territoriality – including our proposal for 'overlapping territorialities' – to close consideration of some examples from Colombia.

Beyond Westphalian territoriality

The territorial state is a highly specific historical and geopolitical entity. It initially arose in Western Europe in the sixteenth and seventeenth centuries. Since that time, political power has come to be seen as inherently territorial because statehood is seen as inherently territorial. From this viewpoint, politics takes place exclusively within 'the institutions and the spatial envelope of the state as the exclusive governor of a definite territory' (Hirst 2005: 27). The process of state formation has always had two crucial attributes. One is *exclusivity*. All of the political entities (the Roman Catholic Church, city-states, etc.) that could not achieve a reasonable semblance of sovereignty over a contiguous territory have been delegitimized as major political actors. The second is *mutual recognition*. The power of states has rested to a considerable extent on the recognition each state receives from the others by means of non-interference in their so-called internal affairs. Together these attributes have created a world in which there can be no territory without a state and vice versa. In this way, territory has come to underpin both nationalism and representative democracy, both of which depend critically on restricting political membership by homeland and address respectively.

From this viewpoint, state sovereignty may be understood as *the absolute territorial organization of political authority*. Most accounts of sovereignty accept its either/or quality: a state either does or does not have sovereignty (Lake 2003). They differ as to whether this is a foundational principle (originating in the seventeenth century) or an emergent social practice. They also vary by acceptance of actors in international politics (such as militarily weaker states, or 'failed states') that are not fully sovereign. But what if the absolute political authority implicit in this story about state sovereignty and its presumed territorial basis has always been problematic?

Territoriality – the use and control of territory for political, social, and economic ends – is in fact a strategy that has developed differentially in specific historical-geographical contexts. The territorial state, as it is known to contemporary political theory, is only one form of territoriality. It developed initially in early modern Europe with the retreat of non-territorial dynastic systems of rule and the transfer of sovereignty from the personhood of monarchs to discrete national populations. That modern state sovereignty did not occur overnight following the Peace of Westphalia in 1648 is now well established. Yet, territory has popularly been associated with the spatiality of the modern state with its claim to absolute control over a population within carefully defined external borders. Indeed, until Sack (1986) extended the understanding of human territoriality as a strategy of individuals and organizations in general, usage of the term territory was largely confined to the spatial organization of states. In the social sciences such as sociology and political science it is still mainly the case that the challenge posed to territory by network forms of organization (associated with globalization) is invariably characterized in totalistic terms as 'the end of geography'. From this viewpoint territory takes on an epistemological centrality that is understood as absolutely fundamental to modernity. But it can also be given an extended meaning to refer to any socially constructed geographical space, not just that resulting from statehood. Especially popular with some French-language geographers, this usage often reflects the need to adopt a term to distinguish the particular and the local from the more general global or national 'space'. It then signifies the 'bottom-up' spatial context for identity and cultural difference (or place) more than the 'top-down' connection between state and territory (Agnew 1987).

From this wider theoretical perspective, territoriality can be judged as having a number of different origins including: (1) as a result of explicit territorial strategizing to devolve administrative functions but maintain central control (Sack 1986); (2) as a secondary result of resolving the dilemmas facing social groups in delivering public goods (as in Michael Mann's sociology of territory); (3) as an expedient facilitating coordination between capitalists who are otherwise in competition with one another (as in Marxist theories of the state); (4) as the focus of one strategy among several of governmentality (as in Michel Foucault's writings); and (5) as a result of defining boundaries between social groups to identify and maintain group cohesion (as in the writings of Georg Simmel and Fredrik Barth, and in more recent sociological theories of political identity). Whatever its specific social origins, however, territoriality is usually put into practice in a number of different if often complementary ways: (1) by popular acceptance of *classifications* of space (e.g. 'ours' versus 'yours'); (2) through *communication* of a sense of place (where territorial markers and borders evoke meanings); and (3) by *enforcing control* over space (by barrier construction, interception, surveillance, policing, and judicial review).

Overlapping territorialities

In many countries a pluralization of meaningful territories is producing what we call 'overlapping territorialities'. Though encapsulated within a given state, they do not need to be mutually exclusive of one another and can be based on different social logics. In many Latin American countries, for example, indigenous and black groups base their collective land claims on social and cultural difference that distinguishes them from the dominant *mestizo* population. To them, their ways of relating to nature and space are quite different from the modern state territorial logic of the conquest of nature. As we discuss in more detail below, their historic territorialities – based on sustainable and magic-religious relations with their surroundings – have existed for hundreds of years, although they have mostly been ignored until recently by political science. The official recognition of indigenous and black territories has now resulted in the legal sanctioning of a differential territoriality at a subnational level, creating territorial authorities other than the national government within the space of the nation-state. State and indigenous territorialities quite literally overlap in these areas and create contested spaces of sovereignty.

The available literature is divided over the contemporary nature and origins of these differential territorialities. One school of thought gives priority to the emergence of regional and local identities in response to the pressures of globalization. From this point of view, the growth of autonomist movements, for example, signifies not so much an ethnicization of identities as a redefinition of 'home'. As people are increasingly exposed to world markets without the same protection once offered by national-state boundaries, they must develop strategies to enhance local competitive advantage in global competition. Rather than signalling the progressive expansion of cosmopolitanism, therefore, globalization represents both a de-territorialization of existing identities and interests and a re-territorialization on the basis of localized cultural identities and economic interests.

Another strand of literature places more emphasis on the 'unfinished' or changing character of many nation-states. In Latin America, this has become most obvious with an array of countries adopting new constitutions over the last two decades that reflect their multicultural and pluriethnic nature. Officially sanctioning indigenous and black groups as 'ethnic minorities' has not only given these groups specific rights, but it more widely critically redefines the meaning of the nation itself. The incorporation of these formerly excluded or marginalized groups into the nation-building narrative signifies a profound change in the ways Latin Americans see themselves and the nature of the nation-state. Although these changes do not take effect overnight, constitutional reforms in Latin America have clearly set the stage for a new socialization in the region. One aspect refers to changes in the territorial constitution of the state, in that new political actors, such as black communities, are entrusted with a certain degree of territorial autonomy and authority that hitherto had been the domain of the state. We will discuss this scenario in more detail below. Suffice it to say here that these changes are related to a wider process of rescaling of state functions through the decentralization of some aspects of the power apparatus. It is also important to stress that they are the result of intense mobilization of indigenous and black social movements that have challenged the status quo and contested the space of the nation-state. In some countries these movements have subsequently evolved into ethnic political parties (Van Cott 2005). The decolonial project in Latin America, while stretching back to indigenous and black resistance against the Hispanic colonizers, is only just beginning to redraw the ontological boundaries of the meaning of the nation-state in this region (Cairo and Mignolo 2008; Moraña, Dussel, Jáuregui 2008).

This mobilization may most clearly be seen in the case of Bolivia, where the government of the first democratically elected indigenous president Evo Morales leads an ongoing struggle against the European-descended elites who see their long-standing domination of the country under threat. The elites' main strategy to rebuff the government consists of claims for autonomy of the wealthy region in the east of the country, which they control. They have used their economic power to coerce workers, organize strikes, and disrupt the national economy. The secessionist threats and the potential fragmentation of the Bolivian state are a very real danger, as the elites use their economic power to push for territorial autonomy. Bolivia's vice-president is clear about the potential disastrous impact that the conflict may have for the national project of indigenous empowerment, but he is also clear that this is a conflict that needs to take place in order to create a more democratic and inclusive government in Bolivia (García Linera 2006).

The tendency among students of contemporary statehood has been to invest an idealized 'nation-state' with quasi-mystical powers irrespective of the actual capacity of existing states to rule their respective territories. Yet, many regions and localities within existing states are only weakly integrated into them. This is most obvious in the case of multinational states such as the former Yugoslavia and the former Soviet Union. Not only did major economic disparities create a popular basis for mobilization against the states that had failed to deal with them, the main territorial divisions within the states ran along officially sanctioned national/ethnic lines, thus creating the clear impression of a pattern of internal colonialism. The 'stories' various groups developed about one another highlight the tendency to transform the inadequacies of the existing state into ethnicized claims about the others. For example, in the former Yugoslavia enemies are invariably likened to nomads, foreign to civilization, preying on the peaceful and prosperous (Bougarel 1999). When central institutions collapse, as in both the Yugoslav and the Soviet cases, regions increasingly replace them as the primary focus of political organization, representing both a reaction to the power vacuum at the centre and the relative ease with which power can be reconstituted regionally.

Cultural divisions of labour within states also often seem to take regional forms, even in cases where states might appear more firmly established. Or, movements can argue that their regions are disadvantaged by structural biases built into existing national economies. A now classic argument about the reality of a cultural division of labour was made by Hechter (1975, reissued 1999), claiming that the Celtic fringe of the British Isles had been underdeveloped to the benefit of England and that the growth of separatist movements in Ireland, Wales, and Scotland was a direct result of popular resentment at this state of affairs. Evidence from such cases as the clusters of Palestinians in economically disadvantaged northern and southern regions of Israel (Yiftachel 1999), the 'ethnocracy' of Ashkenazim dominance of Israeli politics and society through regional settlement and economic policy (Yiftachel 1998), the poverty of minority-occupied regions in western China (Safran 1998), and the use of rural colonization schemes by the Sri Lankan government to displace local Tamils (Manogaran 1999) suggest that there can well be the direct impacts elsewhere that Hechter saw in the British case.

Critical of the veracity of Hechter's specific argument, Keating (1998: 19) points out, however, that proponents of greater regional autonomy often have recourse to claims such as those made by Hechter. Internal colonialism or uneven development, therefore, serve as ideological premises upon which political movements can mobilize popular support. Orthodox Irish nationalism in Northern Ireland often has had recourse to such arguments. Sometimes movements from richer regions can adopt the approach and reverse the logic, often in very reactionary ways. In northern Italy, for example, the federalist/secessionist Northern League berates the Italian state for neglecting the affluent north to favuor the poorer south and in so doing retarding the growth of the north. This case may also serve as a reminder that not all regional (or social) movements are necessarily of a progressive nature, but quite on the contrary may seek to reinforce structures of regional domination and exclusion (Oslender 2004).

Finally, government restructuring can encourage regional identities by partitioning the national space into units that can generate degrees of loyalty/ disloyalty, or by promising devolution of power and then reneging on it, provoking resentment from elements in regional populations. Northern Ireland is one clear example of a region whose very existence as a political entity reflects its incorporation into one state (the United Kingdom), while a large minority of its population has rejected this status and supports its integration into the adjacent Irish Republic. The tortured course of the political movements devoted to either maintaining the status quo or abolishing it, in this case, provides a good opportunity to examine the varied roles of local religious affiliations, landscape images, demographic projections, and competing political languages in creating an intensely regionalized political life. The varied separatist movements in the northeast of India provide a similarly good example of the impact of 'centralizing federalism', in which the promise of devolution has been replaced in a region distinct from the rest of India, across multiple dimensions from religion and literacy to economic and social history, by an increasingly repressive political regime. The promise of a multinational pan-Indianism is thus compromised by the tendency of the central government to usurp ever more power at the expense of local states, justified through national security claims, which is then met by local response (Baruah 1999).

What these examples show is that the nation-state is very much an unfinished, constantly evolving space of contention between different regionally situated social groups. The Eurocentric and US-centric focus in standard political theory may blind us to the fact that throughout the world many different struggles are going on that attempt to redefine the nation, reinterpret its meaning, and even redraw its boundaries. The very dynamic nature of these processes in postcolonial states should give rise to a closer examination of the individual cases and what can be learnt from them for a progressive development of political theory. In this sense, Hansen and Stepputat (2001: 9) propose an exploration of 'the local and historically embedded ideas of normality, order, intelligible authority, and other languages of stateness'.

In the case of Colombia, we will show how 'other languages of stateness' have emerged over the last two decades that have significantly reshaped our understanding of the nature of the nation-state.[1] Crucial in this reshaping has been the mobilization of a range of social movements that have engaged with, contested and undermined the state's apparent power monopoly. As we will show, territory and territoriality have been central as both object of contention and resource of struggle. We should make clear that the Colombian case discussed here is illustrative of a wider trend taking place throughout the Americas. By focusing on developments in Colombia, we move away from this country's usual depiction as the 'poster child for studies of violence' (Davis 2006: 182) and concentrate on its surprisingly progressive constitutional amendments, one of which the anthropologist Michael Taussig (2004: 95) refers to as 'one of the most innovative experiments in political theory this century'.

Colombia's overlapping territorialities

Colombia provides a fascinating case study for the multiple ways in which exclusive state territoriality has been challenged by a range of actors. Social movements, including armed guerrilla groups, have been crucial in these contestations of space. Most observers of the political conflict in Colombia have

---

1   See also Hunt (2006), who has applied the notion of languages of stateness to her study of the rural space of *el pueblo* in the Colombian State.

focused on its violent nature and the increasingly complex interactions between the national government, guerrilla groups and paramilitary organizations, so much so that the social sciences in this country sport a special category of experts on violence – the *violentólogo*. There is a broad consensus that Colombian sovereign authority and its national territory have been fragmented throughout its history (Bushnell 1993; Pécaut 2001; Safford and Palacios 2002). State institutions have been characterized by their weakness and alternative authority regimes have thrived in the face of the state's failure to control large areas of the country, most notably enforced by a range of armed actors (Mason 2005; Pizarro 2004).

The FARC: an alter-state within the state

The history of Latin America's most powerful guerrilla organization, the FARC (Revolutionary Armed Forces of Colombia), for example, has been one of continuous territorial expansion and growth. With its roots in peasant self-defence groups that formed as a direct result of government violence in the 1950s during the partisan conflict known as La Violencia,[2] the FARC evolved from a mobile guerrilla force into a revolutionary movement expanding its armed struggle to most rural regions of the country (Pizarro 1987). In a number of guerrilla conferences the ever-growing movement decided on military strategies, defined new combat zones and drew up recruitment plans. From 1985 onward an accelerated geographical expansion of zones of influence can be observed and FARC's military activities have since targeted over half of the country (Echandía 1999; Sánchez and Chacón 2005).

A clear pattern of engagement became apparent with the FARC quickly establishing footholds in poorer rural regions where the state was effectively absent (Pizarro 2004). There the FARC provided public services to local communities that would normally be expected to be delivered by the government, including education, policing, and jurisdiction. Such a long-term vision for a regional structure of social welfare has been a characteristic of this guerrilla organization, a fact that helps to explain its strong support base and peasants' deep loyalties to the FARC in many regions that were abandoned or neglected by a weak central state.

Colombian sociologist Alfredo Molano (1992, 1994) has examined at length the history of land colonization and violence in Colombia. In *Trochas y Fusiles* (1994) he eloquently describes the culture of the FARC and their interactions with the peasantry, showing the mutually constitutive character of this relationship. On the one hand, the guerrilla controls the management of local economies and imposes taxation as well as its penal and moral codes on the local population. On the other hand, people approach the guerrillas and solicit solutions to everyday issues and quarrels. In order to maintain their authority, the guerrillas need to respond to these demands. Failing to do so, the FARC would lose its legitimacy

---

2   For a concise summary of this period, see Hylton (2003); for more elaborate analysis, see Bergquist et al. (1992), Roldán (2002) and Sánchez and Meertens (2001).

among the local population. The provision of security is one of the central requests and has indeed been the historical raison d'être of the FARC. The guerrillas in fact exploit the state's failure to respond to rural conflicts, and thus fill a hegemonic void left by the state (Richani 2002: 98). In these parts the FARC has become a de facto alter-state within the state.

The most dramatic and visible manifestation of such an alternative territorial regime being officially sanctioned within state boundaries was the demilitarized zone (DMZ) that was established in the southern part of Colombia in 1998 as a pre-condition for peace talks between the guerrillas and the government of President Andrés Pastrana. On request of the FARC, Pastrana ordered the withdrawal of the armed forces from an area of 42,000 square miles in the Meta and Caquetá Departments, so that peace negotiations could take place in a space the guerrillas felt sufficiently safe in. The peace talks never got off the ground in a meaningful way and were beset with accusations on both parts – government and FARC – that the other party did not adhere to previous agreements. Eventually the peace negotiations were declared a failure by the government, and President Pastrana ordered the army to retake the zone on 21 February 2002. Yet for just over three years the FARC constituted the officially sanctioned territorial authority in this demarcated area roughly the size of Switzerland. The guerrillas provided police and judicial powers, set up administrative organizations, and dispensed revolutionary justice. Since the FARC had already established itself as de facto authority in this region, providing everyday protection and services and involving the local population in their project of national revolution, this was neither such a difficult task nor an abnormal scenario. In many ways, the government's decision to grant the DMZ to the FARC merely reflected the real-life situation on the ground. However, in political science terms, this development marked a sharp contestation of space and of exclusive state authority within the nation-state's boundaries.

One may argue that the DMZ is in fact a classroom scenario for what the geographer Robert McColl in the 1960s called the 'territorial imperative' in the creation of an 'insurgent state'. In his comparative analysis of national revolutions he identified a 'commitment to the capture and control of a territorial base within the state' (McColl 1969: 614) as a crucial step in the project of any revolutionary movement that aims at the overthrow of the state:

> Looked at from the viewpoint of internal political developments, the creation of an insurgent state has a number of values to a national revolutionary movement. First, it acts as a physical haven for the security of its leaders and continued development of the movement. Second, it demonstrates the weakness and ineffectiveness of the government to control and protect its own territory and population. Third, such bases provide necessary human and material resources. Finally, the insurgent state and its political administrative organizations provide at least an aura of legitimacy to the movement. It is not a process of state breakdown. The creation of an insurgent state is an effort gradually to replace the

> existing state government. The geopolitical tactic is the attrition of government
> control over specific portions of the state itself. (McColl 1969: 614)

These were, of course, precisely the points that critics of the DMZ voiced from
the outset. Undoubtedly, the FARC used the space as a 'physical haven for the
security of its leaders and continued development of the movement'. And it did
provide the guerrillas with an officially sanctioned legitimacy that put it on a par
with the government at the negotiating table. It was also a massive publicity coup
for the FARC. This eventually disbanded, officially sanctioned experiment of the
insurgent state within the state provides a fascinating example of overlapping
territorialities, which were constantly challenged, fought over, and re-negotiated.

While Colombia's violence and civil war have been prominent in the writings
of political scientists, little attention has so far been given to less belligerent
territorial challenges. This seems strange, perhaps, as Colombia embarked on a
significant decentralization programme of the state apparatus in the mid-1980s
that would bring to the fore and even promote alternative territorialities within
the state. Decentralization was seen as a way out of an institutional crisis that had
brought the country to the 'brink of chaos' (Leal Buitrago and Zamosc 1991).
The closed bipartisan political system, tightly controlled by the Conservative and
Liberal Parties, had ensured the exclusion of broad sectors and movements from
political participation. At the same time, powerful guerrilla movements, right-wing
paramilitary groups (often supported by the armed forces), wide-spread corruption,
and the all-pervasive influence of the illicit drug trade had reduced the legitimacy
of the state in the eyes of many and brought governability to a standstill in the late
1980s. Decentralization was designed to diffuse tensions within a framework of
broader and more inclusive political participation. Above all it aimed to strengthen
democracy at the municipal level and bring government closer to the people. In
other words, a rescaling of state functions took place and new powers were given
to local government, such as the popular election of mayors from 1988 onwards.

The passing of a new Constitution in 1991 marked a further important step in
the spatial restructuring of the state. For the purpose of our principal argument,
we will dedicate the rest of this chapter to examining how state territoriality was
opened up to other non-state actors through a range of constitutional amendments,
and how the state recognized, legitimized and promoted non-state territorialities
within its boundaries. The state's degree of legitimacy and sovereignty may
actually be increased rather than undermined by acknowledging the presence of
other territorial authorities within the space of the nation-state. This, we believe, is
a crucial if often overlooked point.

Constitutional amendments and black territorialities

The new Constitution of 1991 replaced the existing one of 1886. It was drawn up
by a Constituent Assembly, a national public body popularly elected in December
1990 that included independent representatives from ethnic, political, and religious

minorities, as well as from re-incorporated guerrilla movements.[3] Although not directly aimed at the country's ethnic minorities, the Constitution also declared the nation to be multicultural and pluriethnic.[4] In an unprecedented move, it affirmed that 'the state recognizes and protects the ethnic and cultural diversity of the Colombian nation' (Article 7), for the first time officially acknowledging the country's black population as an ethnic minority worthy of special protection. Whereas various articles dealt specifically with Colombia's indigenous populations and outlined their territorial and political rights, only Transitory Article AT-55 made explicit reference to the country's black communities. It marks a watershed in the changing relationship between the Colombian state and the Afro-descendant population. Referring to the 10 million hectares of tropical rainforest in the Pacific coastal region, AT-55 states:

> Within two years of the current Constitution taking effect, Congress will issue ... a law that grants black communities who have been living on *state-owned lands* in the rural riverside areas of the Pacific basin, in accordance with their traditional production practices, the right to collective property over the areas that the law will demarcate. ... The same law will establish mechanisms to protect the cultural identity and the rights of these communities and to foster their economic and social development. (emphasis added)

This constitutionally mandated law was passed in 1993 and became known as Law 70. By 2008 a total of 132 collective land titles had been issued to black communities covering almost five million hectares of tropical rainforest in the Pacific coastal lowlands. As Figure 6.1 illustrates, this marks an impressive 50 per cent of the entire region.[5]

---

3   Guerrilla representatives were from the demobilized movements of the M-19, the People's Liberation Army (EPL) and the indigenous guerrilla group Quintin Lamé. Conspicuously absent from the Constituent Assembly was the FARC, whose previous negative experience of peace negotiations made them more cautious. As a result of a peace treaty between the FARC and the administration of then president Belisario Betancur, the Communist People's Union party UP (Unión Popular) was founded in 1984. Yet, in the following months and years, thousands of its members were killed by right-wing paramilitary groups that were often linked to state institutions. This experience is crucial to understanding today's suspicion and reluctance of the FARC to engage in peace negotiations with the Colombian government.

4   Colombia is not alone in this regard of course. In recent years many Latin American countries have introduced constitutional reforms and opened up official ideologies of nationhood to notions of multiculturalism and pluriethnicity. Van Cott (2000: 17) talks in this regard of an 'emerging regional model of multicultural constitutionalism'.

5   For details on the collective land titling process, see Offen (2003). See Restrepo (2004) for an interpretive ethnography of the articulation of black ethnicity in the process.

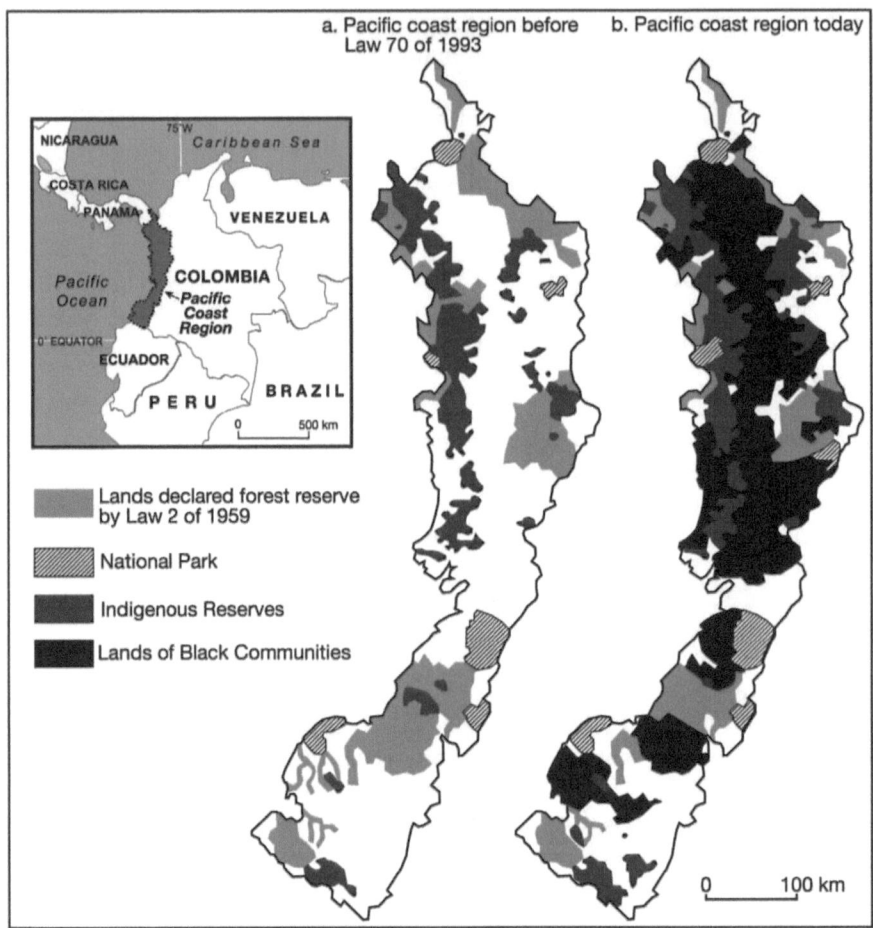

**Figure 6.1   Land distribution in the Pacific coast region before and after
Law 70 of 1993**

Source: adapted from Agudelo (2002: 445).

But why were the Pacific lowlands singled out for this treatment? With 93
per cent of the population, Afro-Colombians constitute the overwhelming majority
in this region, and the black peasantry had increasingly felt threatened by the
accelerated and uncontrolled exploitation of natural resources by external capital.
The often violent predatory extraction practices of national and foreign companies
through logging and mechanized gold mining had led to widespread deforestation
and pollution that threatened the very existence of local populations and their
lifestyles based on subsistence farming, fishing, hunting and gathering. Beginning
in the mid-1980s in the northern Chocó Department, black peasant organizations,

aided by the Catholic Church's Afro-American Pastoral, mobilized around the defence of their lands and the environment. It was here that direct links were first articulated between the notions of a peasant identity and blackness, linked to specific relations to territory. As Afro-Colombian social movement leaders would later point out: 'In fact, the relationships between culture, territory, and natural resources constitute a central axis of discussion and strategy building both within movement organizations and in their dealings with the state' (Grueso et al. 1998: 209).

The Pacific coast region then became the first battlefield around which an emerging social movement of black communities began to mobilize (Escobar 1997; Oslender 2004). Their demand for collective land titles was not just about delimiting their land rights in space. It was first and foremost an affirmation of an ancestral black territoriality that had been exercised for hundreds of years but that had not been recognized and respected as such. The reference in AT-55 to 'state-owned lands' – or *tierras baldías* in Spanish – is a clear indicator of this lack of recognition. Law 70 of 1993 was more specific, referring to 'the lands situated within the limits of the national territory that belong to the state and *have no other owner*' (Diario Oficial 1993: point 4, emphasis added). Thus, the 1993 legislation still did not recognize the existing ancestral black territoriality, or the 'territories of difference', as Escobar (2008) puts it. Law 70 simply makes provision for establishing a collective ownership over lands considered as belonging to the state and as 'empty'. What may appear to be a subtle difference shows in fact the discrepancy between traditionally held and exercised territorialities and the Western territorial (if you want, Westphalian) state logic that had been superimposed. What had in fact existed for centuries were overlapping territorialities.

It was this silent and unarticulated discrepancy that had led to conflict in the region. The Pacific lowlands were first referred to as *tierras baldías* in 1959 legislation. This designation had allowed sawmill owners to appropriate these 'empty' territories for timber extraction through government concessions, mostly without regard to the ways in which local black and indigenous populations used such lands according to their traditional production practices. While these communities had developed a complex socio-ecological relationship with the environment in which the forested hinterland (the *respaldo de monte*) fulfilled material as well as magic-religious functions, to outsiders keen on exploiting the rich resources of timber and alluvial gold deposits, the notion of *tierras baldías* became equivalent to free access to anyone (Taussig 1979: 123). Successive governments granted concessions to entrepreneurs over lands that black communities had made collective use of in ways that did not require delimitations in space in terms of fixed boundaries. Whereas private property amongst rural black populations does exist and is clearly delimited in space – mainly by referring to natural boundaries such as streams, rocks or trees – the *respaldo de monte*, the hinterland is perceived as open collective space. It is the public space per se, which does not require clearly established boundaries. This space, rather, is characterized by fluid boundaries and a powerful interethnic territorial understanding between black and indigenous groups that cohabit in this region. These two groups have in fact shared this space for hundreds of years and have created overlapping territorialities. Afro-Colombians are generally allowed to enter and use what is known to be indigenous collective territory and vice versa, if their respective activities do not infringe upon the other ethnic group's territorial rights. We might also refer to these local epistemologies as a 'tolerated territoriality', a territorial accord that consists of fluid boundaries, which are nevertheless marked and respected in imaginary space. As the Colombian geographer Patricia Vargas comments on these shared understandings among indigenous and black communities in her work on social cartography in the Pacific region:

> Between neighbouring groups there exist *fluid territorial and social boundaries* crossed by relations of co-operation and of commerce. Therefore, the resources or the land that belong to one group can be used by others if the social relations are sufficiently close in order to turn strangers into practical members – yet, without them acquiring rights. (Vargas 1999: 149; our translation and emphasis)

Interestingly enough, it is now the very legislation regarding the creation of collective land titles for black communities that has caused some degree of interethnic conflict. Given the state's demand for clearly defined boundaries that enable the cartographic representation and fixation of the lands to be titled, local communities were required to draw up maps with fixed boundaries in their land rights application to Colombia's Agrarian Reform Institute, INCORA. Local geographical imaginations and mental territorial bounding processes were in fact disciplined by the modern territorial state's spatial logic. The imposition of fixed boundaries onto local epistemologies of 'fluid territorial and social boundaries' forces local communities to translate their territorial aspirations onto maps which Western-style institutions will accept as legitimate documentation accompanying land rights claims.[6]

As a result, on occasion indigenous groups felt their territoriality was violated by borderlines drawn up by black communities that would interfere with their previously established reserves. In order to preclude interethnic conflicts from breaking out, a number of mechanisms were put in place. Interethnic committees were set up, for example, facilitating discussion and negotiation between indigenous groups and black communities, including representatives of the communities involved and government officials. It was these committees that made the final decisions over the delimitation of collective land rights claims. The

---

6   Elsewhere one of us has shown how these two very different ways of knowing and representing the world were reconciled by black communities in heated debates in workshops on social cartography (Oslender 2007a). There, local dwellers created mental maps for the areas that they considered should be titled collectively to their community, with the oral tradition playing a crucial role in eventually 'disciplining' the mental fluid boundaries and rendering them stable as lines drawn on paper.

potential for interethnic conflict is a little discussed but important side effect of the legislation, and one which has to be judged negatively. Yet, given the increasing penetration of external capital in the Pacific coastal region, the drawing of clearly established and fixed boundaries can be seen as protecting the land rights of both indigenous and black communities. Potential territorial conflicts between the two groups may be considered the lesser evil.

To date 132 collective land titles have been issued to black communities in the Pacific coastal lowlands. These titles cover some five million hectares, roughly half the size of the entire region. This is a huge achievement for the social movement of black communities in the country that has mobilized on an unprecedented scale. Afro-Colombians have stepped out of the structural invisibility and marginality in which they had been held for hundreds of years by the dominant nation narrative of *mestizaje*. And although this is only the first step out of the tunnel of racist discrimination, it is a big leap.

The ecological significance of rural black territorialization processes has also been recognized at an international level, for example, with the award of the prestigious Goldman Environmental Prize to a prominent Colombian black activist.[7] Of course, the collective land titling should not be understood as a mere philanthropic gesture by the Colombian state. In fact, the issuing of land rights may be interpreted as a strategy of 'employing' rural black populations as 'guardians' of the fragile ecosystem's biodiversity and its intrinsic future commodity value for pharmaceutical exploitation (Escobar 1997; Wade 1999). However, there is no doubt that this legislation, for the first time in Colombia's history, officially empowers black communities to exercise control over the natural resources of their lands so that enterprises interested in their exploitation will have to deal directly with local communities. That, at least, is the ostensible value of this legislation. In practice, the recent escalation of Colombia's armed conflict into the Pacific region has painfully shown its limitations. Black peasants and fishermen, caught up in the crossfire of the various armed groups, are increasingly forcefully displaced from their lands, as the Pacific region is transformed into geographies of terror and landscapes of fear (Oslender 2007b, 2008).

Conclusions

The case study of Colombia's black communities is but one in a range of alternative territorial authority regimes that have become consolidated in this South American country in recent decades. These territorial regimes may not be

---

7　The Goldman Environmental Prize is considered the Nobel Prize for the Environment. It is given every year to grassroots ecological activists from six geographical world regions. Libia Grueso from Colombia's Process of Black Communities won the prize in April 2004 in the category South/Central America (see http://www.goldmanprize.org/node/106; last accessed 29 November 2010).

spectacular compared to more violent challenges to state territorial authority, such as FARC's temporary creation of an insurgent state within Colombia's boundaries. Yet, the effects of these low-intensity challenges may be more far-reaching. Rather than providing a radical alternative to the current state model, they complement it and may even enhance the legitimacy of the modern territorial state through democratic practice. As Mason (2005: 50) concludes in her observation of a range of authority alternatives in Colombia:

> Paradoxically, the state's legitimacy may be enhanced through challenges to, and the delegation of, its authority, to the extent that alternative social arrangements become a force for progressive reforms, norm observance, and the reconstitution of the state-society relationship.

This development is not restricted to Colombia either. Throughout Latin America constitutional reforms have opened up official ideologies and narratives of nationhood to notions of multiculturalism and pluriethnicity. These are often accompanied by the delegation of a certain degree of territorial power and autonomy to non-state actors. Social movements have played a crucial part in these challenges to exclusive state authority and the contestation of the space of the nation-state. What is at stake for many of the place-based social movements in Latin America is more than mere land rights that provide them with a space to be. It is in fact a radical redefinition of the nature of the state itself. Theirs is a struggle to open up the space of the nation-state for more democratic practices, in which otherness is not only accepted, but recognized as a fundamental part in the constitution of the state itself.

The struggle for alternative territorialities is a crucial point that is often overlooked in debates about state territoriality, both in international relations theory and political geography. Social movements in Latin America do not just carve out a space for themselves within their respective countries. They de facto contribute to increasing the state's degree of legitimacy and sovereignty, as their demands for territorial autonomy are met and they are acknowledged as alternative territorial authorities within the space of the nation-state. The rescaling of state functions by delegating a degree of territorial power to the local scale can thus be argued to extend or deepen democratic practice. Rather than posing a direct challenge to the territorial state, as revolutionary movements like the FARC do, Colombia's black communities show the complex territorial compromises at play that both empower local groups and assert state power in new ways by extending or deepening state institutions, norms, and juridical practices in parts of the country where the state previously had a weak presence. The overlapping territorialities that we propose as an analytical framework for understanding these multiply scaled processes are constitutive of the modern state itself.

References

Agnew, J. (1987) *Place and politics: the geographical mediation of State and society.* London: Allen and Unwin.

—— (2005) 'Sovereignty regimes: territoriality and state authority in contemporary world politics', *Annals of the Association of American Geographers*, 95(2): 437-461.

Agudelo, C. (2002) *Populations noires et politique dans le Pacifique colombien: paradoxes d'une inclusion ambigue.* Unpublished PhD dissertation, Paris: L'Université Paris III – Institut des Hautes Études de l'Amérique Latine, October 2002.

Assies, W., van der Haar, G. and A. Hoekema (eds) (2000), *The challenge of diversity: indigenous peoples and reform of the state in Latin America.* Amsterdam: Thela-Thesis.

Baruah, S. (1999) *India against itself: Assam and the politics of nationality.* Philadelphia: University of Pennsylvania Press.

Bergquist, C., Peñaranda, R. and G. Sánchez (eds) (1992) *Violence in Colombia: the contemporary crisis in historical perspective.* Wilmington, Delaware: Scholarly Resources Inc.

Bougarel, X. (1999) 'Yugoslav wars: the 'revenge of the countryside' between sociological reality and nationalist myth', *East European Quarterly*, 33(2): 157-175.

Bushnell, D. (1993) *The making of modern Colombia: a nation in spite of itself.* Berkeley: University of California Press.

Cairo, H. and W. Mignolo (eds) (2008) *Las vertientes americanas del pensamiento y el proyecto des-colonial: el resurgimiento de los pueblos indígenas y afrolatinos como sujetos políticos.* Madrid: Trama Editorial.

Davis, D. (2006) 'The age of insecurity: violence and social disorder in the new Latin America', *Latin American Research Review*, 41(1): 178-197.

Diario Oficial (1993), *Ley 70 de 1993.* Bogotá: Ministerio de Justicia.

Echandía, C. (1999) *Geografía del conflicto armado y las manifestaciones de la violencia en Colombia.* Bogotá: Vice President's Office.

Escobar, A. (1997) 'Cultural politics and biological diversity: state, capital, and social movements in the Pacific coast of Colombia', in R. Fox and O. Starn (eds) *Between resistance and revolution: cultural politics and social protest.* New Brunswick: Rutgers University Press, 40-64.

—— (2008) *Territories of difference: place, movements, life, redes.* Durham and London: Duke University Press.

García Linera, Á. (2006) 'State crisis and popular power', *New Left Review*, 37: 73-85.

Grueso, L., Rosero, C. and A. Escobar (1998) 'The Process of Black Community organizing in the southern Pacific coast region of Colombia', in S. Alvarez, E. Dagnino and A. Escobar (eds) *Cultures of politics, politics of cultures:*

re-visioning Latin American social movements. Oxford: Westview Press, 196-219.

Hansen, T. B. and F. Stepputat (2001) 'Introduction: states of imagination', in T. B. Hansen and Finn Stepputat (eds) *States of imagination: ethnographic explorations of the postcolonial State*. Durham: Duke University Press, 1-40.

Hechter, M. (1975*) Internal colonialism: the Celtic fringe in British national development, 1536-1966*. Berkeley and Los Angeles: University of California Press.

Hirst, P. (2005) *Space and power: politics, war, and architecture*. Cambridge: Polity Press.

Hunt, S. (2006) 'Languages of stateness: a study of space and el pueblo in the Colombian State', *Latin American Research Review*, 41(3): 88-121.

Hylton, F. (2003) 'An evil hour: Uribe's Colombia in historical perspective', *New Left Review*, 23: 50-93.

Keating, M. (1998) *The new regionalism in Western Europe: territorial restructuring and political change*. Cheltenham: Edward Elgar

Krasner, S. (1995) 'Compromising Westphalia', *International Security*, 20(3): 115-151.

Lake, D. (2003) 'The new sovereignty in international relations', *International Studies Review*, 5: 303-323.

Leal Buitrago, F. and L. Zamosc (eds) (1991) *Al filo del caos: crisis política en la Colombia de los años 80*. Bogotá: Tercer Mundo Editores.

McColl, R. (1969) 'The insurgent state: territorial bases of revolution', *Annals of the Association of American Geographers*, 59(4): 613-631.

Mann, M. (1986) *The sources of social power: a history of power from the beginning to A.D. 1760*, Vol. 1. New York: Cambridge University Press.

Manogaran, C. (1999) 'Space-related identity in Sri Lanka', in Guntram H. Herb and David Kaplan (eds) *Nested identities: nationalism, territory, and scale*. Lanham MD: Rowman and Littlefield.

Mason, A. (2005) 'Constructing authority alternatives on the periphery: vignettes from Colombia', *International Political Science Review*, 26(1): 37-54.

Molano, A. (1992) 'Violence and land colonization', in C. Bergquist, R. Peñaranda and G. Sánchez (eds) *Violence in Colombia: the contemporary crisis in historical perspective*. Delaware: Scholarly Resources Inc., 195-216.

—— (1994) *Trochas y fusiles*. Bogotá: El Ancora Editores.

Moraña, M., Dussel, E. and C. Jáuregui (eds) (2008) *Coloniality at large: Latin America and the postcolonial debate*. London: Duke University Press.

Offen, K. (2003)'The territorial turn: making black territories in Pacific Colombia', *Journal of Latin American Geography*, 2(1): 43-73.

Osiander, A. (2001) 'Sovereignty, International Relations and the Westphalian myth', *International Organization*, 55(2): 251-287.

Oslender, U. (2004) 'Fleshing out the geographies of social movements: black communities on the Colombian Pacific coast and the aquatic space', *Political Geography*, 23(8): 957-985.

—— (2007a) 'Re-visiting the hidden transcript: oral tradition and black cultural politics in the Colombian Pacific coast region', *Environment and Planning D: Society and Space*, 25(6): 1103-1129.

—— (2007b) 'Spaces of terror and fear on Colombia's Pacific coast: the armed conflict and forced displacement among black communities', in D. Gregory and A. Pred (eds) *Violent geographies: fear, terror, and political violence.* New York: Routledge, 111-132.

—— (2008) 'Another history of violence: the production of 'geographies of terror' in Colombia's Pacific coast region,' *Latin American Perspectives*, 35(5): 77-102.

Pécaut, D. (2001) *Guerra contra la sociedad.* Bogotá: Planeta.

Pizarro, E. (1987) *La guerrilla en Colombia.* Bogotá: Cinep.

—— (2004) *Una democracia asediada: balance y perspectivas del conflicto armado en Colombia*, Bogotá: Norma.

Restrepo, E. (2004) 'Ethnicization of blackness in Colombia: toward de-racializing theoretical and political imagination', *Cultural Studies*, 18(5): 698-715.

Richani, N. (2002) *Systems of violence: the political economy of war and peace in Colombia.* Albany, NY: State University of New York Press.

Roldán, M. (2002) *Blood and fire: La Violencia in Antioquia, Colombia, 1946-1953.* Duke University Press.

Sack, R. (1986) *Human territoriality: its theory and history.* Cambridge: Cambridge University Press.

Safford, F. and M. Palacios (2002) *Colombia: fragmented land, divided society.* Oxford: Oxford University Press.

Safran, W. (ed) (1998) *Nationalism and ethnoregional identities in China.* London: Routledge.

Sánchez, F. and M. Chacón (2005) 'Conflict, State and decentralisation: from social progress to an armed dispute for local control, 1974-2002', Crisis States Programme Working Paper No. 70, London: LSE (available at: http://www.crisisstates.com/download/wp/wp70.pdf).

Sánchez, G. and D. Meertens (2001) *Bandits, peasants, and politics: the case of 'la violencia' in Colombia.* Austin: University of Texas Press.

Sieder, R. (ed) (2002) *Multiculturalism in Latin America: indigenous rights, diversity and democracy.* Houndmills: Palgrave Macmillan.

Taussig, M. (1979) *Destrucción y resistencia campesina: el caso del Litoral Pacífico.* Bogotá: Editorial Punta de Lanza.

—— (2004) *My cocaine museum.* London: The University of Chicago Press.

Van Cott, D. L. (2000) *The friendly liquidation of the past: the politics of diversity in Latin America.* Pittsburgh: University of Pittsburgh Press.

—— (2005), *From movements to parties in Latin America: the evolution of ethnic politics*, Cambridge: Cambridge University Press.

Vargas, P. (1999) 'Propuesta metodológica para la investigación participativa de la percepción territorial en el Pacífico', in J. Camacho and E. Restrepo (eds)

De montes, ríos y ciudades: territorios e identidades de la gente negra en Colombia. Bogotá: Ecofondo/ICAN/Fundación Natura, 143-176.

Wade, P. (1999) 'The guardians of power: biodiversity and multiculturality in Colombia', in A. Cheater (eds) *The anthropology of power: empowerment and disempowerment in changing structures*. London: Routledge, 73-87.

Yiftachel, O. (1998) 'Nation-building and the division of space: Ashkenazi domination in the Israeli "ethnocracy"', *Nationalism and Ethnic Politics*, 4(3): 33-58.

——(1999) 'Between nation and state: "fractured" regionalism among Palestinian-Arabs in Israel', *Political Geography*, 18(3), 285-307.

Chapter 7

LimiteLimite: Cracks in the City, Brokering Scales, and Pioneering a New Urbanity

Johan Moyersoen and Erik Swyngedouw

Introduction

In 1998 a rather unusual nine-metre high artistically designed tower (see Figure 7.1) was constructed in one of Brussels' most deprived neighbourhoods, the Brabant neighbourhood (for details, see http://www.citymined.org/projects/limitelimite.php). The building became a landmark construction in the area between 1999 and 2004. This project was an emblematic and highly visible component of a wider and innovative project – called LimiteLimite – of urban re-development in the area.

In August 1998 a small coalition of four individuals decided to initiate the LimiteLimite project. This chapter seeks to unravel the social and institutional dynamics of how a small group of pioneering urban activists engaged with some of the key actors in a deprived neighbourhood of Brussels in this urban renewal process. The LimiteLimite project's most visible realization was the LimiteLimite tower, but it also initiated a chain of spin-off projects and influenced the institutional and associational configuration of urban governance in important ways. Examples of spin-off projects were, among others, a neighbourhood festival, a flower project, a women's project and a variety of artistic interventions. LimiteLimite evolved incrementally by realizing a sequence of small projects through the painstaking crafting of unprecedented partnerships that brought together actors active at the neighbourhood levels (local civil society groups) with agents operating at the metropolitan scale (firms, regional government, higher education institutions, national foundations, among others). The initiating core group included Steven Degraeve, a social worker from RisoBrussel, a non-profit association involved in community work; Jim Segers, working for City Mine(d), a production house for social-artistic projects in the city; Chris Rossaert, an independent architect; and Jacques Lechat, a teacher at APAJ Classe Chantier, an apprenticeship and vocational training school. They had no track record of collaborating and thus very little reputation, financial means or power to start, let alone impose, an urban renewal process. But they shared a common interest in urban life and re-development and, in particular, in the disintegrated urbanity that characterized the Brabant neighbourhood. They saw a need to pioneer socially innovative arrangements to organize and implement urban planning and redevelopment in a socio-spatial environment of institutional disorganization, intense socio-spatial conflict and complete absence of co-ordinated urban interventions from existing institutions such as the municipality or the Brussels Regional Government. Their objective was to initiate a process of urban renewal within this perceived ecology of impossibilities. Consequently, they sought innovative ways to mobilize actors in the neighbourhood, engage with professional urban planners and managers, and gain access to key political and economic power brokers.

To gain social leverage, LimiteLimite effectively appropriated a 'structural hole' in social space. 'Structural hole', or vacant space (see Swyngedouw 2000), is defined here as locations in the city where there is a socio-political power vacuum among contentious actors in situations of open competition (conflict) and/or non-communication (Burt 1992). The project's initiators thus exploited the 'vacant' institutional and socio-political space between the diverse and non-communicating local actors in the Brabant neighbourhood (the intra-urban realm) and the extra-local/metropolitan-global actors (the extra-urban realm). They strategically located and conceptualized the LimiteLimite tower in such a way that the project would appeal on a broader scale that was far too big for the local actors to manage alone. At the same time, the project was too locally embedded for the extra-local actors to claim it as their own realization alone.

In short, the LimiteLimite project mobilized through an emblematic vehicle to bring heterogeneous socio-economic and political actors together and to put the Brabant neighbourhood on the regional agenda. Architect Chris Rossaert designed a highly visible nine-metre high translucent tower that protruded into the street, serving as a meeting and exhibition space. Through Wijkpartenariaat (a local partnership arrangement), local residents were involved in the design and building process. APAJ-Classe Chantier, an apprenticeship training centre that prepares local unemployed for jobs in the construction industry, trained a number of its students through the construction of the tower. The construction and use of the building served as a catalyst to bring together disparate groups from both within and outside the neighbourhood. JP Morgan Guarantee Trust Company financed the structure, but also took responsibility for keeping the new LimiteLimite network together. A number of local higher education establishments – the Vlaamse Economiische Hogeschool (VLEKHO) (economics and business studies), Sint-Lucas Architectuurschool (School of architecture), and the College for Social Work (both higher education establishments) – participated with their students in one or more stages of the project. Local shopkeepers also participated in the network. The tower was dismantled in 2004 to make place for a social housing complex and re-erected in Belfast (for a project called ReLimite) to be part of a similar project in an attempt to 'globalize' the initiative (see http://www.citymined.org/projects/relimite.php), but the organization, LimiteLimite vzw, continued to work in the area until 2008. After that period, several of the projects initiated by LimiteLimite continued under a variety of different institutional and organizational forms.

Figure 7.1   LimiteLimite

The tower's construction and the chain of spin-off initiatives it propelled fostered the formation of an inclusive and multi-scalar partnership that was, at the time, unparalleled in Brussels. LimiteLimite was a reaction against a governance impasse provoked by the disjunction between the deprived and disintegrated reality of the neighbourhood on the one hand, and the extra-local pressure to accommodate in the neighbourhood companies serving the global service economy, on the other. The project's initiators were able to gain significant social leverage by assuming – in social space as well as in physical place – the role of the 'tertius', that is, a brokerage role between the local and extra-local governance realms of the Brabant neighbourhood (Simmel 1955; Burt 1992, 2005).

In this chapter, we explore how a broker or 'tertius,' occupying a 'structural hole' (Burt 1992), organized hitherto unrelated heterogeneous actors and mobilized alliances across scales, producing 'glocal' networks of association to initiate projects of urban change and redevelopment, despite frequently contentious interests and entrenched practices of non-communication. We explore how local actors can find mobilization opportunities in contexts of disarticulated inter-scalar governance, and how such opportunities are related to the ability to facilitate and control information flows among actors positioned at different scales of governance. We shall also show how the 'tertius' position enables actors to define terms of innovation and development and how the continued power of local actors depends on their ability to continuously control information. In the process, power relations may change as new actors enter alliance networks.

The project initiators, through their unique positioning as the tertius between the local and extra-local scale, were able to mobilize the diverse positive potentialities in the cosmopolitan city of Brussels, and thus produce a process of 'glocal' (local and extra-local) empowerment. Initially bound by the uncertain, fuzzy and disjointed socio-spatial conditions characteristic of the Brabant neighbourhood, the project was unable to mobilize extensive support from government or other dominant market, state or civil society institutions in Brussels. This led the project's initiators to experiment and engage with alternative *path-dependent* development strategies, notably through processes of effectuation and complex good provision, that could effectively bring together – in each phase of the project – actors from the local and extra-local spheres. These glocal empowering processes re-configured, rescaled and dramatically re-ordered the social geometry of power in the neighbourhood in an incremental and *socially innovative* manner. However, this re-ordering also had socially undesirable, disempowering side-effects. The project's initiators could only realize the full potentiality of the tertius-role when they could monopolize their position. To maintain their socially unique positioning, the project's initiators were not always inclined to promote consensus between the fragmented groups, but rather had paradoxically strategic incentives to maintain division.

Context

LimiteLimite emerged in response to the fragmented, diffuse and overlapping governance structures that were operative in the Brabant neighbourhood (see Figure 7.2) and negatively affecting neighbourhood participation. Two distinctive inter-scalar processes can be identified. On the one hand, a series of urban redevelopment projects with global reach and ambition, i.e., the Manhattan Project, had been implemented to the west of the Brabant neighbourhood. The global service and administrative functions and activities of the Manhattan Project clashed with the needs of local residents. On the other hand, the multi-scaled governance structure, including restrictions on voting rights for foreigners, entrenched clientelistic practices by municipal political clans, and political apathy on the part of the most disenfranchised residents. In this way, inter-scalar processes produced powerful grievances but also closed off political space for the expression of discontent. Despite the culturally rich and attractive amenities in the neighbourhood, such as the multi-cultural atmosphere, the vibrant commercial street specializing in Middle-Eastern and south Mediterranean goods, the variety of higher-education colleges and the proximity of Brussels' sole big business district, deeply entrenched antagonisms and competition between different municipalities and between local and regional authorities, combined with a heterogeneous local public, rendered any citizen action in the neighbourhood virtually impossible (Degraeve 1999) and stalled any attempt to initiate a concerted process of urban renewal.

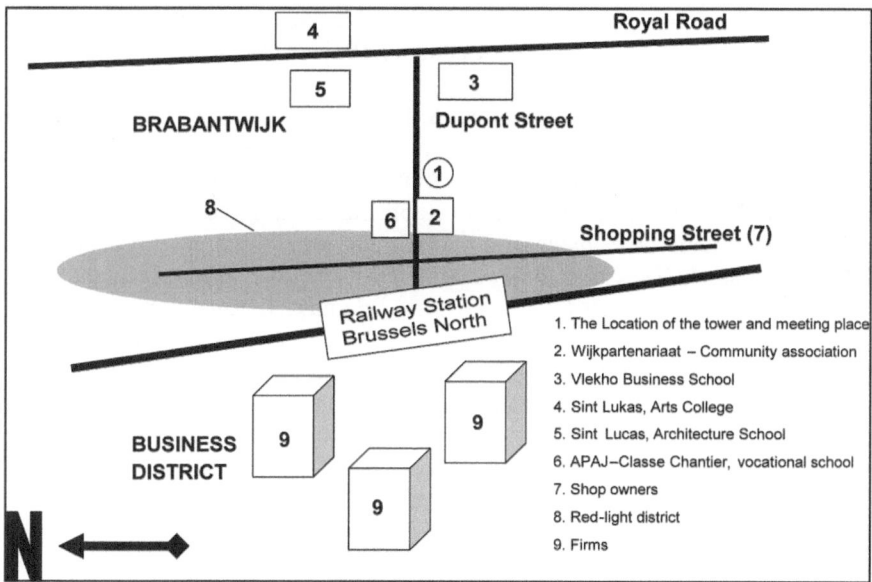

Figure 7.2   The Brabant neighbourhood and LimiteLimite

The Brabant neighbourhood was essentially squeezed between two 'governance realms' operating on radically different scales. Over the past thirty years, on the western side of Brussels North railway station, as sketched in Figure 7.2, the economic and political elites of the city have been developing the city's new business district, accommodating headquarters of major companies as well as key administrative services. The Brabant neighbourhood is located on the other (that is, eastern) side of the railway station. The neighbourhood is one of Brussels' most densely populated areas. Originally white working class, the social composition of the area has, over the years, become more mixed as migrants (mainly from Turkey and Morocco) moved in. The area is also host to a thriving red-light district and street prostitution. These two distinct 'realms', on either side of Brussels North station, embodied radically different tensions in the city. The first extra-urban (or global) realm is where Brussels was competing with other cities in an attempt to attract part of the global service economy. The second intra-urban realm is the everyday and finely-textured neighbourhood living space of a highly diverse, but poor and socially excluded, local community.

The extra-urban realm or spatializing the global aspirations of Brussels

From the sixties onwards, the Brussels North area came to be emblematic of the pursuits of Brussels' political and economic elites to transform the city from a provincial town into a modern and globally competitive city. The densely populated working class area on the western side of the Brussels North railway station was bulldozed to make way for a one-kilometre long business district. These new developments became known as the 'Manhattan Project' (Martens 1994), reflective of the Brussels' elites' aspirations to turn Brussels into the European counterpart of New York's Manhattan (Papadopoulos 1996). It was widely anticipated that the Manhattan Project would expand into and replace surrounding popular neighbourhoods, including the Brabant neighbourhood. The latter, consequently, became a key target for speculative development, as suggested by its increasing vacancy rates, underinvestment, and overall deterioration of the physical fabric and housing stock (De Corte 2001, 2002). This negative spiral accentuated the neighbourhood's reputation for low-cost housing attracting new immigrants, on the one hand, and the further expansion of a thriving low-end sex economy, on the other. The disinvestment drove the Brabant neighbourhood further downwards on the slippery slope towards social deprivation and marginalization.

The new urban policy of the late nineties characterized by public–private partnerships, narrow economic policy targets and city marketing, was highly influenced by the aspirations and desiderata of the entrepreneurial elite. Their agenda, however, was often incompatible with the particular demands for mediation and adjustment raised from the diverse, heterogeneous, and multicultural – but non-communicative – space that typified the Brabant neighbourhood (Moulaert et al. 2003). The economic requirements for enhanced business-related services clashed with the everyday demands (e.g., jobs, childcare, community centres, green spaces and recreation areas) of the local inhabitants. Company headquarters like those of Banksys, Belgacom and Euroclear, as well as administrative services of the Federal Government and Flemish Commission, moved into the newly built Manhattan Project. The newly created jobs, however, did not match the locally available labour supply, but relied on higher-skilled commuters from the more affluent suburbs. As mentioned above, the Brabant neighbourhood was primarily a working class neighbourhood, with 58 per cent of employment in blue-collar activities in 1997, and only 25 per cent of the active population in white-collar employment. By 1997, the unemployment rate in the Brabant neighbourhood had risen to 25 per cent (BRES 1997) and in 2001 the average household income in the Brabant neighbourhood was 40 per cent below the average for the Brussels Capital Region (De Corte 2001). Thus, the protagonists of the new urban policy perceived local conditions and demands as dead weight obstructing their extra-local aspirations.

Intra-urban realm or the contradictions of multi-scalar governance in Brussels

The systematic neglect of the Brabant neighbourhood and the absence of any coherent urban redevelopment initiative for the area was not only caused by extra-local adjacent developments like the Manhattan Project, but also by the complex multi-scalar governance dynamics of Brussels (Kesteloot and Mistiaen 2002). In 1993, Belgium became a federal state, constituted by two Community Commissions (Flemish and French) and three Regional Governments (Flanders, Wallonia, and Brussels), in addition to the national scale government (Swyngedouw and Moyersoen 2006). In Brussels, this led to the intriguing situation of the French and Flemish Community Commissions overlapping. The most important competences of the Community Commissions are education, health, and cultural policy; their policies differ for French and Flemish residents. The governance of the Brussels Region, in contrast, is bilingual and responsible for 'territorial' matters such as public transport, economy, environment, urban planning and public works. However, the creation of these new layers and scales of governing did not displace the old, incumbent structure of local governance in the Brussels Region. That is, in addition to the Community Commissions and the Brussels Region, the city-region also consists of 19 autonomous municipalities, each with an elected council. They tend to be highly parochial, organized strongly around local party bosses and with strong inter-municipal competition. The Brabant neighbourhood is partly located in the municipality of Sint-Joost Ten Noode and partly in Schaarbeek. The resulting institutional configuration – complex, fuzzy, contradictory, often competing, and little understood – has created a deep rift between local residents and the institutions of governance (Swyngedouw and Baeten 2001). This alienation is further aggravated by the fact that non-European citizens are not allowed to vote, not even at the municipal level. This means that roughly 30 percent of the population in the Brabant neighbourhood cannot participate in the formal political process. Not surprisingly, politicians have tended to neglect the area and focus instead on more secure constituencies with bigger electorates. The neighbourhood consequently has been serviced very poorly.

The friction between the non-residential functions in the neighbourhood and the demands of the inhabitants has further distorted the social network structures. Approximately 3,000 students attending the numerous higher education institutions located in the proximity of the Brabant neighbourhood, as well as a large number of employees working in the service economy, pass through the area every day. This is in addition to the various clients frequenting the red light district and its amenities. The daily morning-evening-night routine has created enormous tensions in the neighbourhood, giving rise to an upsurge of petty theft and criminality (Degraeve 1999). These incidents have further fuelled an already strong, if subjective, sense of insecurity. They obliged, for example, firms and schools to take initiatives to 'protect' their employees and students against the neighbourhood.

These tensions and frictions, as summarized in Table 7.1, led to a situation whereby each action in the area taken by one group was seen and experienced as an imposition upon the other. Residents and user groups saw each other's actions as threats to their social, economic, or cultural identity, or to their security and sense of fairness and justice (Swyngedouw and Moyersoen 2006). Years of fragmentation and neglect led to the disappearance of community networks and organizations from which socially innovative projects could emerge. Instead, agents were acting in an autonomous, fragmented manner, myopic (or even hostile) to other initiatives and interests in the neighbourhood. It is precisely this space of friction and non-communication, this 'vacant' interstitial space, where the initiators of LimiteLimite inserted themselves and created a fecund arena for new, innovative, and inclusive urban collective action.

Table 7.1   The governance impasse between the intra- and extra-urban realm in the Brabant neighbourhood

INTRA-URBAN Local (neighbourhood)	EXTRA-URBAN Metropolitan (global)
Friction 1 Alienation Complex government structure Contentious Everyday life	**Friction 1** Degeneration of the neighbourhood Real-Estate Speculation Disinvestment
Friction 2 Isolation Voting right Non-EU Migrant Community	**Friction 2** Governance polarization Transparent business-climate Heterogeneous city
Friction 3 Disjunction Residential functions Non-residential functions	**Friction 3** Unemployment Growth service economy Decline labour intensive sector
FRAGMENTATION AND POLARIZATION	**DISJOINTED AND TRANSIENT NEIGHBOURHOOD**

LimiteLimite and the pioneering role of the *tertius*

The construction of the LimiteLimite tower functioned as a physical and social intermediary to bridge, in a unique way, what we defined above as the intra- and extra-urban socio-spatial scales. Filling in a 'structural hole' critically enabled the project's initiators to adopt the role of the *tertius*, or mediator/broker, among groups in conflict or in a position of non-communication. Simmel (1904 (1955)) has argued that the position of the tertius in the social 'gestalt' of conflict is one of the most catalysing and vivid forms of interaction and is in essence a way to achieve (positive or negative) unity by resolving or mediating divergent interests and dualisms. Conflict is thus an unstable state of social disunity that strives for convergence by mutual cooperation, dominance or annihilation (Simmel 1904 (1955)). Taking up the unique position of the tertius in the vacant space between the local and the metropolitan essentially allowed the project's initiators to take advantage of the contentious and unstable dynamics of actors in conflict and initiate a collective process of innovative networking, articulated through the LimiteLimite tower project. The tertius takes a unique and monopolist position. This exclusive position accrues from a situation of scarcity in cross-scalar social relations. In the networking process, two distinct strategies pursued by the project's initiators, linked to the tertius position, can be identified: *Tertius Gaudens* (i.e., the third which rejoices) and *Divide et Impera* (i.e., divide and rule) (Simmel 1902; Burt 1992).

The first strategy of *Tertius Gaudens* relates to the non-partisan aspect of the tertius position, enabling the tertius to function as an information bridge between diverse and competing actors that do not have strong relationships with one another and are mainly related through a third party (Simmel 1902; Simmel and Wolff 1950). The LimiteLimite tower was built in an area of stalemate, where very few channels of communication were available among the groups using the area. The project's initiators thus took up the strategically advantageous role of 'information bridge' (Granovetter 1983) and catalyst to generate communication among local and metropolitan groups. The LimiteLimite process was characterized by an array and sequence of different working groups, mostly composed of local and metropolitan actors, whereby a member of the core group acted as a bridge among different actors and working groups. This position of information conveyor also mitigated possible free-riding problems. Indeed, if one of the user groups decided to free-ride, it removed itself from the communication benefits that the project initiators provided. The threat of being deprived or excluded from the communication link managed by the project initiators gave the latter considerable leverage, including the power to hold groups together in collective action processes.

The second strategy of *Divide et Impera* refers to a situation whereby a third party actively separates participating parties in order to attain or maintain power (Simmel 1902). With the local and metropolitan groups uncertain about whose preferences would dominate the relationship, the project's initiators found themselves in a broker position, able to pit the divergent demands of the groups against one another. This divisive strategy enabled the core project group to maintain and monopolize its position as tertius among the various participating groups and partners. The interest of the core project group was of course articulated around a desire to pioneer urban re-development by fostering new associational arrangements that bring together unlikely partners in a process of collective action. By doing so, a new inter-scalar 'gestalt', bringing together local and global actors, has been actively constructed.

Although the role of tertius empowered the core group to mobilize a wide range of diverse user groups under the umbrella of LimiteLimite, it also had disempowering side-effects. The tertius role can be usefully conceived of as a positional good, that is, a role that is sensitive to social congestion. Social congestion refers to the tendency of those who do not possess positional goods (status goods) to aspire to acquire them (Hirsch 1976). The tertius only achieves its full potentiality when the position can be monopolized in an exclusive manner. Consequently, any actor in the neighbourhood that challenges the project's initiators in their tertius role may undermine the latter's power position. Therefore, in order to maintain their socially unique positioning, the project's initiators were not always inclined to promote consensus within highly fragmented networks. We can usefully understand the processes underlying the governance effects of LimiteLimite as a dialectical relationship between the competitive claims of the project's initiators in their tertius role, and their endeavours to initiate inclusive collective action. This dialectical relationship, as we shall discuss in the following section, is also present in the two main socially innovative processes that the project has fostered: effectuation and complex good provision.

The drivers of social innovation

The positioning of the LimiteLimite project in a 'structural hole' (i.e. a socio-spatial configuration characterized by conflicting positions and non-communicative relations) set the initiative in an 'institutional void', that is to say a space characterized by both absence of codified and coherent institutional arrangements and the lack of agreement on the rules and norms that should conduct or guide the policy making process (Hajer 2003). Indeed, the 'structural hole' in-between the local and extra-local actors was a space where the codified framework and institutional order, which traditionally provide the stage for organizing and implementing urban planning, had fallen apart. Policy actions became caught in a 'double dynamic' of negotiation: negotiation not only about which action to undertake, but also, and perhaps more importantly, which norms, rules and institutional framework should guide the policy-making process. In other words, the fuzziness and distrust that characterize multiply contested areas impeded traditional consensus building, co-ordination and governance processes around aims, procedures, phasing and objectives. LimiteLimite thus needed a fundamentally different strategic vision and innovative coordination processes to successfully achieve collective action.

Below, we advance two alternative strategies for social innovation with which the LimiteLimite core group experimented: effectuation and complex good provision.

Effectuation

Sarasvathy (2001) defines 'effectuation' as a path-dependent and means-driven strategy to develop action in an environment of radical uncertainty. Radical uncertainty is a condition under which no one can anticipate the outcome of a particular course of action. Sarasvathy (2001) illustrates the difference between predictable strategies, where one can anticipate the future, and strategies in situations of radical uncertainty, where the future of a process is unpredictable, by drawing a parallel between two different ways to cook a meal:

> Imagine a chef assigned the task of cooking dinner. There are two ways the task can be organized. In the first, the host or client picks out the meal in advance. All the chef needs to do is list the ingredients needed, shop for them, and actually cook the meal. This is a process of causation. It begins with a given menu and focuses on selecting between effective ways to prepare the meal. In the second case, the host asks the chef to look through the cupboards in the kitchen for possible ingredients and utensils and then cook a meal. This is a process of effectuation. It begins with given ingredients and utensils and focuses on preparing one of many desirable meals with them (Sarasvathy, 2001).

Since the project's initiators had to rely each time on the commitment of financial, human and social resources of the partners to realize their goals, they were forced to negotiate, on an ongoing basis, the objectives and targets (policy) of the project and the norms and rules (polity) that would guide the project. They deployed the tertius instruments, *Tertius Gaudens* and *Divide et Impera*, to encourage participants to commit financial, human and social capital. From the perspective of the project's initiators, LimiteLimite became a means-driven rather than a goal-driven project. The capacities, commitment and desires of each of the partners drove the project rather than a priori agreed targets or objectives. The project's initiators behaved, much like the cook looking in his cupboard for possible ingredients, as a mediator mobilizing resources and means to hedge or assure the future dynamics of LimiteLimite. Hence, each step in LimiteLimite's development process was the product of a deliberation process among different groups linked through the tertius, each with their own interests, resources and expertise. In this way, the process of effectuation became an empowering and bottom-up participatory process.

Of course the various interacting social networks were not static but rather highly dynamic and erratic. The contentious choreographies of uneven socio-spatial power relations and, once started, the processes of LimiteLimite itself have reshaped and reconfigured the social networks in the neighbourhood. To hold on to the tertius role, the project's initiators had to constantly reposition themselves in line with the ongoing and fluid social and embryonic institutional dynamics. As a result, LimiteLimite has evolved as a highly path-dependent process, involving a sequence of small collective actions, each time with another partnership. Figure 7.3 summarizes this process. In each collective action, the project's initiators repositioned themselves and worked further with the actors dependent on the particular social dynamics, available expertise, and commitment present at that moment. The necessity to continually initiate cross-boundary partnerships among local and extra-local actors made LimiteLimite a highly socially inclusive process. In addition to the partners mentioned above, the process of LimiteLimite engaged with a variety of other local groups such as VIVA (a women's organization), Wijkpartenariaat (a community work association) and Sports Hall 58 (a sports association). Also, the project mobilized prominent extra-local actors such as the RISO (Regional Institute for Community Development), private companies, and The Koning Boudewijn Stichting (a private foundation supporting social, cultural and art projects in Belgium) which hitherto had not taken any active role in participating in the governance dynamics of the neighbourhood. This particular process of effectuation, however, made it difficult for the project's initiators – at the onset of the project – to develop a structural relationship with government bodies. Government institutions unfamiliar with the process, and certainly hesitant to accept that they had lost the capacity to control the agenda and regulate the area in a coherent way, preferred to steer away from LimiteLimite. It was only when the project needed authorization and sought financial support that ad hoc arrangements and relations with government bodies were initiated. In the main, the stakeholders were relatively autonomously positioned vis-à-vis the state and, much like a social movement, dictated the agenda, the norms, rules and timing of the LimiteLimite project in a tactical innovative way (McAdam et al. 2001).

The process of effectuation resulted in LimiteLimite becoming a sequence of successive small projects (see Figure 7.3 for an overview of the projects). There were two main reasons this happened. Firstly, once a project started, it was difficult for new actors to join the already launched collective action as the development strategy was grounded in the stakeholders' *ex ante* commitment. Hence, new actors were invited to start spin-off projects. Secondly, each stage of the project was limited in its objectives as these were the outcome of prior negotiations to mobilize resources. The creation of a spin-off project was an effective way for new actors to participate and for committed new activists to push forward their agenda for the locality. These two dynamics generated a chain reaction of spin-off projects that would be brought together under the umbrella network of LimiteLimite. The institutional architecture of this process of participation took the form of a 'modular system'. Each LimiteLimite spin-off initiative, as highlighted in Figure 7.3, can be usefully thought of as a module within a larger system (Simon 1981; Sarasvathy et al. 2003). Such a module essentially consisted of a partnership between local and metropolitan actors, initiated by the tertius to realize a specific sub-project for the neighbourhood. This process of modularity created a particular dynamic form for the overall collective action (Simon 1981). Negotiations between the diverse participants in each module set the agenda for the overall process. Participants in each module interacted much more intensely and easily amongst each other than with participants from other modules (or spin-off initiatives). In the short run, the high interaction between stakeholders in one spin-off initiative was more or less independent of other initiatives. In the long run, the sporadic interaction across the various projects affected the overall process in a positive aggregate manner. As a result, LimiteLimite became a self-directed and self-organized bottom-up process regulated by stakeholder negotiations.

Possible objectives of each stage of the project were open for negotiation, providing a basis for discussion and interaction. The core group sought to hedge the future by mobilizing and engaging stakeholders' commitment through negotiations for the required resources, prior to the start of each project. For example, building the LimiteLimite tower was not a pre-planned objective of the core group. It emerged as the outcome of a deliberation (negotiation) process among the network participants. In this way, the process of effectuation became an empowering and participatory process open to new possibilities, ideas, and alternatives. As a result, effectuation enabled LimiteLimite to emerge as a spontaneous and creative (even artistic) process of, and for, different user groups.

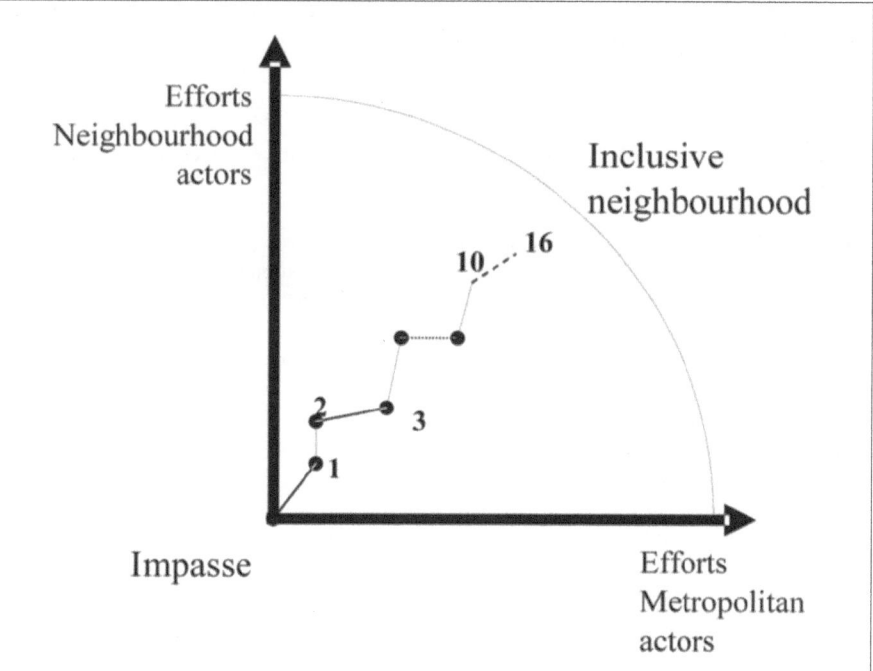

1. (31.10.1997) Submission of an application for funding in the program 'Extraordinary neighbourhood' to the King Boudewijn Foundation, a private foundation sponsoring cultural, social, and related projects. Partners: Wijkparternariaat, City Mine(d), Euroclear NV, Municipality Schaarbeek.

2. (15.11.1997-01.04.1998) Formation of core-group Partners: Steven Degraeve – a social worker of RisoBrussel, a non-profit association involved in community work – Jim Segers – working for City Mine(d), a production house for social artistic projects in the city; Chris Rossaert, an independent architect, and Jacques Lechat, a teacher of APAJ – Classe Chantier – an apprenticeship school.

3. (01.04.1998-01.08.1998) Discussion of idea of constructing The Tower + preparation of sponsorship dossier. Partners: Core-group + Ondex.

4. (01.08.1998) Submission of an application for funding for the building of the tower in the program 'Extraordinary neighbourhood' to the King Boudewijn Foundation. Partners: Core-group; JP Morgan Bank; King Boudewijn Foundation; Municipality Schaarbeek; Neighbourhood committee Brabantwijk.

5. (01.08.1998-01.10.1998) Formation of supportive networks at the neighbourhood level. Partners: The neighbourhood committee Dupont Street, three primary schools, VIVA-a women organization, Wijkpartenariaat – a community work association, and sports hall 58 – a sport association.

6. (01.08.1998-01.10.1998) Formation of supportive networks at the metropolitan level. Partners: VLEKHO Business School; IRIS-hogeschool – A School for Social Works; St Lucas, A School for architecture; JP Morgan Bank, Koning Boudewijn Stichting.

7.(01.01.1999) Formation of technical working group. Partners: Jacques Lechat, Jim Segers, Chris Rossaert and Steven Degraeve.

8. (05.1999) Organization of a neighbourhood festival and a flower project for the neighbourhood. Partners: Wijkpartenariaat; neighbourhood committee Dupont Street, three primary schools; sports hall 58 – a sport association.

9. (01.03.99-01.04.1999) Dupont Bizarre – cooperative project between neighbourhood partners and students of art school. Partners: St-Lucas, City Mine(d), Wijkpartenariaat.

10. (1999-2001) Participation in international exchange program and workshop on socio-economic urban development between network LimiteLimite and initiatives in Amsterdam, London and Paris. Partners: Brusselse Raad voor het Leefmilieu, City Mine(d), Katholieke Universiteit Leuven, neighbourhood committee Brabantwijk.

11. (05.1999) Theatre group DitoDito in Tower. Partners: Theatre group DitoDito, Wijkpartenariaat, City Mine(d).

12. (31.12.1999) Approval of two 'neighbourhood contracts' between municipality and Brussels Capital Region for the regeneration of the Brabantwijk. Partners: Brussels Capital Region, Municipality Schaarbeek.

13. (2000) Brussels European Cultural Capital, 2000 – Artistic intervention in Brabantwijk and interventions in the tunnels bridging Brabantwijk with Business District. Partners: City Mine(d); Committee of retailers in the Brabantwijk.

14. (2000) Monthly breakfast for the women in the neighbourhood. Partners: VIVA – a women's organization; Wijkpartenariaat.

15. (2000) Approval of European Structural Fund Objective II grant for the neighbourhood.

16. (01.12.2001) Formation of LimiteLimite as a non-profit association. Partners: Frank Pottie (president), vice-manaager of Euroclear; Micheline Goossens, director of VLEKHO Business School; Nancy Van Espen, co-coordinator UNIZO-Brussel, the organization of the self-employed; Inge Maes, coordinator of student facilities VLEKHO Business School; Kris Audenaert, president of Wijkpartnerariaat, an organization involved in community work; Seven Degraeve; staff member of Centrum voor Sociale Stadsontwikkeling, an NGO in support of Flemish organizations involved in urban development and community work; Alain Storme, staff member Riso-Brussel, an organization involved in community work; Lahcen Hammou, manager of the multicultural shopping street Brabantstraat.

* Note that the listed incremental steps are not always direct initiatives of the network LimiteLimite. For example steps 10, 12 and 13 are significant projects in the neighbourhood. Although LimiteLimite was not directly responsible for their success, the dynamics generated by LimiteLimite played an important part in their realization.

Figure 7.3   LimiteLimite, a complex and multi-layered process of small cooperative steps

Complex goods provision

The term 'complex good' here relates to the multi-layered characteristics of LimiteLimite's goal-setting process amongst diverse actors with divergent needs and mentalities (Bessen 2001; Rossi 2003). A complex good consists of many components that have multiple interactions and have different uses for different people. Most urban amenities (for example, safety, environmental quality, infrastructure, housing provision and so on) are complex goods. Complex goods provision is engendered through the divergent demands of actors in a socially heterogeneous and fragmented social fabric. As LimiteLimite had to respond to the heterogeneous – and not always complementary – social needs and objectives of the participants, the core group facilitated the making of heterogeneous or complex goods. The diversity of project modules produced communication among heterogeneous user groups in the city with actors contributing divergent views, interests and perspectives in the partnership. The heterogeneous composition of the partnership allowed the participants to deliberate creatively and imaginatively (Perry-Smith and Shalley 2003).

Given the socially complex, diverse and fragmented neighbourhood and the inclusive nature of the project, the initiators were continually confronted in their mediation for each spin-off initiative with different (and often incompatible) demands. As a result, each spin-off initiative became a creative discursive action to agree on the goal and the particular polity wherein projects would be realized. Moreover, the frictional social space of the project's location engendered a multitude of crosscutting issues. The small-scaled and seemingly 'single issue' project of building the tower on a physically and institutionally vacant plot of land in an underprivileged urban neighbourhood raised many issues related to, among others, security, gender, mobility, environment, poverty, and local governance. The project's initiators had to draw on a variety of people (school teachers, local neighbourhood committees, women's groups, local government, architects, social workers, representatives of various ethnic groups, etc.) for their expertise in order to emerge with an adequate response to these diverse development facets and needs (Lowndes and Skelcher 1998). In order to mobilize this expertise, the core group created an umbrella 'reflective' network for innovation, involving all project modules. Since each module had its own motivation and goals, the exchange across modules could and did lead to a clash of interests, but this was mitigated and constrained by the organizational autonomy of each individual module that pursued its own goals and objectives. LimiteLimite neither embodied nor fostered consensus on the set of goals and targets to be achieved; rather, each module evolved its own understanding of what should be realized in the neighbourhood, based on its own and distinct history and motivation. Precisely for this reason, the project's initiators started to experiment with different layers of consensus on the nature of the collective action itself. Some groups of actors aimed to construct a better image of the neighbourhood, others focused on jobs, and still others concentrated on obtaining housing renovation subsidies. Indeed, it is the modularity of the emerging networks that ensured that any change of or addition to the objectives or motives in one module would have only limited or no reverberations on the rest of the system. Modularity allowed different partnerships to work simultaneously without fear of direct interference or hijacking of objectives and targets. The result was a complex, hybrid, multi-layered 'good' that reflected the diversity and richness of approaches in the neighbourhood.

Although the process of 'complex good provision' generated a stimulating framework to promote the revitalization of the Brabant neighbourhood, the process also showed, perhaps paradoxically so, a tendency to become increasingly self-selective. The complexity of the neighbourhood motivated the formation of expertise-based partnerships, actively recruiting specific actors with specific knowledge and resources. As a result, a self-selective matching process of expertise preceded the actual involvement of actors and, consequently, group negotiations often took place among self-selected partners. For example, the vocational training school, APAJ Classe Chantier, was a prominent actor in the building of the tower as it had the expertise and labour power to actually construct the tower. Once built, the role of APAJ Classe Chantier diminished as its expertise was no longer immediately relevant. Table 7.2 gives an overview of the self-managed and self-selective processes that the two strategies – effectuation and complex good provision – respectively gave rise to.

Table 7.2   Processes of effectuation and complex good provision in the case of LimiteLimite

PROCESS Triggered by the role position of the core group – the tertius – and the location of the project in an 'institutional void'		
EFFECTUATION		COMPLEX GOOD PROVISION
Uncertainty		Heterogeneity
Hedging strategy based on stakeholders' commitment Means-driven strategy instead of goal-driven		Diverse needs Creative production
Incremental modular partnership – shaped by sequence of small joint actions Modular partnership		Diversity as trigger for expertise-based partnership Cross-Boundary partnerships (scale, sector, cultural ...)
SELF-MANAGED PROCESS		SELF-SELECTIVE PROCESS
INCLUSIVE COLLECTIVE ACTION		

Conclusion

The initial driver of the LimiteLimite dynamic was directly related to its strategic positioning in a socio-spatial environment characterized by friction and non-communication among actors operating on the local and metropolitan scales. In a context of uncertainty, fuzziness, conflict and heterogeneity, urban redevelopment demanded cognitive and coordination processes fundamentally different from those traditionally deployed in state-led urban regeneration policies. The policy impasse in the Brabant neighbourhood enabled the LimiteLimite core group to occupy the position of the 'tertius', which, in turn, permitted forging cross-scalar partnerships. During the LimiteLimite process, alternative mechanisms for engendering urban renewal were experimented with, namely effectuation and complex good provision.

Effectuation emerged as an incremental process organized as a sequence of small joint cooperative projects between local and metropolitan groups. The uncertain, fuzzy and disjointed socio-spatial conditions in the Brabant neighbourhood forced the core group to avoid long-term planning and to develop the project gradually and incrementally. Hence, the core group, using its tertius power, negotiated a new partnership between local and metropolitan groups in each stage of the project. Moreover, as each stage was a product of stakeholders' financial, human, and organizational resource commitments, the project was primarily means driven rather than goal oriented. The goals of LimiteLimite were continuously open for re-negotiation in each partnership, making LimiteLimite a self-directive process led by the participating actors themselves.

Complex goods provision was engendered through the divergent demands of the actors, differentiation that arose from the socially heterogeneous and fragmented location of the project. The production process of complex goods provision generated, in turn, a contradictory effect: the requirement to produce a complex good was accompanied by a process of self-selection. Despite the meta-goal of LimiteLimite to foster an inclusive partnership, this process of self-selection led to a socially exclusive dynamic.

The experiences of LimiteLimite in Brussels show a 'glocal' empowering process at work, one that was conducive to generating socially innovative governance arrangements. The project and its numerous spin-off initiatives has forged networked interactions between actors who were active at different scales and aimed to (re)gain control of the regulatory dynamics of, and in, their locality. LimiteLimite took as its starting point the governance impasse in the Brabant neighbourhood. This apparent deadlock situation created an institutional void, which gave the project's initiators the political opportunity to claim the role of the tertius. The radical uncertainty that characterized this institutional void, however, required the project's initiators and their collaborators to engage in so-called 'double dynamic' negotiations: with respect to the policy processes (i.e., negotiations regarding the nature of the envisioned initiative) as well as to the structural polity processes (i.e. negotiations regarding the decision-making rules). The simultaneous occurrence of these two types of negotiation at the local and extra-local levels allowed the project's initiators and interlocutors to fundamentally change the geometry of power in the neighbourhood in an inclusive way.

In 2000, local and extra-local members of the LimiteLimite partnerships established a formal non-profit association, LimietLimite.[1] Today, this institution is still a privileged point of contact for government bodies, community organizations and development agencies wishing to deliberate on urban revitalization initiatives in the Brabant neighbourhood. The non-profit association has been a structural partner in two neighbourhood contracts (Brabant-Groen and Aarschot-Vooruitgangstraat (2001-2005)), which were funded programmes of the Brussels Capital Region to revitalize the most disadvantaged neighbourhoods in Brussels. LimiteLimite has thus enabled local, disenfranchised and disempowered actors in the Brabant neighbourhood to jump scales and to claim participation in the future decision-making of their neighbourhood. It has also allowed extra-local actors to be more embedded in the local social fabric. These processes of 'glocal empowerment' have promoted a form of 'rooted cosmopolitanism,' that is to say, policy and polity actions which are embedded in the particularities of a locality, yet aim to bridge the inter-scalar gap between the local everyday life spaces and extra-local cultural, political and economic processes (Tarrow 2005). At the same time, they have also, and perhaps paradoxically so, generated counter-productive dynamics of self-selection, which risk leading to disempowerment.

Acknowledgements

We are grateful to the European Union's Framework VI project *SINGOCOM* for the funding that permitted the field-work for this project. We thank Byron Miller for his careful editing and the referees for their constructive comments. With thanks also to Frank Moulaert, Jim Segers and the great people of CityMine(d) for their input and support. Any remaining errors of fact or reasoning are, of course, entirely of our own making.

---

1   The founding members of the non-profit association are Frank Pottie (president), vice-manager of EUROCLEAR; Micheline Goossens, director of VLEKHO, the business school; Nancy Van Espen, co-coordinator of UNIZO-Brussel, a lobbying organization for the self-employed; Inge Maes, coordinator of student facilities at VLEKHO; Kris Audenaert, president of Wijkpartnerariaat, a community work organization; Seven Degraeve, staff member of the Centrum voor Sociale Stadsontwikkeling, an NGO in support of Flemish organizations involved in urban development and community work; Alain Storme, staff member of RisoBrussel, a regional community development in community work; Lahcen Hammou, manager of the multicultural shopping street Brabantstraat.

References

Bessen, J. (2001) 'Open source software: Free provision of complex public goods', Research on Innovation 2001 <www.researchoninnovation/opensrc.pdf>.

BRES (1997) *Indicateurs statistiques bruxellois 1997*. Brussels: Iris.

Burt, R. S. (1992) *Structural holes: The social structure of competition*. Cambridge: Harvard University Press.

—— (2005) *Brokerage and closure. An introduction to social capital*. Oxford: Oxford University Press.

De Corte, S. (2001) *Tien Brusselse wijkfiches, in opdracht van de Vlaamse Gemeenschapscommissie, welzijn en gezondheid, sif-projecten, brussel*. Brussel: Cosmopolis.

De Corte, S. (2002) *Vijf Brusselse wijkfiches, in opdracht van de Vlaamse Gemeenschapscommissie, welzijn en gezondheid, sif-projecten, brussel*. Brussel: Cosmopolis.

Degraeve, S. (1999) 'Limite/limite een leefbaarheidsproject in de brabantwijk – schaarbeek', *Opbouwwerk Brussel*, 17-20.

Granovetter, M. (1983) 'The Strength of Weak Ties: a Network Theory Revisited', *Sociological Theory*, 1: 201-233.

Hajer, M. (2003) 'Policy without polity? Policy analysis and the institutional void', *Policy Sciences*, 36: 175-195.

Hirsch, F. (1976) *Social Limits to Growth*. London: Routledge and Kegan Paul.

Kesteloot, C. and P. Mistiaen (2002) 'The Brussels case: Institutional complexity', in C. Kesteloot (ed.) *The spatial dimensions of urban social exclusion and integration*. Amsterdam: Urbex.

Lowndes, V., and C. Skelcher (1998) 'The dynamics of multi-organizational partnerships: An analysis of changing modes of governance', *Public Administration*, 76: 313-333.

McAdam, D., Tarrow, S. and C. Tilly (2001) *Dynamics of Contention*. Cambridge: Cambridge University Press.

Martens, A. (1994) *De Noordwijk. Slopen of wonen*. EPO: Berchem.

Moulaert, F., Rodriguez, A., and E. Swyngedouw (eds) (2003) *The Globalized City*. Oxford: Oxford University Press.

Papadopoulos A.G. (1996) *Urban regimes and strategies. Building Europe's central executive business district in Brussels*. Chicago: Chicago University Press.

Perry-Smith, J. and C. Shalley (2003) 'The social side of creativity: A static and dynamic social network perspective', *Academy of Management Review*, 28 (1): 89-106.

Rossi, M. A. (2003) 'Decoding the "open source puzzle" a survey of theoretical and empirical contributions', *Workingpaper*, Dipartimento di Economia Politica, Università di Siena.

Sarasvathy, S. (2001) 'Effectual reasoning in entrepreneurial decision making: Existence and bounds', *Academy of Management Best Paper Proceedings*, Society for Effectual Action.

Sarasvathy, S. D. et al. (2003) 'Accounting for the future: Psychological elements of effectual entrepreneurship', under review at *Journal of Applied Psychology*, 2003 <http://www.effectuation.org/Topics.htm#Effectuation>.

Simmel, G. (1902) 'The Number of Members as Determining the Sociological Form of the Group', *American Journal of Sociology*, 8: 1-46, 158-196.

—— (1904 (1955)) *Conflict*. Glencoe, Ill: Free Press.

Simmel, G. and K. Wolff (1950). *The sociology of Georg Simmel*. Glencoe, Ill: Free Press.

Simon, H. A. (1981) *The sciences of the artificial*. Cambridge, Mass.: MIT Press.

Swyngedouw E., (2000) 'The Mont des Arts as a Ruin in the Revanchist City', in De Meulder B., Van Herck K. (eds) *Vacant City – Brussels' Mont des Arts Reconsidered*. Rotterdam: NAI-Publishers, pp. 267-281.

Swyngedouw, E., and G. Baeten (2001) 'Scaling the city: The political economy of "glocal" development – Brussels' conundrum', *European Planning Studies*, 9 (7): 823-849.

Swyngedouw, E., and J. Moeyersoen (2006) 'Reluctant Globalisers: The Paradoxes of "Glocal" Development in Brussels', in M. Amen, K. Archer, and M. Bosman (eds) *Relocating Global Cities – From the Center to the Margins*. Lanham, MD: Rowman and Littlefield Publishers Inc., pp. 155-177.

Tarrow, S. (2005) *The New Transnational Activism*. Cambridge: Cambridge University Press.

Chapter 8

Multiscalar Mobilization for the Just City: New Spatial Politics of Urban Movements

Margit Mayer

Introduction

The space of contention addressed in this chapter is the city. Contemporary social movements that contest and resist the neoliberalization of cities have generated new spatial forms and practices (as have other contemporary movements), including the use of new network and scalar strategies transcending their locality. Like many other place-based movements in response to globalization, urban movements have expanded their action repertoires and differentiated the spatial dimensions of their activities in terms of place, scale, and network. While these emerging spatialities of social movements are gradually getting charted (see e.g. Miller 2000; Leitner et al. 2008; Nicholls 2009), less attention is as yet paid to the concomitant substantive shifts in the critique of neoliberalism and shifts in the demands and goals of the movements. The growing prominence, for example, of a rights discourse, which appears to go hand in hand with an NGOization trend of social movements, is barely discussed.[1] Even where the effects of entanglements between local and global activism are the object of inquiry (for example as in Cumbers et al. 2008), the operational logics of global NGOs and local social movements are barely distinguished.[2] Asking whether the expansion and selective use of movements' scalar strategies have implications for their chances of challenging neoliberal power relations, this chapter concentrates on one particular place-specific movement. It explores the ways movements for a more just city articulate the 'global' in struggles that take place 'locally' and vice versa. Drawing on cases and examples from the Euro-American zone, the chapter finds that urban movements encounter, on each and every scale, specific fields of social and power relations, which need to be realized and confronted if cooptation or neutralization are to be prevented.

It is no news that urban politics occurs not only in cities. Even before political relations between cities, nation states and supranational institutions became significantly reconfigured during the phase of neoliberal globalization, urban

---

1   Harvey (2006: 51) sees, however, a connection between the rise of NGOs and the rise of rights discourses, both of which have accompanied the neoliberal turn.

2   For the difference between social movements and NGOs see Demirovic (2000).

politics occurred not only in municipalities, but also through regional and national governments as these set up frameworks and parameters, stimulate renewal or infrastructure programmes at one point, housing development at another, or cut back and cancel such programmes in favour of revitalization programmes or plans to enhance locational economic competitiveness. Trans-border networks and partnerships which cities have been building across national boundaries are also no novelty. What is new, though, is the re-scaling[3] of decision-making processes about many contested issues: which issues get down-scaled to local and urban authorities, which get up-scaled or externalized to supra- or extra-national bodies, and how these shifts affect power relations between ruling and challenging groups is of crucial importance for the opportunities and potentials of social movements.

This shifting scalar organization of statehood is reflected in grassroots movements' similar reorganization in multi-scalar ways. Responding to where contested decision-making takes place, movements mobilize not only locally, but also regionally, supra-regionally, nationally, and in new ways globally, making use of technological developments as well as the emergence of global publics. Urban social movements fight for 'the right to the city' not merely on site, but worldwide, at global summits as well as with regional, continental and larger campaigns. Identifying the institutions and actors of corporate globalization as responsible for the degradation and polarization of their – and not only their – city, they push for an urbanization that respects the urban dwellers' desires and claims as opposed to those of corporate investors or developers. Contesting and challenging the re-scaled architecture of both state and corporate power, social movements have been developing new scalar practices, which imply new possibilities as well as new risks and problems, which have however hardly been explored and are thus not yet very well understood.

Urban movements have also made increasing use of another sociospatial dimension: networks, which span space and frequently cross hierarchical scales. They build links between local struggles and transnational organizations, between place-based contestation and global NGOs, thereby bringing into contact vastly different movement lexicons and political cultures. Researchers are still in the initial phases of trying to conceptualize the resulting multitude of forms of connectivity between urban movements around the world.

The literatures on transnational social movements tend to distinguish between two types of practices: On the one hand, they identify transnational (advocacy) networks that push for the democratization of international institutions and

---

3   Jessop et al. (2008) remind us that the concept of (re)scaling emerged at a time when inherited global, national, regional, and local relations have become recalibrated through post-Cold War capitalist restructuring and state retrenchment, propelling particularly supra-national and subnational arenas to the fore of sociospatial regulation. Capturing the rescaling of state activities since the crisis of Fordism, it brought into focus the mutable, contested, but very real hierarchical structures of neoliberalism in which actors act (Mayer 2008: 417).

agreements, organize campaigns to that end, or draft alternative charters; to the extent these networks pursue urban agendas, they do so with the goal of strengthening the participation and access rights of city dwellers, especially of the most vulnerable ones. Distinct from these transnational NGO-type organizations and networks, they see local initiatives that experiment with and implement direct-democratic projects committed to social justice principles, from participatory environmental policy to participatory budgeting that have proliferated throughout the world to form a source of counter-hegemonic politics (Santos and Rodriguez-Garavito 2005; Smith 2008; Tarrow 2005b).

A broader definition of emancipatory political practices at the local scale would also include the sundry movements challenging global corporate domination in the city in more indirect ways, for example as migrant movements do when they deplore their disenfranchised situation in the first world's metropoles as an outcome of the dynamics of neoliberal globalization; or the anti-gentrification movements when they attack the upgrading of downtowns and of hip residential districts as well as the mega-projects pushed by city governments in the name of global locational competition and carried out by global investors, for displacing and marginalizing disadvantaged groups and spaces. These urban movements insert themselves in various ways into supra-local and supra-national politics: through legal practices, by, for example, invoking EU law against national jurisdiction, or through discursive practices, by invoking the international human rights discourse and the respective UN institutions as tools in their struggle.

Similarly, the definition of transnational protest needs to be broader – and more heterogeneous – than is generally assumed in the literatures. It includes not just the transnational advocacy networks that have launched campaigns to democratize international institutions such as the World Trade Organization, World Bank, or International Monetary Fund, but also those activists' networks that organize the events and campaigns critically accompanying the summits of representatives of the neoliberal world order. These counter summits provide a space not just for challenging and publicizing the social and ecological destructiveness of corporate globalization, but also to exchange insights and experiences with other (local) activists from around the world, and to plan and coordinate upcoming joint civil disobedience and other actions in the common struggle for the right to the city. The way this occurs is, invariably, strongly influenced by the respective local movement milieu (of, for example, Seattle, Genoa, Quebec, Prague or Rostock).

Since the first meeting of the World Social Forum in 2001 and thanks to the many regional and national Social Fora meetings around the world, an alternative, counter-hegemonic form of globalization has been taking shape. Its activists and organizations have, over the last 10 years, been taking the core issues of the alter-globalization movement, the struggles against privatization, dispossession, and eviction, from the global summits back to their cities – of the first as well as second and third worlds.

This multi-scalar jumble of interurban movement connectivity does not appear to be structured by any causal relationships between, for example, movement strength and location in the scalar hierarchy. Neither can we observe that up-scaling struggles brings more effective results, nor can one say the closer to the grassroots, the more authentic or radical the movements will be (nor is the inverse invariably the case: the more inter- or transnational a campaign, the more aloof or easily co-opted its activists). Are movements more successful if they employ a bigger repertoire of spatial strategies? Do place-based movements overcome their localism only if they embrace additional spatial strategies? Looking at examples from these new local/global arenas and their relationships with each other, this chapter finds that questions such as the ones just listed are not answerable by abstract accounting of the various spatial strategies employed in contentious urban politics. If we are to uncover the new opportunities as well as the new challenges urban movements encounter in this emerging multi-scalar architecture, we need to look closely – not just at the spatial forms, but *at the social actors who act*, embedded in particular (multidimensional) spatial forms and making use of particular 'glocal' scales and networks. The chapter is thus an attempt to 'politicize' the space/scale/network concepts suggested by Jessop et al. (2008), i.e. to bring politics into their multidimensional, polymorphic framework of socio-spatial analysis. That means, the differentiation and transformation processes that have taken place amongst urban contestants during the phase of roll-out neoliberalism have to be accounted for. As state actors from the municipal to the UN level have cultivated partnerships with grassroots organizations in their search for best practices, the struggles within and against the neoliberal city have taken on many faces. In this situation, we need an acute awareness of the political meaning of sociospatial categories as well as of traditional movement vocabularies, as their substance is transforming in front of our eyes. Categories like 'empowerment' or the 'right to the city' have come to mean rather different things in different contexts. Depending on context, some distinctions become even more important than in the past: for example the distinction between NGOs and social movements matters a great deal in terms of political substance, but is not always clearly made (e.g. Sikkink 2005; Cumbers et al. 2008: 190).

In order to make these arguments, the chapter first looks at the ways in which classical urban-based movements have come to refer to and make use of the global scale, either by protesting the negative effects of globalization in the city, or by harnessing scales of discourse and politics beyond the local, thus diversifying their strategies and activities in a multi-scalar way (*Local Protest*). Second, it looks at the ways in which transnational networks and campaigns have taken up the claims and demands of urban movements and thus impact on the local movement milieus (*Transnational Protest*). The interplay of these different clusters of 'glocal' movements creates a novel multiscalar architecture of urban protest, in which impetus from supra-national scales may boost and strengthen local grassroots movements (as when *Anti-globalization Movements Discover the Local*), and particular urban initiatives may turn into beacons of the global social justice movement. Building on these cases, the last section traces some of the tensions and conflicts that have emerged between mobile transnational activism and the locally or community-based groups, between activists from resource-rich first world countries and those representing urban struggles in the global South, and between radical calls for a fundamental transformation of cities as they exist versus demands for 'good urban governance' and rights to the city. Exploring the differences between these goals, the concluding section reveals problems with assumptions underlying the literatures on and policies toward a transnational civil society. It problematizes the apparent antagonism between 'good' urban residents and their advocates on the one hand and 'threatening' but intangible global market forces on the other, which underlies observable trends of cooptation and de-radicalization in the new global right to the city movement, both in the circles of the World Charter network where movements work hand in hand with UN institutions and municipalities, as well as on the local scale of first world cities, where newly celebrated rights to the city coalitions are easily hijacked by neoliberal city marketing and entrepreneurial creative city strategies. These trends put the question and necessity of building activist alliances with urban movements in developing and emerging countries firmly on the agenda of the privileged countries' movements.

Local protest: urban movements scaling upwards

In the course of the last few decades' globalization, cities and local politics have acquired new significance in two ways, and as a consequence, urban movements have taken on new roles:[4] First, the shift towards decentralization in a series of policy fields (particularly in labour market and social policy fields) has expanded municipalities' pertinent powers and functions and has simultaneously entailed an opening of local governance arrangements, through which municipalities now seek to benefit from the expertise and experience of local NGOs and social movement organizations. Secondly, the upgrading of cities as engines of economic growth has enhanced their autonomy vis-à-vis national governments and turned them into important nodes and actors within worldwide networks. Since this upgrading of the role of cities has not been accompanied by any improvement in the living conditions or increased access to decision-making processes for the majority of urban dwellers, local struggles over social justice and the fight for the right to the city have become more pronounced. Both activists and supportive academics realize more and more that their struggles and projects are hampered if they remain locally confined, and that they need to connect with similar struggles in other places or with movement actors on other, 'higher' scales. '[To] build effective social justice strategies ... efforts must emphasize the formation and mobilization of a consciousness of justice and a multi-scalar understanding of place that can be

---

4   More detail on these connections is provided in Hamel et al. (2000) and Mayer (2007a and b).

linked with other local scale ventures and/or larger scale actors' (Pendras 2002: 831).

Globalization in the city

Local struggles against corporate urban development and the 'entrepreneurial city', which challenge the make-over of central business districts (CBDs) and concomitant gentrification processes, increasingly confront global investors and developers. Ever since (the third wave of) gentrification is being pushed as an instrument of intra-urban competition, global finance capital has been flowing into large-scale development projects in CBDs (Smith and DeFilippis 1999), justified by local politicians pointing to job creation, tax revenues and gains from tourism. While North American cities have been leading this development, Europeans have also advanced gentrification as a tool in the global urban competition with 'urban regeneration' programmes designed to help cities make themselves attractive and remove 'irritants' such as the homeless or other potentially blemishing factors. Local governments in this situation often strive to be global capital's active partner, no longer even pretending to regulate or control development (Smith 2002: 443). Movements challenging this form of urban development have galvanized across Europe, as for example in coordinated 'Downtown Action Weeks' that took place simultaneously in 20 German and Swiss cities to protest not only the intensifying global location competition and the corresponding image marketing campaigns, but also the 'cleansing' of the downtown citadels, the marginalization of whole neighbourhoods and groups of residents that do not 'fit' into this entrepreneurial city. Mobilizations against privatization and commercialization of public space and against the deregulation and deterioration of public services in European as well as North American cities are well documented (Hamel et al. 2000; Mitchell 2003; Low and Smith 2006; Leitner et al. 2007), struggles against gentrification less so.

Also, many North American as well as European cities are sites of protests by 'undocumented' migrants who contest the role and position forced on them within northern nations, which treat them as outsiders while crucially integrating their labour power into the expanding economies of the urban centres of the global political economy (Rosewarne 2001). The Sans-Papiers, for example, have been claiming, through demonstrations and occupations, that their presence in first world metropoles is legitimate and thus their right to membership is valid before the granting of any citizenship rights (McNevin 2006: 143-144). With actions such as the occupation of the headquarters of the French Construction Confederation they emphasize their prominent role in meeting labour shortages in key urban industries as well as the tight connection between formal and informal economic sectors. Such migrant organizations of 'immanent outsiders' – integrated into the urban economy but excluded from social and political participation – have formed networks, together with advocacy groups, across Europe and beyond (e.g. NoBorder Network; No-one is illegal).[5] They also have a strong presence at meetings of the European Social Fora (see below) as well as at the first US social forum meetings in Atlanta 2007 and Detroit 2010, which indicates that they are making use of the transnational organizational structures of the global justice movement, parallel to their local actions.

Supra-local scales of politics

Another strand of local protest is made up by residents of abandoned and deindustrialized districts who contest the effects of neoliberal globalization with strategies as varied as squatting vacant buildings or developing self-help structures. These neighbourhoods represent the obverse of the booming CBDs; their (unemployed, welfare dependent, or otherwise precarious) residents are targeted, in this era of roll-out neoliberalism (Brenner and Theodore 2002), with a novel type of socio-spatial programme designed to 'stop the downward spiral' presumably characteristic of blighted areas. In Europe, the community organizations active in such districts have increasingly become integrated into (often EU-supported) neighbourhood development programmes, which address social exclusion and marginalization with territorially targeted empowerment approaches, social capital building, and even microcredit programmes, i.e. with instruments that were originally developed to combat urban poverty in the global South (Mayer and Rankin 2002). Involving community-based organizations in policies to combat urban decline and poverty has a longer tradition in North America. Policy diffusion in this realm is as transnational as community-based movements seeking to learn from so-called best practice in 'successful' cases of harnessing civil society energies to stop decline.

When such movements shift in scale to the international level, they do not necessarily become transnational movements. As Tarrow and McAdam (2005) have found, transposition of part of the movement's activities (rather than its transformation) is the more common pattern. This pattern can be observed when urban (anti-poverty) movements forge transnational networks of support as an operational strategy to improve their constituencies' lot. An illustration of this is presented by the Kensington Welfare Rights Union (KWRU) in Philadelphia, for example, which has harnessed global human rights discourses and organizations for their concerns. The Kensington neighbourhood, which lost most of its local businesses and jobs in the course of the 1970s and 80s, had the highest poverty concentration in Pennsylvania by the beginning of the 1990s. In 1991 six women on welfare took over an abandoned welfare office, set up a community centre, and began to mobilize for better social services (Ford Foundation 2004: 50).

---

5   See http://www.noborder.org/ and http://en.wikipedia.org/wiki/No_Border_network for the former, http://noii-van.resist.ca/ or http://www.noii.org.uk/ for no-one is illegal campaigns in Canada and Britain (last accessed February 20, 2009).

> Through direct action campaigns, including the takeover of empty HUD housing
> by homeless families, we have housed more than 500 families, fed and obtained
> utility services for thousands, and educated on the streets for basic skills (Babtist
> and Bricker-Jenkins 2002: 204).

After passage of Clinton's welfare reform in 1996, the KWRU organized, together
with other anti-poverty-groups, a 140-mile march to the state capital Harrisburg
to protest the loss of benefits. But its members soon realized that even on the state
level their struggle remained constrained and therefore broadened their framework
from a civil rights to a human rights focus, and linked up with anti-poverty groups
internationally, especially with groups active in third world countries. With their
'March for Our Lives' from Philadelphia's Liberty Bell to New York City in 1997
they initiated a Poor People's Economic Human Rights Campaign (PPEHRC)
claiming the rights defined in the UN Universal Declaration of Human Rights
(particularly articles 23, 25, and 26, i.e. the right to work for just pay, housing
and education). Thus, they not only demanded that international human rights
conventions apply in the US, but they simultaneously internationalized the
struggle of US poverty groups (Smith 2008). At two Poor People's Summits,
which KWRU convened 1998 in Philadelphia and 2000 in New York, activists
from 30 different countries gathered for workshops and 'reality tours', in which
KWRU activists showed their international visitors the devastating impacts of
neoliberal globalization on US inner cities. Similarly, with their 1998 month-long
'March of the Americas' from Washington, DC to the United Nations in New York
they involved international activists, especially from Latin America, in order to
highlight that US poverty needs to be understood in a global context. Because
Washington 'could not be counted on to protect and promote the economic
human rights of its residents, ... we turned toward the court of world opinion
symbolized by the UN Social and Economic Council (Babtist and Bricker-Jenkins
2002). In 1999, KWRU globalized its struggle further by taking the PPEHR
Campaign to the Hague Appeal for Peace (which marked the 150th anniversary
of the intergovernmental Hague Peace Conference), where nearly 10,000 people
working in peace and human rights groups from more than 100 countries were
gathered. KWRU activists also participated in the 'People's Tribunal on Corporate
Crimes Against Humanity' held in preparation of the World Trade Organization
ministerial in 1999. And since 2001, they have regularly participated in the World
Social Forum (Smith 2008). Together with the Center for Economic and Social
Rights (Ford Foundation 2004: 26f.) they have been pressuring international
human rights institutions such as the UN Human Rights Commission in Geneva
(where they raised their concerns about widespread US poverty to government
delegates) and its regional counterpart in the Organization of American States
(OAS), the Inter-American Human Rights Committee (which they petitioned in
2003 to rule against the 1996 US welfare reform as a violation of the OAS human
rights standards, Smith 2008: 207).

The dual strategy which KWRU has been applying – on the one hand addressing, with community action and programmes, the concrete problems of urban poverty, and on the other developing and using international contacts with NGOs and human rights institutions as well as, through the WSF process, similar movement groups active elsewhere – has helped to publicize the concerns of anti-poverty groups and also to make visible the connections between struggles against the marginalization of 'redundant' people and neighbourhoods around the globe.

Yet another field of urban contestation in the struggle over the neoliberal city is that of community participation, often framed as 'practical utopia', where urban residents seek to expand their input in municipal planning and decision-making processes. A variety of such projects and initiatives expanding participatory democracy and citizenship rights – whether through participatory budgeting, innovative forms of involving citizens in planning, or policy-specific round tables – have been diffusing and propagating via transnational networks. UN Habitat and the World Urban Forum offer opportunities for disseminating and sharing such experiences and best-practice models, and also for persuading local policy makers of the advantages of such participatory programmes and direct-democratic procedures. Networks and cooperative linkages have also emerged between cities – to further sustainable development, employment, or civic engagement. Such partnerships and networks are globally supported by the World Bank, through its various *good governance* programmes, or nationally, as by the German Bertelsmann Foundation (which initiated a so-called *Civitas Network* of civil society-oriented municipalities).[6]

The turn towards transnational arenas and organizations that local activists have embarked on as they seek to strengthen the input and participatory claims of active citizens is owed to the same logic as KWRU's turn towards international human rights forums. As this local welfare rights organization learned not to expect a correction of the violation of poor people's rights from the city of Philadelphia or from the state of Pennsylvania, it reached to higher scales for backing of its demands. In spite of the growing salience of cities in the global context, it appears pointless to restrict movement claims to the local/urban scale, where locational competition, financial crisis, and privatization measures have robbed municipalities of resources and latitude. Even before the impact of the 2008 recession made its impact felt, urban movement groups have begun to wonder whether local authorities are still an appropriate target or addressee for them, as their capacity for public provisions has diminished more and more. This is certainly an additional reason for the growing supra-local orientation of urban social movements.

------

6 After organizing a competition, this network elaborated quality criteria for citizen-oriented municipalities (Pröhl et al. 2002), which detail desired forms of civic engagement in a variety of municipal policy fields, from budget to neighbourhood development (Roth 2003).

Transnational protest: the (right to the) city in the global justice movement

The mobilizations labelled by Europeans as 'alter-' or 'anti-globalization movements' and by North Americans as 'global justice movements' are most manifest in the protests against supra-national organizations such as the WTO and IMF and against summit meetings (e.g. of the G8),[7] as they seize on the political opportunities and public attention which these meetings create. They are also manifest in the 'open space' of the World Social Forum and in the national, regional and local Social Fora, which have created novel transnational spaces of activism, within which urban protest claims and milieus are equally present (cf. previous section). But the global justice movement itself also does not only home in on (counter)hegemonic globalization, it directly addresses the city as well, and not only in symbolic actions at urban sites where the command and control functions of the global economy are concentrated (such as headquarters of multinational corporations, banks and investment firms, stock exchanges etc.). Over the last few years the anti-globalization movement has explicitly identified the city as the place where the negative effects of the global neoliberal project become tangible for many different groups, and where it therefore makes sense to organize the resistance against the neoliberal project (Köhler and Wissen 2003).

Summit protests

Since 1994 international mega-events have been staged where thousands of people rally against the multilateral economic and financial institutions and their political representations. Whether at G8 summits (Cologne 1999, Gleneagles 2005, Heiligendamm 2007) or at meetings of global institutions such as WTO, IMF, World Bank, or FTAA, movements organize and coordinate across national boundaries and political differences in order to articulate broad, plural resistance against neoliberal globalization, to disturb or disrupt the meetings, and to network and exchange with each other at parallel 'alternative summits' (cf. Ainger 2009). In preparations well in advance of the events they develop formal as well as informal coalitions and networks that build on existing local, regional, national and global organizations, which then cooperate – more or less successfully – during the days of the protest, in demonstrations, blockades, assemblies, workshops and cultural events, always creating unique happenings.[8]

---

7   Mobilization against globalization can be dated back to 1986, when over 80,000 people protested against an IMF meeting in Berlin (Gerhards and Rucht 1992). In the early 1990s, struggles emerged in the US against GATT as well as protests by movements such as Reclaim the Streets in Britain (Routledge 2003: 347).

8   The events around Heiligendamm 2007, for example, were summed up like this: 'The kaleidoscope of emotions and inspirations swirling in Rostock, in its demos, actions, camps, media and art centers, cannot be easily described. It was a manic rush, an incredible show of radical strength and post-national solidarity' (Foti 2007).

While the transnational network organizations Attac,[9] People's Global Action,[10] Reclaim the Streets,[11] and various European and North American anarchist and autonomous groups play central roles in carrying out these events, the local movement organizations of the summit hosting city are equally crucial.[12] At the Seattle protest (WTO 1999) it was not just the old leftist and anarchist groups and the new environmental, women's, and faith-based organizations, but also the longshoremen, Teamsters, and other union groups who had already carried out the local strikes of 1919 and 1934 (Levi and Olson 2000); in Genoa (G8 2001) it was the *Centri Sociali Autogestiti*, the self-managed social centres of the local alternative milieu (Piazza 2007). And it was local movement organizations from Berlin, Hamburg, and Hannover that mobilized in 2007 for the protest against the G8 in Heiligendamm and the counter-summit in Rostock, together with

---

9   Attac (Association pour la Taxation des Transactions pour l'aide aux Citoyens), founded in 1998 in order to implement the Tobin tax worldwide, today constitutes a network of professionalized NGOs that is particularly well grounded in France, Germany, and Switzerland with hundreds of local affiliate groups (cf. http://www.attac.org; Eskola and Kolb 2002).

10   People's Global Action (PGA) owes its genesis to an encounter between activists and intellectuals that was organized by the Zapatistas in Chiapas in 1996 which called for an intercontinental network of resistance against neoliberalism. In 1997, the idea of a network between different resistance formations was launched by 10 social movements including Movimento Sem Terra (Landless peasants movement) of Brazil and the Karnataka State Farmers Union of India. Since its official birth in 1998 PGA consists of leftist groups both from the global South as well as North to inspire people to resist corporate domination through civil disobedience and people-oriented constructive action.

Since the groups have had difficulties functioning at a distance (due to access to information problems as well as language and cultural problems), much of the organizational work has been conducted by Support Groups of activists who are mostly based in Europe (cf. Routledge 2003; Tarrow and McAdam 2005: 143; cf. http://www.nadir.org/nadir/initiativ/agp/en/index.html and for original documents www.nadir.org/nadir/initiativ/agp/en/PGAInfos).

11   Reclaim the Streets (RTS) formed in Britain in the mid-1990s, initially as a radical environmental movement that transformed urban industrial landscapes into eco-friendly oases, occupied city streets and concrete-covered squares to build gardens and celebrate parties. Later RTS activists began to disrupt business-as-usual in public spaces with provocative interventions and protest parties: globally on June 18, 1999 simultaneously in 40 cities around the globe with street parties and happenings to protest the World Economic Summit taking place in Cologne (J18); locally they would attack, on days of action announced as 'carnival', the centres of global finance as, for example, in the hermetically controlled square mile of the City of London, where they would symbolically 're-appropriate' the representations of global corporate power (http://en.wikipedia.org/wiki/Reclaim_the_Streets; http://rts.gn.apc.org/).

12   With regard to the protest events in Seattle, on which most research so far has focused, Hadden and Tarrow (2007) note that the majority of the participating organizations and individuals were of local origin.

supra-regional groups of the 'Block G8' Campaign,[13] the Interventionist Left,[14] Dissent!,[15] and Attac.

Building on the legacy of prior summit protests,[16] the actions around Heiligendamm and Rostock reveal the urban edge of the anti-globalization movement in all its diversity. In the demonstrations and blockades of access routes to Heiligendamm and in the counter summit activities in Rostock and the nearby camps in Reddelich, 'plurality' was realized, and collective and joint actions as well as marked differences were visible.[17] A form of 'urban consciousness' beyond political and tactical differences appeared to be prevalent among the participants:

> the black sweatshirt has become a universal symbol of anti-capitalist self-identification, even among people that would never throw a bottle: it simply means you're on the side of Ungdomshuset, Mehringhof, Rote Flora, Köpi, and other nodal European social squats currently threatened with eviction and persecution. Urban rebellion is spreading across many European cities because there is a widespread feeling that the whole anarcho-punk, radical-autonomist, pink-queer way of life could be wiped out if we don't put up major resistance against police repression and the assorted forces of bourgeois and clerical respectability (Foti 2007).

---

13   'Block G8' constituted a broad coalition including large organizations such as Attac and the Left Party, but the Hannover Coordinating Committee played a key coordinating role, and also initiated the Rostock action conferences.

14   The Interventionist Left (http://www.g8-2007) spearheaded the block 'Make Capitalism History', but was also part of the 'Block G8' coalition.

15   The 'anarcho-globalists' of the Dissent! network (dissentnetzwerk.org) mobilized autonomous and anarchist activists from Poland, Denmark, Holland, Britain, Greece, Italy, Sweden, France, and Switzerland to Rostock (cf. Holloway and Sergi 2007).

16   The traditional dual strategy, to be both on the street and in the hallways where negotiation takes place, works less and less well at G8 meetings than at WTO, IMF or World Bank summits. This is due to the gradual transformation of the function of G8 meetings: while they initially served to mediate between competing state and capital interests, their task has shifted to legitimating global rule as benevolent. The summits are supposed to show that global problems (debt at the Cologne summit, poverty/Africa at Gleneagles, climate change at Heiligendamm) can be resolved through these meetings. At Gleneagles the British government even presented itself as an extension of the legitimate claims of social movements, at Heiligendamm the impression was created that the leaders of the industrial nations were actually going to do something about climate change (Müller and Sol 2007).

17   In contrast to the summit protests in Gleneagles, where the movement was fractured and divided and little willingness for cooperation or even coordination existed among the three different mobilizations (Make Poverty History, G8 Alternatives, Dissent!), the 2007 G8 protest was far better coordinated and marked by higher levels of cooperation, even though it, too, saw intense conflict, especially around the issue of militancy.

The themes of urban social movements were also explicitly present: workshops offered discussions on how local activism might internationalize the struggle against displacement, eviction, speculation, privatization, and for the right to housing, city, water and land;[18] one workshop sought to create a better understanding of the relationship between global finance capital and urban restructuring: under the heading of 'Global financial markets, privatization, and investments' European and North American urban activists and researchers shared information and experiences with Asian and Latin American movements,[19] mediated by INURA[20] and the Habitat International Coalition.[21] At a 'Gathering of homeless, marginalized and tenants', representatives of community and homeless organizations from Brazil, Japan, France and Germany discussed the globalization of the struggles for the right to housing. In spite of the interactions and information sharing at such events, however, the follow-up evaluations noted self-critically that 'we succeeded only in very rudimentary fashion to build bridges to local struggles, whether to self-organized groups of unemployed, to homeless initiatives, or to the residual activists from the strikes at Gate Gourmet, Siemens or Opel' (Samsa 2007).

The (World) Social Forum, World Urban Forum, and Social Urban Forum

Emerging out of the massive protests against free trade policies and the global financial institutions, the World Social Forum was initially designed as a counter summit to the World Economic Forum in Davos, and reflected a desire to shift energies from street protests toward generating alternatives to neoliberal globalization.[22] The first meeting in Porto Alegre, Brazil, 2001 drew many more

---

18   Cf. the website of Habitat (http://www.habitants.de/en/campaigns/g807/) which describes urban movements at G8 Heiligendamm.

19   <http://www.habitants.de/en/campaigns/g807/agenda/index.php/art_00000030>

20   INURA (International Network of Urban Research and Action), founded in 1991, is a transnational mix of movement organization, alternative professional association, and network (Lehrer and Keil 2007).

21   Habitat International Coalition (HIC) is an alliance with worldwide membership for the right to housing, dedicated to advocacy for the poor. It seeks to derive practical and strategic lessons from experiences around the world (http://www.hic-net.org/).

22   Many people saw the WSF as an intentional response to counter-summit protests such as J18, Seattle, Prague. But at its outset, there was little connection to the alter-globalization movement. The WSF was tied to oppose the World Economic Forum, but rather than protest, it was to do so by being a space to share and discuss positive proposals. It would not have had as much resonance beyond the typical NGO and UN crowd, had it not been for the mobilizations of the alter-globalization movement, and it would not have transformed as it did without the kinds of knowledge, experiences, critiques contributed by various areas of the alter-globalization movement: it evolved in crucial ways thanks to the conflicts, demands and self-organized responses contesting the centralized and hierarchical form of the first few forums, the presence of corporate products, the use of non-ecological materials, etc. (Osterweil 2008: 85).

participants than the French and Brazilian organizers had anticipated. At the second WSF, participants were called on to organize similar processes in their own places and at whatever scale made sense to them. Since then, this novel form of alternative globalization has not only convened annually on the world scale,[23] but also in regional as well as national and local gatherings.[24] The WSF has evolved into a completely novel socio-spatial and political praxis, a globally diffused political form and methodology that is regularly enacted on different scales, and unites a broad spectrum of distinct movements and organizations, from all over the world,[25] in the struggle against neoliberal globalization, oppression, and discrimination. The large variety of different, and not only western, currents, and the broad scope of political positions and – insurgent as well as institutionalized – action repertoires, scarcely provide a basis for consensus, except with regard to nonviolence, but even this is contested.[26] With its slogan 'Another world is possible!' the WSF articulates a radical critique of contemporary neoliberal realities and declares a better society as possible, without however specifying its features in much detail. The diversity and variety of participating movements are, in fact, highlighted as positive, valorized as an important source for a progressive

---

23   The world event took place four times in Porto Alegre, 2004 in Mumbai, 2007 in Nairobi, 2009 again in Brazil, but for the first time in the equatorial city of Belém. The vast array and richness of events at any one iteration of the WSF and its mutations across time and space are impossible to capture; each WSF reflects both the global historical conjuncture and the particular conjuncture and social struggles of the host country and region (Conway 2008, 2009).

24   Self-organized Social Fora appeared on every continent: After the violent repression of the anti-G8 protest in Genoa in 2001 local Social Fora sprang up all over Europe. Acknowledging the strength of the Italian Social Fora, the coalition of European associations at the WSF in Porto Alegre decided to hold the first European Social Forum meeting in Italy (which met 2002 in Florence, to be followed by meetings 2003 in Paris, 2004 in London, 2006 in Athens, 2008 in Malmö, and 2010 in Istanbul). In the Americas, hemispheric Social Fora took place in Quito, Ecuador in 2004 and Guatemala City in 2008. In 2006 the WSF was polycentric, i.e. dispersed over three sites (Caracas, Bamako, Karachi). In North America, the social forum process had a later start and is not as widely anchored as elsewhere (cf. Smith and Reese 2008). Groups that were involved in the Seattle protest also participated in the WSF in 2001 and 2002, and in 2004 tried to peg down 'global justice' by organizing a 'North West Social Forum' (which failed, cf. Mayer 2007a: 107); 9/11 seemed to encourage a shift away from the anti-globalization towards the peace movement. So it was not until 2007 that a first US-wide Social Forum took place in Atlanta (Albert 2007; Ponniah 2007), to be followed by a second iteration in Detroit in 2010 (cf. http://www.ussf2010.org/).

25   As with the earlier events, Belém attracted masses of participants – 130,000 from 142 countries – but the majority come from the region, in this case 'the Forum remained an overwhelmingly light-skinned, young, urban, Brazilian and Portuguese-speaking space' (Conway 2009).

26   The WSF Charter outlines the goals and political character of the forum, cf. <http://www.forumsocialmundi.org.br>.

societal transformation, because the global movement arena constructed by the WSF allows a multiplicity of emancipatory practices to flourish next to and engage with each other, which helps to make differences visible and thus more acceptable. This allows groups to form alliances on the basis of commonalities – as well as on the basis of differences.[27] Santos captured this feature of the WSF process with a vignette from Nairobi 2007:

> Urban dwellers from different cities of the planet were planning collective actions against forcible evictions and the privatization of water supply; community leaders from all over Africa were setting up the Africa Water Network and, together with NGOs and human rights and health movements and organizations from all over the world, were planning the most comprehensive campaign against HIV/Aids (Santos 2007: 24).

Besides the human rights, environmental, climate change, women's, unions, indigenous and other movements, urban and community-based organizations also use the open space of the WSF to encounter and learn from 'others' and to share information across places and scales. While local groups that visit and participate in WSF events still remain local movements (as Tarrow and McAdam insist), they become not only more eager to embrace transnational commitments (without abandoning their domestic ones), but they can also use access to this transnational space to support each other and to spread.

Compared with counter summits, WSF events are more dominated by (international) NGOs and formal organizations than (local) social movement groups. While local movement activists obviously participate – often in large numbers – they partake here in an arena significantly shaped by transnational networks and organizations (many established or funded by UN programmes) dedicated to urban problems, such as for example the World Urban Forum.[28] The thousands of WUF participants who congregated 2002 in Nairobi, 2004 in Barcelona, 2006 in Vancouver, 2008 in Nanjing, and 2010 in Rio de Janeiro encompass the spectrum of relevant urban actors: government leaders, mayors, and members of national, regional, and international associations of local governments, non-governmental and community organizations and international associations of local governments, professionals, academics, young people, women and slum dwellers groups – but have been accompanied and challenged by

---

27 Even though differences are explicitly not perceived as destabilizing per se, some of the academic literature still assumes them to be an obstacle that needs to be overcome, e.g. Nicholls 2009.

28 The World Urban Forum was established by the United Nations to examine the problems of urbanization by bringing together government leaders and mayors with members of national, regional, and international associations of local governments, as well as non-governmental and community organizations, professionals, academics, and grassroots organizations (see http://www.unhabitat.org/categories.asp?catid=535).

simultaneous Social Urban Forums[29] which are much smaller but bring together more grassroots and social movement organizations. The World Urban Fora held so far were used to dialogue and network as well as to draw up reports and appeals. Activist networks such as the International Alliance of Inhabitants (IAI)[30] and the Habitat International Coalition (HIC)[31] make use of such UN-sponsored fora as well as of the WSF process to launch massive campaigns against the effects of neoliberal globalization in (not only third world) cities. They use these arenas for global campaigns (e.g. *Housing and Land Rights*) and appeals for a new *Urban Social Pact*, and also for drafting charters for the 'right to the city', which, over the last few years, has become a joint focus of international human rights groups, UNESCO institutions, and these urban-oriented NGOs.

A 2002 'World Seminar for the Human Right to the City' sponsored by the WSF led to the draft of a 'World Charter for the Human Right to the City' in 2003.[32] In order to harmonize the different charters, UN-Habitat and UNESCO

---

29   The Social Urban Forum in Rio was called by social movements and organizations of Rio de Janeiro including favela-based groups (see Marcuse 2010).

30   Founded in Madrid in 2003, the International Alliance of Inhabitants (IAI) has brought together a large network of grassroots associations of urban inhabitants from many parts of the world. It seeks to coordinate actions to 'jointly stand against the perverse effects of exclusion, poverty, environmental degradation, exploitation, violence, and problems related to transportation, housing and urban governance produced by the neoliberal globalization' (http://eng.habitants.org/who_we_are/). It has concentrated its international campaigns on evictions, demanding unrestricted rights to housing for marginalized groups, and has lent support to squatters and other movements defending residential neighbourhoods against demolition or commercialization (cf. Holm 2009).

31   See http://www.hic-net.org/.

32   The original proposal for a 'Charter for Human Rights in Cities' was presented by a Brazilian NGO (FASE) at the 6th Brazilian Conference on Human Rights in 2001, shortly before the Brazilian City Statute came into force. It was inspired by the 'European Charter for the Safeguarding of Human Rights in the City' of 2000, adopted by more than 200 European cities. As a result of the recognition of the right to the city by the 2001 City Statute and the presentation of the initial FASE document during the WSF in 2002, several NGOs and urban social movements, especially from Brazil and other Latin American countries, started drafting the text later called the 'World Charter on the Right to the City'. This draft was subsequently discussed and expanded by Habitat International Coalition and others in 2004, at the America's Social Forum in Quito and the 2nd World Urban Forum in Barcelona. At the 2005 WSF in Porto Alegre, a 'Workshop on the Right to the City' was attended by hundreds of people, and the Brazilian Minister of Cities formally subscribed to the process of discussion and implementation of the 'World Charter on the Right to the City' (Fernandes 2007: 215-216). Following the WSF 2005, growing interest in and mobilization around the proposed World Charter spread, increasingly beyond Latin America. The theme was also picked up at the 2006 World Urban Forum, and at Rio in 2010 became its theme 'The Right to the City – Bridging the Urban Divide'.

have sought, since 2005,[33] together with the International Social Science Council (ISSC) and various international NGOs (such as AIVE, Metropolis, and others) to generate consensus among central actors, especially local politicians, about policies that are to guarantee sustainable, just and democratic cities. Towards this end and to further internationally comparative research on such policies, an international working group and a UNESCO-chair were established, which also play an active role in publicizing the international debates on urban policies and the right to the city.

The organizations pushing this type of a 'right to the city' agenda see some of its elements as already implemented, as with the participatory municipal budget – not only in Porto Alegre, but in many cities around the world;[34] further, they point to digital democracy as implemented in the city of Bologna, which provides free internet access to its residents; or to youth governments that have been installed in the Latin American and Caribbean region with the help of UN-Habitat.

All of these statutes and charters seek to influence public policy and legislation in a way that combines urban development with social equity and justice. They strive to put 'our most vulnerable urban residents' rather than investors and developers at the centre of public policy, and in this effort enumerate specific rights which a progressive urban politics should particularly protect. Thus, contrary to Lefebvre's definition of the right to the city,[35] which these documents frequently refer to,[36] in fact they invoke specific struggles for particular rights (not *the* right to the city), and combine 'a bundle of already-existing human rights and related State obligations, to which, by extension, local authorities are also party' (paragraph 7). The right to the city in these declarations entails the human rights to housing and work, food and clean water, health, security, access to public infrastructure, participation in decision-making, and many more. These rights are supposed to hold for all 'urban inhabitants', both as individuals and as a collective, but some groups are mentioned as deserving particular protection (the poor, ill, handicapped, and migrants get mentioned). In practice, these charters are not binding, globally enforceable guidelines, but rather are proposed to work

---

33   In March 2005 UN-Habitat and UNESCO signed a Memorandum of Understanding to confirm their cooperation towards the 'right to the city'. Together with ISSC, they launched a joint initiative on 'Urban Policies and the Right to the City' at UNESCO headquarters in Paris in 2005. The second meeting of the UNESCO/UN-Habitat/IAEC working group on 'Urban Policies and the Right to the City' took place at the municipality of Barcelona in early 2006, the third in late 2006 again in Paris, and the fourth in February 2008 was hosted by the city of Montreal.

34   Cf. http:///www.participatorybudgeting.org.

35   Born out of the context of May 68 in Paris, the right to the city for Lefebvre meant the 'creative surplus of the city, which points beyond the rationality of economics and state planning, as well as the right to participate in urban centrality' (Lefebvre 1968/1996; cf. Mayer 2007c).

36   Cf. UNESCO-UN-Habitat Discussion Paper of March 2005 <http://www.hic-net. org/articles.asp?PID=229>.

as blueprints for municipalities and NGOs interested in *good urban governance*. Their goal is to establish effective legal monitoring mechanisms and instruments to ensure the enforcement of recognized human, social, and citizenship rights. Towards this end, UN-Habitat campaigns such as the 'Global Campaign on Urban Governance' proselytize the charters, using toolkits on participatory decision-making, transparency in local governance, and participatory budgeting to demonstrate how these principles can be implemented in practice.

Contrary to Lefebvre's right to the city, which builds on a class-based concept of difference, these charters and declarations as well as their repertoire build on a more general concept of diversity and heterogeneity, in which civil society as a whole appears as worthy of protection from (destructive) global forces – as if it did not itself harbour economic and political actors who participate in and profit from the production of poverty, discrimination, and racism. It thus obfuscates the fact that civil society, and 'the city', are themselves deeply divided by class and power, and it remains up to social movements to challenge those reproducing exclusion and exploitation.

Yet, one might still argue that, once fully realized, these enumerated rights to access all that the existing city has to offer might spell a significant improvement. For one, the public recognition through governmental and UN institutions certainly lends added weight and legitimacy to movement demands and enhances the influence and status of the groups articulating them. But these charters and the coalitions devising and promoting them often tend, in the process, to modify the political content and meaning of the original movement demands. The laundry-list of rights boils down to claims for inclusion in the current system as it exists; it does not aim at transforming the existing system – and in that process ourselves. This type of rights discourse merely targets particular aspects of neoliberal policy, e.g. in combating poverty, but not the underlying economic policies which systematically produce poverty and exclusion. It shifts the nuances of the political stances and tends to dilute some of the radical demands of transformative movements. Both concepts of the right to the city are actually present at WSF meetings, but some of the documents published[37] tend to reflect the institutionalized version of a 'top down agenda agreed on by some NGO networks who already know what the rights are, but want to build a larger alliance … for which they need a name and branding' (Unger 2009).

The transnational forms of cooperation with NGOs and UN institutions do provide urban movements with rich opportunities for exchanging and networking with multiple emancipatory struggles, and to link up with grassroots mobilizations in cities of the global South and North – as do the camps at counter summits such as Heiligendamm. For example, at the WSF 2007 a variety of urban networks

---

37   For example, the 'Urban Movements Building Convergences at the World Social Forum, WSF 2009', see http://www.hic-net.org/content/convergencies-wsf2009.pdf.

met[38] and drafted a call for a 'Global Housing Campaign' that put the struggle against evictions and forcible displacement worldwide on the agenda. This call was supported and affirmed by the meeting of the WSF International Council in Berlin a few months later, and ratified at the alternative counter summit in Rostock. In 2009 'urban movements and networks against the crisis' met at Belém and passed a declaration to promote the right to the city, democratize the World Urban Forum, and prepare a 'Popular Urban Forum ... leading up to WSF 2011'.[39] The boundaries between local and global movements, as well as those between NGOs and direct action groups appear as even more fuzzy at the local, regional and national fora held simultaneously in 2008 and 2010, and the decentral 'action weeks' established in the years between global WSF meetings (deliberately designed to give more space to and strengthen local and regional structures within the WSF process).[40]

Anti-globalization movements discover the local/the urban

The experiences gained at summit protests as well as those from World Social Forum meetings are brought home. Conference delegates report back to their home organizations, and counter summit participants often attempt to import inspirations gained and lessons learned. Lesley Wood described how activists from New York City and Toronto tried to transfer the 'Seattle model' to their home cities. After participating in the G8 protests in Seattle they experimented in their local movement activities with some of the hallmarks of transnational summit protest such as black bloc street fighting tactics, blockade strategies, affinity groups, spokes councils, and radical puppet theatre (cf. Tarrow 2005a: 63). Importing these forms of organizing and of fighting turned out to be more difficult in Toronto than in New York, and implanting action repertoires seems to be harder than to transfer the themes of neoliberal globalization from the global to the local setting (Diani 2004). Still, 'report backs' have become an important tool also for the participants of the US Social Forum in Atlanta 2007 and Detroit 2010. These back home gatherings, where the activists share their experiences and observations at the Forum with members of their local community, also reflect the process orientation of the Social Forums (Smith, Kutz-Flamenbaum and Hausmann 2008: 46).

---

38   200 representatives of urban movements ('for housing and city rights') from 20 countries (from Africa, South and North America, Asia and Europe) as well as of municipal governments, cf. <http://www.hic-net.org/news.asp?PID=395>.

39   See <http://www.hic-net.org/articles.asp?PID=921>.

40   See the call of the International Council of the WSF for the Global Day of Action January 28, 2008 <http://www.forumsocialmundial.org.br/dinamic.php?pagina= chamada2008_eng>.

The localization and urbanization of global-scale movements occur in a variety of ways and with varying impact.[41] Especially since transnational networks such as Attac, Global Action Network, and Reclaim the Streets have become aware that free trade and market deregulation not only wreck sustainable production structures in the global South, but also threaten unions and consumers in North America and Europe, they have refocused their activities onto national and local scales in these first world regions as well. In addition to their efforts to democratize international institutions, they now also emphasize the local impacts of global neoliberal restructuring in their home cities. Many have moved the defence of public urban services and infrastructures to the top of their agendas, some (like Reclaim the Streets) zero in on the detrimental effects of corporate-driven urban restructuring. In Germany, the more than 200 local Attac groups have, since 2003, been turning towards whatever the local impositions of neoliberal restructuring happen to be, whether the privatization of public utilities or the dismantling of social services. One year they held actions at 50 cities' train stations against the privatization of the German railway company, in another they rallied at the local offices of the Social Democratic Party protesting its dismantling the German welfare net; after the financial crisis they staged actions – including blockades and other civil disobedience – in front of banks in 75 cities demanding that the government redistribute the bailout monies.

In similar fashion, the Social Fora that have sprouted since 2003 in Germany have been bundling together the work of local progressive initiatives, rank and file union groups, autonomous as well as church and charitable organizations. They align themselves with the WSF Charter of Principles (stressing cooperation in non-hierarchical networks), in order to 'locally benefit from the dynamics of the WSF process'.[42] This local-scale instantiation of the anti-globalization movement emphasizes alliance building and networking just like the global enactment does. German Social Forum groups have, in the face of mounting attacks on social rights, sought to bring together the splintered protests of students and childcare workers, jobless and handicapped, migrants and welfare recipients, arguing that their fragmented protest does not find resonance with left parties, church representatives, or other established organizations. With their political networks they seek not only to coordinate and support the activities of these diverse and often isolated groups, but especially to break up the climate of resignation and to instigate a public critique of neoliberal social policies.[43]

---

41   Because movement activists confront different situations as they translate or import experiences and insights gained on global scales into their distinct national and local environments, this may lead to significant differences in the 'down-scaling' of global issues and strategies (cf. Tarrow 2005a: 60, 2005b: Chapter 3).

42   Leitlinien der Zusammenarbeit im Berlin Social Forum (Guidelines of cooperation in the Social Forum Berlin, which were adopted May 2003, cf. <www.wikiservice.at/esf/wiki.cgi?BSF-Leitlinien>.

43   Cf. <http://germany.indymedia.org/2003/03/44293.shtml>.

Together with Attac, the Social Fora have also carried the global campaign against the General Agreement against Trade and Services (GATS)[44] into the cities: they organized protests and referenda against the privatization of public goods such as municipal utilities, and petitioned for plebiscites against cross-border leasing of public facilities to US finance trusts (a form of selling off public infrastructure that was wildly popular in German municipalities before the global financial meltdown, cf. Rügemer 2004, 2008). In many cities the protests succeeded in stopping or preventing these deals, as for example the plan to lease out the Frankfurt subway grid.

In 2006 the German Social Forum groups congregated for the first time for a 'Social Forum in Germany' (SFiD) in Cottbus; a second national meeting was again staged in this Eastern city, in the autumn of 2007, deliberately to strengthen local movement milieus. Because 2008 saw a (decentral) Global Action Day (instead of a central WSF meeting), the third SFiD meeting took place over four days in 2009 in Hitzacker/ Wendland.[45] The multi-scalar work of the local social fora is vividly demonstrated in an interactive film DVD presenting the initiatives and groups making up the Berlin Social Forum: video clips showing various Berlin activities are linked to presentations of groups that were present at the German Social Forum gathering, as well as to clips on international social movement groups shaping the WSF process. The referendum against the privatization of water provision in Berlin, for example, is linked to the campaign against privatization against water in India.

It is clear that anti-globalization movements which have newly entered the urban stage have brought fresh momentum to the local movements, helping them overcome their fragmentation, and supporting their consolidation as well as their professionalization. In 'localizing' the issues of the transnational movements, they contributed to the transfer of repertoires associated with the work of transnationally oriented organizations, such as professional PR work, conscious media orientation, and a flexible action repertoire utilizing pragmatic as well as militant action forms. Especially the dual tactics proven and tested at summit meetings, where activists operate both in the negotiation arenas (as representatives or advocates for various disadvantaged groups, and as partners of business and state actors) as well as in the streets (with demonstrations, blockades, creative spectacles, clowns and puppets)

---

44   The GATS extends liberalization efforts from the sphere of goods to the one of services, i.e. to areas that had previously been considered outside the remit of trade talks. How services are provided and by whom has impacts on basic human rights (to health, to education) and development (e.g. access to water, public transport, financial services, etc., i.e. GATS links economic processes directly to social and human rights (cf. UN High Commissioner for Human Rights, 2001).

45   Cf. <http://www.sfid.info/d_sf/index.html>. However, this third SFiD meeting showed, like some of the other European ones, fewer attendees, as parallel progressive alliances triggered by the economic crisis absorbed much of the protest potential (Wallrodt 2009).

have found their way into the urban movement practices. In both dimensions, as negotiation partner and as movement activist, urban movements have since enhanced their organizational and professional skills. Whether engaging in theatre actions against workfare jobs in front of employment offices, or warding off evictions of squatted buildings or social centres, their action forms increasingly tend towards media savvy, professionally managed events, and slick websites report on their action, link to related ones carried out by others, and help spread the message and build the networks. The shift of the anti-globalization movements toward the city has thus given a boost to place-based movements, and global/ local connections and mutual learning processes have frequently politicized local projects and initiatives. These trans-scalar diffusions of the anti-globalization movement have also put the city, and the struggle to reclaim the city, on top of the agenda of transnational struggles, emphasizing its global significance (cf. Portaliou 2007).

The diffusion of knowledge and experience goes both ways; not only have the anti-globalization movements and WSF events impacted on local struggles, but particular local/national instantiations of the Forum Process are also impacting on the WSF. A good example of this has been the US Social Forum which has challenged existing notions of open space so far characteristic of the WSF by fusing the culture of the broader WSF process with US-specific movement dynamics, especially that of base-building organizations concentrated in poor and minority communities. They prefer to see the Forum as an 'intentional' rather than as an open space,[46] arguing that the latter is easily dominated by the resource-rich – in the US, white middle-class males. Thus, the USSF organizers made deliberate efforts to include Blacks, Latinas, migrants, indigenous and other marginalized groups. And, in recognition of the need for political intervention by a unified movement, they provided a 'People's Movement Assembly' as locus for coordinated political action, in an effort to move the WSF beyond the open space model (Smith, Juris and the Social Forum Research Collective 2008).

This last example of local-national-global connectivities and diffusion processes reconnects the circle these last three sections have been describing: we are back to 'urban movements scaling upwards'. All the cases sketched in these sections illustrate that every one of the scales in which the (right to) the city is being fought over is shaped by specific power relations and conflict structures. The political and discursive context, the particular features of the actors, and the opportunity structures are particular to each case. But the various instances also reveal that the struggle for a better city takes not only different scale-differentiated, but also distinct place-based, territorially anchored, and network-specific forms. All of these socio-spatial dimensions would need to be accounted for in order to adequately understand the shifts and reconfigurations in the contemporary social and political power relations (cf. Jessop 2008; Brenner 2009). The brief

---

46   How 'intentionality' reflects US political and cultural legacies and how it differs from the conception and practice of open space are elaborated in Juris 2008.

survey presented in this chapter provides but an initial impulse; more systematic empirical research on the practices of movements on the various scales of urban politics and their interactions is necessary in order to identify the opportunities and possibilities set up by these shifts. A few preliminary clues may, however, be gathered from our initial synopsis, which the final section will present.

Opportunities and problems of multi-scalar struggles for the city

The battle for another, better city is fought in a variety of differently scaled, partially overlapping arenas simultaneously, constituting a complex set of relations of global and local engagement: alliances and networks forged at WSF meetings or alternative counter summits may reverberate as productively on local struggles as do the global activities of human rights and anti-poverty groups; simultaneous actions at different sites around the world (as with J18 in 40 cities across continents against the Cologne World Economic Summit 1999, or with the demonstrations in many different countries against the World Bank and IMF meeting in Prague in 2000, or with the decentral WSF global days of action) signal the global connectedness of local movements, thus politicizing them as constitutive elements of counter-hegemonic globalization. Also, local struggles defending social rights or pushing for direct-democratic practices transcend their local limitation (and potential regressive bent) through transnational interconnectedness. But to realize these potentials requires specific preconditions: there is no automatic mechanism intensifying or enhancing the movements' strength and mobilizing power. For 'glocal protest' to be effective – this much has become clear – the movement milieus of the host locality play a crucial role (and the relocation, after Genoa and Seattle, of summits to remote luxury resorts has robbed protests of this on-site base). Since then, the success of summit protests has depended more on the movement groups of nearby cities and on supra-local coalitions and networks. These conditions have made it more difficult to create bridges to local struggles than was the case in Seattle and Genoa, where thick protest milieus rooted in local everyday relations provided a fertile organizational base.

The connection between local movement milieus and transnational urban politics has become more complicated in other ways as well. When an online discussion forum on social movements raised the question of how best to describe the process 'whereby social movement leaders ... gravitate away from small and local actions to more diffuse national or international ones',[47] replies ranged from neutrally descriptive terms such as 'institutionalization' and 'action diversification' to pejorative classifications such as 'cooptation' and 'bureaucratization'. The

---

47   SOCIAL-MOVEMENT@LISTSERVE.HEANET.IE in September 2007: '... to put it in more practical terms: avoid leafleting on a Saturday morning outside windswept shopping centres, in order to attend 'important' international meetings with fellow leading activists?'

cases of local/global activism presented in this chapter reveal, however, that the transnationalization or up-scaling of urban movement activities does not always and without fail imply simultaneous processes of depoliticization or NGOization. Rather, we find the whole spectrum of political positions and demands on every scale, from radical anti-neoliberal (frequently in tandem with direct action and blockade type of repertoires) to narrower, more pragmatic demands (targeting select aspects of neoliberal politics, as, for example, combating poverty, without challenging the underlying economic system) usually put forth by NGOs and movement entrepreneurs, often in collaboration with multilateral organizations (cf. Wallace 2003). While the splits and divisions emerging within the urban movement landscapes and the dilemmas they create for emancipatory struggles are addressed in some initial research (cf. Mayer 2007b; Twickel 2010), the activities, collaborations and alliances that have formed in the global arenas appear as so complex, variegated, and ambiguous that accompanying fragmentations or shifts in political orientation and direction of the contestations have yet to be systematically investigated.

'Glocal' collaboration around summit events, WSF as well as ESF and other regional/transnational gatherings has in many ways invigorated movement activism, but also created new and problematic dilemmas. On the positive side, the interchange taking place here between activists from different countries and between particular local and transnational movements can occur only via horizontal, democratic communication structures, within which multiple acts of translation have to take place, which aids and supports the emergence of pragmatic and tolerant positions (cf. Doerr 2009). The effects of practising respect for national, cultural as well as ideological differences are visible in the establishment of democratic local-transnational structures. And since networking and coalition-building (including with partners from outside the movement scene) is so crucial to organizing the transnational events, these strategies have broadly publicized the movements' agendas and expanded the front of resistance against neoliberal strategies of privatization and social dismantling.

But success or effectiveness of social movements is not easy to define or measure – and social movement theorists disagree on the criteria: some highlight their contribution to boosting emancipatory efforts, others measure success in terms of political leverage gained. There have always been movements that, while failing in terms of emancipatory and transformative criteria, continue with some success in terms of power politics. The networks of NGOs and advocacy organizations that have been working on drafting a World Charter on the Right to the City and are designing policies that seek to guarantee sustainable, just, and democratic cities might be viewed as exemplary cases of urban movements having gained political leverage. On the other hand, the novel spaces created by the anti-globalization movement in the social forum process (on global as well as sub-global scales) provide modalities of transformational politics, the effectiveness of which – while hardly measurable with categories and criteria developed from traditional definitions of political power – may still be significant. Osterweil (2008) for example points to the ways in which conflicts between different movement groups were resolved at the US Social Forum in Atlanta as positive and transformative in that they made previously unarticulated conflicts between different political cultures visible and helped produce less dogmatic and less formulaic knowledge as well as political actors more capable of reflexive practice. Such learning processes and experiences, with potentially long-term transformative possibilities, have been allowed to occur at the scale and in the (open) space created by the social forum process, which brings into face-to-face contact the needs and cultures of different local struggles. Their clashes are either worked out in a context of mutual recognition and respect and in knowing that what joins them is the struggle against neoliberalization, which harms them all – or they do not get worked out.

The (narrower) successes in terms of gaining political leverage, however, often come at a price. One price is the emerging cleavage between a small elite of transnationally mobile activists who dominate the flow of information and monopolize decision-making, and the home-based rank and file of the movements (cf. Cumbers et al. 2008). While urban movements in poor and developing countries also send their representatives and spokespeople to global gatherings and networks, their numbers are comparatively small due to fewer resources. Their issues and viewpoints tend to be represented instead by advocacy networks such as Habitat International Coalition (HIC), who may or may not authentically speak for the positions of shack dwellers' movements of South Africa's cities, or the self-organized struggles of Latin American favelas or of South Asian informals (cf. Bayat 2004); authors such as Pithouse emphatically deny that they can (Pithouse 2009).

Another price for the gain in visibility and influence thanks to participation in the transnational arena, as was shown for the networks engaged in the charter movement, is a watering down of the movement agenda. Responsible for such dilution effects is not the 'up-scaling' from local to higher dimensions of activism, but rather the specific composition of civic alliances 'from below'. As illustrated with the broad movements towards Right to the City charters, such statutes and the associated discourses assume and advance civil society alliances between urban inhabitants, municipal governments, and NGOs. These alliances and their participating organizations provide arenas where the content of 'good urban governance' or direct-democratic procedures may be open to the input and radical demands of activist mobilization and political pressure – if activists choose to mobilize such pressure. But there is a world of difference between the agenda of the (UN-sponsored) World Urban Forum and that of the Social Urban Forum, both of which met at Rio de Janeiro in March 2010 on the theme of the Right to the City:

> the desirability/inevitability of capitalism was a foundational belief at the WUF; not so at the SUF, where it was frequently called into question … At the WUF, the poor were dealt with as the objects, the beneficiaries of the policies there debated … At the SUF, the poor and their movements were the subjects of

concern … Bridging the Gap, in the call for the WUF, was there seen as moving
the poor a little closer to those above them; in the SUF, it was rather eradicating
the distinction between above and below (Marcuse 2010: 31-32).

In spite of such contrasts, cooperation on some immediate actions appears as
possible, even if with different long-term perspectives. As Sikkink (2005) pointed
out, some international institutions provide opportunity structures, even arenas
for social movements, not just threats. But where the goals are too divergent,
confrontation may be more appropriate. Confrontation and critique may be the
only viable option with regard to the programmes and initiatives sponsored by the
World Bank or WTO, for which strengthening civil society networks has become
a means to increase efficiency; urban poverty, which is here seen as resulting from
inefficient local government, is combated by prescribing more businesslike public
management. The NGOs partnering in such World Bank or WTO programmes tend
to comply with, rather than challenge, the standards and definitions prevalent there
(Fox and Brown 1998), and frequently grassroots groups fall prey to such views
as well. Underlying these views is the assumption that, when urban inhabitants get
together with municipalities to develop endogenous potential and local growth,
structural contradictions between local autonomy and international competition,
or between sustainability and economic growth, can be harmonized (cf. Jessop
2000).

   The problem with such thinking is that it strips the various scales of contestation
of their social content and of the tensions and conflicts residing within them,
making them appear either as homogeneous and worth protecting (as with the
local or urban scale), or as threatening but inevitable (as with the global scale).
In this perspective, the different scales become reified, and the political and
organizing processes, the clashes and changing relations which take place within
all of them disappear from view. In reality, of course, the scales themselves remain
contested, power is unevenly distributed within each of them. If, as is suggested
here, the success of collective action is measured by the movements' capacity to
transform, in the course of struggle, the terrain and the constraints of conflict (cf.
Santos 2008), then the goal is always the fundamental transformation of existing
structures, and the struggle for a better city is always also a struggle about power,
which can neither be left to international NGOs (even well-meaning ones) nor to
local governments (even social-democratic ones).

*North-South dialogue: new relations between urban movements in the Global
North and South?*

Some of the effects of the world's uneven development on urban movements
have already been mentioned above: the issues and demands of slum dwellers and
other local movements in the sprawling mega-cities of emerging and developing
countries are not as easily (re)presented in the upper scales of the urban resistance
networks as those of their first world counterparts. Even though global gatherings

and counter summits have helped to detect similarities and to build on linkages in the common struggle against neoliberal globalization, the differences go far beyond unequal access to resources. There are also structural differences stemming from the changing global division of labour, in which more and more manufacturing is outsourced to the global South – leaving first world cities as sites of privileged 'global suburbs', hosting the global economy's command and control centres with its FIRE sectors,[48] vying for globally mobile creative workers (both in technology as well as culture) and for tourists, who are in turn serviced by downgraded, increasingly migrant, labourers.

While the new global proletariat assembles at the production platforms in the global South, the class composition of first world cities is now marked by a new antagonism (and hence new types of conflict) between top-end users and 'creative classes' on the one hand and low-wage and precarious groups of excluded and marginalized 'urban outcasts' on the other. That is not to say that the struggles over this antagonism are not relevant. Resistance against the neoliberal restructuring of these cities, and struggles in defence of urbanity as well as of alternative spaces and lifestyles or those in support of social economy are obviously relevant to emancipatory change, but they do not really threaten the structures of domination and exploitation of the neoliberal system. Contestations along the new antagonism between privileged urban users and advanced marginality are, indeed, of global significance, because low waged work has expanded to include more women and more immigrants, making first world cities important sites of anti-colonial struggles as well as struggles against racism and sexism. But many of the first world urban struggles are primarily defensive, and much of the anti-gentrification movement seeks to save a piece of urbanity or protect alternative lifestyles. The risk of co-optation and (partial) integration into an urban model in the image of corporate and financial interests is immense. Among former squatters and newly engaged cultural activists, many have become more interested in projecting a city where their own – self-determined and politically-correct – liberated space is guaranteed, and less concerned with the exclusion and repression of less fortunate ones. Such activists increasingly succeed today in securing their own survival by buying into the new, entrepreneurial 'creative city' policies that exploit their vibrant cultural scenes for branding, as a locational asset in the intensifying interurban competition.

Under these conditions, where urban resistance is often limited to defending 'alternative biotopes' and/or easily incorporated into creative city policies, we need to ask whether first world cities still bring forth the social forces with an interest in transforming them into more equitable and attractive living environments for all. In whose interest would it be to form alliances that would challenge the structures of *global* inequality?

At the same time, new movements have emerged in cities of the global South – especially in struggles with local governments and local elites that act as stooges

---

48   Finance, insurance, real estate as well as other advance services.

for global corporations and global institutions – that have developed organizational structures and forms of protest of their own and do not always find the support of western NGOs and leftist movements as helpful. The struggles of the pavement dwellers in India, the favela residents of Latin American cities, the slum residents of the rapidly urbanizing Asian tiger countries, or of the shack dwellers of the urban peripheries of Capetown, Durban, and Johannesburg all demonstrate that the urban poor, in resisting dispossession, eviction, police violence and repression, have organized themselves into independent structures, developed their own local protest cultures, and have achieved – through mass mobilization, occupations, and political protest – improvements in their living conditions.

These new local movements – their repertoires, demands, and organizational forms – do not fit neatly into the models of transnational social movement research. The enormous material and political difficulties they confront prevent them from directly participating in global organizing processes, and they frequently find the support offered by transnational NGO networks less effective than that provided by local churches or volunteers from the global North who spend some months with them, sharing their daily lives and struggles. It is also striking that concepts that do not loom large in most western movement milieus, such as the dignity of the individual, here are an essential part of the movement vocabulary: ' ... it was the traditional language of the dignity of each person, reworked into a cosmopolitan form appropriate for urban life, that was ... given primary consideration ahead of any of the more explicitly political languages', Pithouse (2009: 246-7) wrote of the Abahlali Basemjondolo activists in Durban. While this emphasis on the role of human dignity as part of the political struggle of the movements of the landless and poor in the global South distinguishes them from traditional progressive movements in the West, it does play a role in the poor people's movements of the US, which play such a leading role in the practice and coalitions of the American social forum process. Similarly, the continuities between everyday life and protest action are not only characteristic of resistance movements in the global South (Bayat 2004), but also of the political mobilization of migrant informal workers in the 'first world' (cf. Boudreau 2009). Thus, we increasingly find the orientations and protest forms of the movements of the urban poor of the South also in the disenfranchised and marginalized areas of northern metropoles.

Even though this is indicative of how the global North and South do intersect, the creation of linkages between the different struggles for the right to the city – that of the leftist, alternative, and 'creative' challengers of neoliberal urban politics on the one hand, and that of the global poor on the other – is not always easy, both because of the uneven development and because of the different movement cultures. But the fight, even though one is far more existential than the other, is the same struggle for a liveable, sustainable city for all. That is why, in practice, building linkages and coalitions, on all scales, has become so important; and why, in theory, it has become so important to find analytical ways to discern the old and new scales of mobilization against the neoliberal project not as reified but themselves shaped by conflicts of interest between social forces. All scales are themselves politicized institutional spaces for carrying out the struggle over (the right to) the city. Locally as well as globally the claims and demands of urban movements have already been picked up and incorporated into national as well as international institutions' strategies to develop a 'softer, kinder' neoliberalism. Scholars and activists may or may not want to prefer a reformist route to alleviate specific or particularly problematic effects of neoliberalization, but at least they should distinguish clearly between problems within and of neoliberalism, i.e. to distinguish between a critique and an agenda that remain immanent to the regime of neoliberal capitalism, and positions that point beyond such a regime – even if multiscalar deflections now obfuscate the realities of social and power relations.

References

Ainger, K. (2009) 'Once beaten for stating the obvious, our time has come', in *The Guardian*, 26 March 2009, http://www.guardian.co.uk/commentisfree/2009/mar/26/anticapitalism-protest-recession-g20/.

Albert, M. (2007) 'USSF – 2007 and After …' in *ZNet* (July 12), http://www.zmag.org/znet/viewArticle/14970.

Babtist, W., and M. Bricker-Jenkins (2002) 'A View from the Bottom: Poor People and Their Allies Respond to Welfare Reform', in R. Albelda, A. Withorn (eds) *Lost Ground. Welfare Reform, Poverty and Beyond*. Cambridge, MA: South End Press, 195-210.

Bandler, M. (2007) 'Faster, higher, stronger: Expansion and boundaries of transnational activism', Paper ECPR Conference Pisa 2007. (http://www.ecpr.visionmd.co.uk/paper_info.asp?paperNumber=PP535)

Bayat, A. (2004) 'Globalization and the Politics of the Informals in the Global South', in A. Roy, and N. AlSayyad (eds) *Urban Informality: Transnational Perspectives from the Middle East, Latin America, and South Asia*. Lanham: Lexington Books, 79-102.

Boudreau, J.A. (2009) 'Taking the bus daily and demonstrating on Sunday: Reflections on the formation of political subjectivity in an urban world', *City*, 13 (2-3): 336-346.

Brenner, N. (2000) 'The Urban Question as a Scale Question: Reflections on Henri Lefebvre, Urban Theory and the Politics of Scale', *International Journal of Urban and Regional Research*, 24 (2): 361-78.

—— (2009) 'A thousand leaves. Notes on the geographies of uneven spatial development', in R. Keil and R. Mahon (eds) *Leviathan Undone? Towards a Political Economy of Scale*. Vancouver: University of British Columbia Press.

Brenner, N. and N. Theodore (2002) 'Cities and the Geographies of "Actually Existing Neoliberalism"', *Antipode*, 34 (3): 349-379.

Conway, J. (2008) 'The empire, the movement, and the politics of scale: considering the World Social Forum', in R. Keil and R. Mahon (eds) *Leviathan*

Undone? Towards a Political Economy of Scale. Vancouver: University of British Columbia Press.

—— (2009) 'Belém 2009: Indigenizing the Global at the World Social Forum', http://unialter.wordpress.com/2009/03/02/belem-2009-indigenizing-the-global-at-the-world-social-forum/.

Cumbers, A., Routledge, P., and C. Nativel (2008) 'The entangled geographies of global justice networks', *Progress in Human Geography*, 32 (2): 183-201.

della Porta, D. (2005) 'Multiple Belongings, Tolerant Identities, and the Construction of "Another Politics": Between the European Social Forum and the Local Social Fora', in D. della Porta and S. Tarrow (eds) *Transnational Protest and Global Activism*. Lanham, UK: Rowman and Littlefield, pp. 175-202.

Demirovic, A. (2000) 'NGOs and Social Movements: A Study in Contrasts', *Capitalism Nature Socialism*, 11 (4): 131-140.

Diani, M. (2004) 'Cities in the World: Local Civil Society and Global Issues in Britain', in D. della Porta and S. Tarrow (eds) *Transnational Protest and Global Activism*. Lanham, UK: Rowman and Littlefield.

Doerr, N. (2009) 'Language and Democracy in Movement. Multilingualism and the case of the European Social Forum process', *Social Movement Studies*, 8(2): 149-165.

Eskola, K. and F. Kolb (2002) 'Attac: Entstehung und Profil einer globalisierungskritischen Bewegungsorganisation', in H. Walk and N. Boehme (eds) *Globaler Widerstand. Internationale Netzwerke auf der Suche nach Alternativen im globalen Kapitalismus*. Münster: Westfälisches Dampfboot, pp. 157-167.

Fernandes, E. (2007) 'Constructing the "Right to the City" in Brazil', *Social and Legal Studies*, 16 (2): 201-219.

Ford Foundation (2004) 'Close to Home. Case Studies of Human Rights Work in the United States', New York: Ford Foundation (http://www.fordfound.org).

Foti, A. (2007) 'Pink, Black, Pirate: Taking Stock of Rostock', http://transform. eipcp.net/correspondence/1182944688.

Fox, J. and L. D. Brown (eds) (1998) *The Struggle for Accountability: The World Bank, NGOs, and Grassroots Movements*. Cambridge, MA: The MIT Press.

Gerhards, J. and D. Rucht (1992) 'Mesomobilization: Organizing and Framing in Two Protest Campaigns in West Germany', *American Journal of Sociology*, 98: 555-596.

Glassman, J. (2001) 'From Seattle (and Ubon) to Bangkok: the Scales of Resistance to Corporate Globalization', *Environment and Planning D: Society and Space*, 19: 513-533.

Hadden, J. and S. Tarrow (2007) 'The Global Justice Movement in the US since Seattle', in D. della Porta (ed) *The Global Justice Movement*. Boulder: Paradigm Publishers, pp. 210-231.

Hamel, P., Lustiger-Thaler, H. and M. Mayer (eds) (2000) *Urban Movements in a Globalizing World.* London, New York: Routledge.

Harvey, D. (2006) *Spaces of global capitalism. Towards a theory of uneven geographical development*. London and New York: Verso.

Holloway, J. and V. Sergi (2007) 'Of Stones and Flowers – Dialogue between John Holloway and Vittorio Sergi' (http://gipfelsoli.org/Home/Heiligendamm_2007/Heiligendamm_2007_english/Discussion_and_Evaluation/4087.html).

Holm, A. (2009) 'Recht auf Stadt – soziale Kämpfe in der neoliberalen Stadt', RLS-Thüringen (ed.) *Die Stadt im Neoliberalismus*. Erfurt: RLS / Gesellschaftsanalyse, pp. 27-37.

Jessop, B. (2000) 'Good Governance and the Urban Question: On Managing the Contradictions of Neoliberalism', *Mieterecho*.

—— (2008) *State Power*. Cambridge, UK: Polity Press.

Jessop, B., Brenner, N. and M. Jones (2008) 'Theorizing Sociospatial Relations', *Environment and Planning D: Society and Space*, 26: 389-401.

Juris, J. (2008) 'Spaces of Intentionality: Race, Class, and Horizontality at the US Social Forum', *Mobilization*, 13 (4): 353-372.

Köhler, B. and M. Wissen (2003) 'Glocalizing Protest: Urban Conflicts and Global Social Movements', *International Journal of Urban and Regional Research*, 24 (4): 942-951.

Lefebvre, H. (1968/1996) 'The Right to the City', in E. Kofman and E. Lebas (eds) *Writings on Cities*. Oxford: Blackwell, pp. 63-181.

Lehrer, U. and R. Keil (2007) 'From Possible Urban Worlds to the Contested Metropolis: Urban Research and Activism in the Age of Neoliberalism', in H. Leitner, J. Peck and E. Sheppard (eds) *Contesting Neoliberalism: Urban Frontiers*. New York: Guilford, pp. 291-310.

Leitner, H., Peck, J., and E. Sheppard (eds) (2007) *Contesting Neoliberalism: Urban Frontiers*. New York: Guilford.

Leitner, H., Shepperd, E., and Sziarto, M. (2008) 'The spatialities of contentious politics', *Transactions of the Institute of British Geographers*, (33) 2: 157-172.

Levi, M. And D. Olson (2000) 'Strikes: Past and Present – and the Battles of Seattle', *Politics and Society*, 28 (3): 309-329.

Lichbach, M. and P. Almeida (2001) 'Global Order and Local Resistance: The Neoliberal Institutional Trilemma and the Battle of Seattle', Ms., Dept of Sociology, University of California Riverside.

Low, S. and N. Smith (2006) *The Politics of Public Space*. New York: Routledge.

McNevin, A. (2006) 'Political Belonging in a Neoliberal Era: The Struggle of the Sans-Papiers', *Citizenship Studies*, 10 (2): 135-151.

Marcuse, P. (2010) 'Two World Urban Forums, Two Worlds Apart', *Progressive Planning*, 183: 30-32.

Mayer, M. (2007a) 'Contesting the Neoliberalization of Urban Governance', in H. Leitner, J. Peck, and E. Sheppard (eds) *Contested Urban Futures: Neoliberalisms and their Discontents*. New York: Guilford Press, pp. 90-115.

—— (2007b) 'Städtische soziale Bewegungen', in R. Roth and D. Rucht (eds) *Handbuch Soziale Bewegungen*. Frankfurt: Campus.

—— (2007c): 'Recht auf Stadt', in Uli Brand, Bettina Lösch, Stefan Thimmel (eds), *ABC der Globalisierung*. Hamburg: VSA-Verlag.

—— (2008) 'To what end do we theorize sociospatial relations?', *Environment and Planning D: Society and Space*, 26: 414-419.

Mayer, M. and K. Rankin (2002) 'Social Capital and (Community) Development: a North/South Perspective', *Antipode*, 34 (4): 804-808.

Miller, B. (2000) *Geography and Social Movements*. Minneapolis: University of Minnesota Press.

Mitchell, D. (2003) *The Right to the City. Social Justice and the Fight for Public Space*. New York: Guilford.

Mudu, P. (2004) 'Resisting and challenging neoliberalism: The development of Italian social centers', *Antipode*, 36: 917-941.

Müller, T. and K. Sol (2007) 'Zwei Siege auf einmal? Das geht nun wirklich nicht!' (http://transform.eipcp.net/correspondence/1183042751).

Nicholls, W. (2009) 'Place, Networks, Space: Theorising the Geographies of Social Movements', *Transactions of the Institute of British Geographers*, 34: 78-93.

Nichols, J. (2005) 'Urban Archipelago', *The Nation*, June 20 (http://www.thenation.com/docprint.mhtml?i=20050620&s=nichols).

Osterweil, M. (2008) 'A Different (Kind of) Politics is Possible: Conflict and Problem(s) at the USSF', J. Blau and M. Karides (eds) *The World and US Social Forums*. Leiden and Boston: Brill, pp. 71-89.

Pendras, M. (2002) 'From Local Consciousness to Global Change: Asserting Power at the Local Scale', *International Journal of Urban and Regional Research*, 26 (4): 823-833.

Piazza, G. (2007) 'Inside the radical left of the global justice movement: The squatted self-managed social centres in Italy', Paper ECPR (European Consortium for Political Research) Conference Pisa (http://www.ecpr.visionmd.co.uk/paper_info.asp?paperNumber=PP556).

Pithouse, R. (2009) 'Abahlali Basemjondolo and the Struggle for the City in Durban, South Africa', CIDADES 6 (9): 256-257.

Ponniah, T. (2007) 'The Contribution of the U.S. Social Forum: a Reply to Whitaker and Bello's Debate on the Open Space' (http://www.lfsc.org/wsf/ussf_contribution_thomas.pdf).

Portaliou, E. (2007) 'Anti-global movements reclaim the city', *City*, 11 (2): 165-175.

Pröhl, M., Sinning, H. and S. Nährlich (eds) (2002) *Bürgerorientierte Kommunen inDeutschland–Anforderungen und Qualitätsbausteine*. Projektdokumentation Bd. 3: Ergebnisse und Perspektiven des Netzwerkes CIVITAS. Gütersloh: Bertelsmann.

Rosewarne, S. (2001) 'Globalization, Migration, and Labor Market Formation – Labor's Challenge?', *Capitalism Nature Socialism*, 12 (3): 71-84.

Roth, R. (2003) 'Bürgerkommune – ein Reformprojekt mit Hindernissen,' in H.J. Dahme (ed.) *Soziale Arbeit für den aktivierenden Staat*. Opladen: Leske + Budrich.

Routledge, P. (2003) 'Convergence Space: Process Geographies of Grassroots Globalization Networks', *Transactions of the Institute of British Geographers*, 28: 333-349.

Rügemer, W. (2004) 'Cross Border Leasing', *Ein Lehrstück zur globalen Enteignung der Städte*. Münster: Westfälisches Dampfboot.

—— (2008) *Privatisierung in Deutschland: Eine Bilanz*. Münster: Westfälisches Dampfboot.

Samsa, G. (2007): *Mythos Heiligendamm*, http://gipfelsoli.org/Texte/Gipfelpro test/4074.html.

Santos, B. S. (2004) 'Das Weltsozialforum: Für eine gegenhegemoniale Globalisierung', in *Utopie kreativ*, 169: 1004-1016.

—— (2007) 'The World Social Forum and the Global Left', Ms University of Coimbar/University of Wisconsin Madison/University of Warwick (http://www.openspaceforum.net/twiki/tiki-read_article.php?articleId=492).

—— (2008) 'Pluralidades despolarizadas: una izquierda con futuro', D. Chavez, C. Rodríguez Garavito and P. Barrett (eds) *La nueva izquierda en América Latina*. Madrid: Los Libros de la Catarata, pp. 359-376.

Santos, B. S. and C. Rodriguez-Garavito (2005) 'Law, Politics, and the Subaltern in Counter-Hegemonic Globalization', in B. Santos and C. Rodriguez-Garavito (eds) *Law and Globalization from Below*. New York, Cambridge University Press, pp. 1-26.

Sikkink, K. (2005) 'Patterns of Dynamic Multilevel Governance and the Insider-Outsider Coalition', in D. della Porta and S. Tarrow (eds) *Transnational Protest and Global Activism*. Lanham: Rowman and Littlefield, pp. 151-173.

Smith, J. (2008) *Global Visions/Rival Networks. Social Movements for Global Democracy*. Baltimore: Johns Hopkins University Press.

Smith, J. and E. Reese (eds) (2008) 'The World Social Forum Process', Special Focus Issue, *Mobilization*, 13 (4).

Smith, J., Juris, J. and the Social Forum Research Collective (2008) '"We are the ones we have been waiting for": The U.S. Social Forum in Context', *Mobilization*, 13(4): 373-394.

Smith, J., Kutz-Flamenbaum, R. and C. Hausmann (2008) 'New Politics Emerging at the US Social Forum', in J. Blau and M. Karides (eds) *The World and US Social Forums*. Leiden and Boston: Brill, pp. 41-56.

Smith, N. (2002) 'New globalism, new urbanism: Gentrification as global urban strategy', *Antipode*, 34 (3): 427-450.

Smith, N. and J. DeFillipis (1999) 'The reassertion of economics: 1990s gentrification in the Lower East Side', *International Journal of Urban and Regional Research*, 23(4): 638-653.

Swyngedouw, E. (1997) 'Neither global nor local: "glocalization" and the politics of scale', in K. Cox (ed.) *Spaces of Globalization: Reasserting the Power of the Local*. New York: Guildford.

Tarrow, S. (2005a) 'The Dualities of Transnational Contention: "Two Activist Solitudes" or a New World Altogether?', *Mobilization*, 10 (1): 53-72.

—— (2005b) *The New Transnational Activism*. Cambridge: Cambridge University Press.

Tarrow, S. and D. McAdam (2005) 'Scale Shift in Transnational Contention', in D. della Porta and S. Tarrow (eds) *Transnational Protest and Global Activism*. Lanham: Rowman and Littlefield, pp. 121-147.

Twickel, C. (2010) *Gentrifizierungsdingsbums oder Eine Stadt für alle*. Hamburg: Nautilus.

Unger, K. (2009) '"Right to the City" as a Response to the Crisis: "Convergence" or Divergence of Urban Social Movements?', http://www.reclaiming-spaces. org/crisis/archives/266.

United Nations High Commissioner for Human Rights (2001) 'Liberalisation of trade in services, and human rights', Sub-Commission on Human Rights Resolution 4.

Wallace, T. (2003) 'NGO Dilemmas: Trojan Horses for Global Neoliberalism?', in L. Panitch and C. Leys (eds) *Socialist Register 2004. The Imperial Challenge*. London: Merlin Press.

Wallrodt, I. (2009) 'Bewegungen im Wartestand', in *Neues Deutschland* October 19, 2009, p. 2 <http://www.neues-deutschland.de/artikel/157659.bewegungen-im-wartestand.html>.

PART III
Networks: Connecting Actors and Resources Across Space

Chapter 9

The Built Environment and Organization in Anti-US Protest Mobilization after the 1999 Belgrade Embassy Bombing[1]

Dingxin Zhao

Introduction

Based on a case study of the 1999 anti-US Chinese student protest triggered by the Belgrade embassy bombing, this chapter explores the relationship between organization and the spatial configuration of the built environment in social movement mobilization. The concept of the built environment includes any human-made surroundings that shape human activity.[2] The built environment can range in scale from personal shelter to neighbourhoods and schools, to large-scale civic spaces. Its spatial configuration can have tremendous impacts on human activity in multiple ways.[3]

The role of the spatial configuration of the built environment (built environment hereafter) in movement mobilization has received attention elsewhere (e.g., Gieryn 2000; Gould 1995; Traugott 1995; Wolford 2003; Zhao 1998). In his study of nineteenth-century Paris uprisings, Gould (1991, 1993, 1995) shows how

1   I thank Lulu Li for his assistance with data collection in China. Thanks are also extended to William Parish, Andrew Abbot, Doug McAdam, John D. McCarthy, and the editors of this book for their very helpful criticisms and suggestions in the process of revising the chapter. Funding for this project was provided by a University of Chicago social sciences faculty grant.

2   Built environment is an important aspect of place, a concept that also contains other meanings, such as geographic location, human territoriality and political geography, and attributed meanings of a place that forms the base for the formation of feelings as well as for memory, identity and ideology construction. See Agnew (1987), Cresswell (2004), Entrikin (1991), Gieryn (2000), Gregory and Pred (2007), and Tuan (1977), among others, for the developments and uses of the concepts in analysing the social processes. See also Earle (1993), Hedstrom (1994), Martin and Miller (2003), Miller (2000), Routledge (1997), Sewell (2001) and Tilly (2000), among others, for studies on the role of place in social movements.

3   See Zhao (1998) for a review of the studies on the impact that the spatial configuration of the built environment has on human activities in the fields of psychology, geography and sociology.

Haussmann's rebuilding of Paris altered the residential structure and consequently changed the basis of insurgent mobilization from working class consciousness during the 1848 June insurrection to neighbourhood solidarity during the 1871 Paris Commune. Bayat (1997) analyses how the active street life of unemployed or poor new immigrants in major Iranian cities gave rise to what he calls 'passive networks', which greatly facilitated poor people's resistance to state control in their collective fight for survival and betterment. Zhao (1998) argues that student mobilization during the 1989 Prodemocracy Movement in China depended more on the dense built environment of Beijing campuses and the associated rhythmic spatial activities of the students than on movement organizations and dissident networks and, moreover, that the impact of the campus environment on student mobilization could not be simply reduced to network-based mobilization. Stillerman (2003) shows how the characteristics of the built environment and everyday spatial routines of the metal workers and coal miners in Chile influenced the tactical repertoires and mobilizing structures of strikers.

To emphasize the role of built environment, however, is not to deny the classic understanding on the importance of organizations and networks in movement mobilization (e.g., Fernandez and McAdam 1988; Klandermans and Oegema 1987; Marwell and Oliver 1993; McAdam 1986; McCarthy 1996; Snow, Zurcher and Ekland-Olson 1980). We know that organizations and networks always operate in a specific built environment. We also know that organizations, networks, and the built environment play their roles in a concrete process of social movement mobilization. What this chapter aims to show, however, is that the built environment plays very different roles in social movement mobilization depending on the existence or absence of organizations in mobilization. More specifically, this chapter shows that spontaneously initiated participant mobilization is more constrained to use whatever tactics are available (which means built-environment-based recruitment in the case examined here). In other words, other factors being equal, the importance of the built environment in participant mobilization tends to increase with a decrease in organizational involvement.

Organization-based, network-based and built-environment-based mobilization can be viewed as three different but related aspects of participant mobilization; I seek to understand how they function in a real life mobilization process. Relationships among organization, networks and the built environment are complex. Organizations need, and try to generate, grassroots networks to expand their mobilization potential. Simultaneously, the built environment shapes people's spatial routines and generates propinquity-based networks that can facilitate participant mobilization. Since both organization and the built environment generate social networks that can be appropriated in participant mobilization, it is difficult to separate the independent functions of networks from those of organization and the built-environment. Therefore, as a starting point, this chapter only explores relationships between organization and the built environment in a protest event.

To examine how two independent variables shape the outcome of a dependent variable, we need at least to be able to alter the values of one of the independent variables. In other words, to understand the relationships between the role of organization and the built environment in participant mobilization, we need at least to be able to alter either the built environment or the levels of organizational involvement in a particular protest mobilization. The point made here is simple enough, but ideal cases that satisfy such conditions are difficult to establish in a real world.

In 1999, anti-US demonstrations broke out in China after four US missiles hit China's Belgrade embassy. In Beijing, the government allowed the students to demonstrate, but asked the government-sponsored graduate and undergraduate student unions (student unions hereafter) in the universities to organize the demonstrations. Yet for reasons that will become clear later, the student unions' involvement in protests at different universities was quite different. In many universities, most students joined the anti-US protest organized by the student unions, but in a few other universities a substantial number of students resisted the government's attempted manipulation and initiated demonstrations of their own. Since the built environment of most Beijing campuses is broadly similar in terms of their impact on the spatial routines of students (Zhao 1998), the different levels of student unions' involvement in the protest created an ideal situation to explore the relationships between the built environment and organization in participant mobilization. In this study I select several otherwise similar universities, but with different levels of organizational involvement in the anti-US protest, to examine how levels of organizational involvement shaped the role of the built environment in the protest.

The background of the anti-US protest

On the night of May 7, 1999, more than 40 days after the NATO bombing of Yugoslavia began, a B-2 stealth bomber that flew directly from the United States fired five missiles at China's Belgrade embassy (four hit and one missed). Three were killed and over 20 other Chinese diplomats were wounded during the incident. As the news reached China on May 8, anti-US protest surged. In Beijing alone, between May 8 and 11, thousands of students protested in front of the US embassy. They attacked US businesses, in particular the McDonald's and KFC restaurants. They also burned American flags, threw rocks, tomatoes, and inkbottles onto the American embassy building and besieged the embassy for four days.

Before the embassy bombing, US-China relations had already experienced a major change. In the United States, the collapse of the Soviet Union in conjunction with the Chinese government's brutal repression of the 1989 Prodemocracy Movement fundamentally changed China's geopolitical importance to the United States as well as Americans' view of the Chinese regime. US-China relations became problem-ridden and the US media's China coverage turned negative in the 1990s (Fewsmith 2001: 2). In China, to boost its declining legitimacy after crushing the 1989 Movement, the Chinese government staged Patriotic Education Campaigns (Zhao S. 1998) and deepened market-oriented reform. The combined effect of the Patriotic Education Campaigns, the success of the reform, and the worsening of US-China relations induced the rise of anti-US nationalism. Although the extent of this anti-US nationalism is often exaggerated, it is certainly true that most Chinese no longer held as rosy a picture of the United States as they had during the 1980s (Zhao 2002).

Perhaps because of the above developments in US-China relations, the 1999 protest came to be seen in the US media as an act that was instigated by the Chinese government. The reality, however, was more complicated. Although the Chinese government has staged patriotic campaigns to boost its eroding legitimacy, their key purpose is always to teach the students to identify the communist government as the vanguard of national interests (Zhao S. 1998; Wang 2000). Spontaneous mass-based nationalistic activities are never encouraged in those campaigns because they pose potential challenges to authoritarian rule.[4] During the anti-US protest triggered by the embassy bombing, however, the Chinese government not only allowed the students to demonstrate but also asked the student unions to organize the protests.

A senior official of a major university provided me with his account of how the government decision on student protests was made (informant no. 62). This official had participated in an emergency meeting held by the central government discussing the government strategies toward the bombing incident. According to him, the most common concern expressed by the meeting participants was the possibility of losing control once a large-scale anti-US protest broke out. Nonetheless, most meeting participants also argued that the students must be allowed to demonstrate this time for two reasons. First, due to the availability of the Internet, many students knew about the bombing soon after it occurred and small-scale protests had already started on campuses, without government permission.[5] Second, less than a year before, in 1998, the Chinese government

---

4 For example, on July 14, 1996, some rightist Japanese students built a lighthouse on the disputed Diaoyutai islands. Following the incident, while the Chinese government condemned the action and reaffirmed China's sovereignty over the islands, it also tried hard to prevent anti-Japanese protest from happening (Zheng 1999: 134). During the 1998 Indonesia riot, many ethnic Chinese women were sexually assaulted and raped. When the news reached China, students organized protests, but the protests were repressed by the government. In 2000, Chen Shuibian's election as Taiwan's new president caused the Chinese government great unhappiness because of his pro-independence stance. Nevertheless, the Chinese government deterred several spontaneous anti-Chen demonstrations initiated by the students after the Taiwan election. A year later, after the crash of a US spy plane with a Chinese fighter jet near Hainan Island, the Chinese government also tried to prevent any large-scale anti-US student demonstrations from happening in universities.

5 At 6:24 a.m., sina.com, a Chinese commercial website, posted the news (Liu 2000: 349). Around the same time, a few of my informants also heard the news from the Voice of America.

had repressed a student protest triggered by the rape of ethnic Chinese women during riots in Indonesia. Many meeting participants warned the top decision-makers that students in some major universities were still very upset about the repression, and that to most Chinese students the embassy bombing was a much more serious incident.[6] Chances of students turning against the government would be very high if the students were not permitted to demonstrate. In the end, the central government decided to allow the students to protest, but at the same time to control the development of the protest events.

Two other possible motives could also be behind the Chinese government's decision to allow managed anti-US protests. First, as the 10th anniversary of the 1989 Prodemocracy Movement drew near, the government was concerned that the protest would turn against the government. Indeed, as several informants remembered (nos. 39, 41, 47), during the demonstrations some people brought up the issue of the 1989 Movement and others attacked various government policies. Second, the government was also worried that the demonstrators would use excessive violence, which could hurt China's economy. During the incident, protests organized by the student unions were much more peaceful than those that were spontaneously initiated.[7] Much of the government's efforts were aimed at preventing radical actions such as attacks on American nationals and businesses (no. 61).

Against this background, on the morning of May 8, the student unions in many universities started to prepare for the protest. In the same afternoon, several thousand students from 11 Beijing universities took the buses prepared by the student unions and demonstrated outside the US and British embassies. By evening, increasingly more students and Beijing residents arrived at the embassy area.

What is rarely known to the outside world is that in some universities the student unions' mobilization efforts actually met with strong resistance. During the interviews, I found that, due to the students' past experiences with the state (such as their experiences during the 1989 Movement and more recently during the 1998 protest triggered by the Indonesian riots), many students were very suspicious of the student unions' involvement in the protest (Zhao 2002, 2003). They consciously avoided joining any protest that they saw as controlled by the school authorities and initiated independent protests of their own. The student

---

6   Under Chinese law, ethnic Chinese who have acquired citizenships of other countries are no longer Chinese nationals. Therefore, while the Chinese people were angry at the rapes, many of them also considered it as Indonesia's internal affair. On the other hand, because the Chinese government opposed the NATO bombing of Yugoslavia, most Chinese believed that the embassy bombing was a deliberate act of the US government or military aimed at punishing China for the opposition.

7   For example, informant 19 is a former university student. He went to the US embassy alone and acted violently. His action was condemned by the students in the student union organized demonstration.

unions' efforts were constantly challenged, which led to frequent conflicts between the students and government.

What is also important for this analysis is that the student unions' ability to control the students varied from university to university. In Qinghua University, the student unions had a good control of the situation, and as a result almost all the protests were organized by the unions, while in Beijing University many protests were spontaneously initiated by students themselves.[8] People's University exhibited the third pattern, where, while the protests were often initiated by students, the student unions always stepped forward trying to gain some control, and their manipulation often became a source of conflict. Since these three universities have similar campus layouts and student residential patterns, the different levels of student unions' involvement in the anti-US protest create an ideal situation for us to examine the relationships between the built environment and organization in participant mobilization.

Background information and logic of analysis

The campus built environment of the three universities is similar, but not identical. Could such differences in any way have an impact on the findings of this study? All three universities considered here have similarly structured student dormitory clusters in one or two places (southwest for Beijing University, north for Qinghua University, and two clusters in the southeast and southwest for People's University). These dense dormitory compounds comprise the centres of students' daily spatial interactions: the Triangle in Beijing University, the 10th dining hall in Qinghua University and the third dining hall in People's University. Yet in comparison with the Triangle, the central locations in Qinghua and People's University are less central. The Triangle, with several campus poster boards on the spot, is where shops such as the post office, the bookstore and the grocery store are located. It is also situated between student dormitories, the library, instruction buildings and several dining halls (Figure 9.1). Whenever students go to classrooms, the library, dining halls, the post office or back to their dormitories, they have to pass the Triangle area. In contrast, the dormitories in People's University cluster in two places, and the 10th dining hall in Qinghua University, although huge and centrally located, does not have many shops and other service facilities. These spatial differences, however, do not change the fact that in all three universities the students are densely concentrated in a small area.[9]

---

8   A later report by Beijing University's graduate students unions reveals that, of the 11 major demonstrations initiated by their students, only five were legally registered and organized by the student unions (no. 7).

9   Qinghua University has a bigger campus than the other two universities. However, this should not undermine my findings because the student dormitory quarter of Qinghua is clustered in the north-central part of the university, and in that area, the student density is very high, perhaps even higher than the other two universities.

The following two pieces of evidence indicate that it was the different levels of organizational involvement, rather than the minor differences in the built environment, that contributed to different patterns of protest mobilization at the three universities. First, because the Chinese government opposed the 1989 Prodemocracy Movement and no other organizations are known to have mobilized protest, the 1989 student protest at all three universities can be considered as largely spontaneously initiated. That is, the student mobilization at all three universities heavily relied on everyday patterns of spatial interaction, structured by the built environment, despite some differences in campus layout (Zhao 2001: ch. 8). Second, during the 1999 anti-US protest, although both People's and Qinghua University had no central place as significant as the Triangle at Beijing University, only Qinghua students were almost exclusively mobilized by the student unions. The student mobilization at People's University was very similar to that at Beijing University, both being based in everyday patterns of spatial interaction, structured by the built environment. Clearly, minor differences in the built environment did not have a significant impact on the patterns of mobilization at the three universities.

The organizations that mobilized student protests in 1999 were not typical social movement organizations but government-sponsored student unions. It might be asked to what extent the findings here could shed light on social movement mobilization in general. The answer lies in the nature of student unions in China and why student unions can be treated as functionally equivalent to movement organizations for the purpose of this study. In today's Chinese universities, student unions are run by the students themselves. Since the 1990s, many student unions have even introduced competitive elections. The unions organize sports tournaments, debate competitions, lecture series, film/theatre events, best teacher awards, and spring/autumn outings as part of their regular activities. They also organize special activities as situations and opportunities arise. The anti-US protest is an example of such a special activity.

The student unions at all the three universities have an identical bureaucratic structure. At the university level, each university has various functional branches including the Offices of Culture, Sports, Female Students, Public Relations, and more. The structure of the departmental level student unions is less elaborate, but similar. When student unions organize an event, they usually post announcements on bulletin boards, including an electronic bulletin board. If the unions need to mobilize more students or, if they organize an event at short notice, they will send cadres to inform students class by class or even dormitory by dormitory.

The student unions' bureaucratic structure provides them with a certain efficiency, but whether a student participates in a union sponsored activity is all up to each student. Student unions in today's China bear no resemblance to Leninist party structures and commanding authorities. In fact, similar to the student unions in the US, Chinese student unions have very limited impact on students' lives. Student unions are 'government sponsored' only in the following senses. First, university authorities usually impose a political clause in the unions' leadership qualification by-laws. In Beijing University, this clause reads: 'Love the country, uphold the four cardinal principles, have clear and correct political views'.[10] The clause is seldom seriously followed in student union elections, but its existence gives the school authorities a veto power to exclude 'trouble makers' from the leadership. Second, even though student unions today increasingly rely on private donations, university funding is still their key resource, creating political dependence. Third, leaders of student unions have higher chances of being recruited by government agencies after graduation if they have demonstrated good leadership. Those who are interested in getting such jobs naturally want to maintain good relations with university authorities for the purpose of obtaining favourable recommendations.

In this research student unions are treated as movement organizations for two reasons. First, as discussed above, student unions are similar to volunteer organizations in the sense that, while the unions can use their infrastructure to mobilize the students, they cannot coerce students. For the purpose of studying participant mobilization, student unions can be considered as functionally equivalent to movement organizations. Second, there is precedent for treating volunteer and interest-based organizations as functionally equivalent in contentious mobilization. In McCarthy and Zald's (1973) classic resource mobilization article, interest group organizations such as the Consumer Rights Organization and Welfare Rights Organization are treated as movement organizations. Volunteer organizations, after all, share great similarities with movement organizations in terms of the means available to mobilize participants.

In this study I found that, while the structure of student unions in the three universities was almost identical, the student unions of the three universities had different levels of involvement in student mobilization. To better situate this finding, I here analyse factors behind the student unions' differentiated involvement in the protest. Extensive interviews with protest participants show that the student unions' differentiated involvement in the protest was mainly a result of two conditions.[11] First, while both Beijing University and People's University

---

10 The four cardinal principles are: adherence to socialism, adherence to the leadership of the communist party, adherence to Marxism-Leninism and Mao Zedong thought, and adherence to the dictatorship of the proletariat.

11 I interviewed 62 individuals in 1999. Among them, 60 were from three major Beijing universities (27 for Beijing University, 21 for Qinghua University, and 12 for People's University). Of the 60 students, 39 are males and 21 females, and 43 are undergraduate students while 17 are graduate students. The three universities were selected because, of over 60 institutions of higher learning in Beijing, they are among the most prestigious large universities that have exerted a great impact on students in other universities and on Chinese politics. The informants were recruited through a snowball method: after each interview, the informant was asked to suggest friends who might be willing to participate. To ensure representativeness, I would decide whether a suggested candidate would be contacted. I started the interview by asking the informant to recall as detailed as possible his/her personal experiences during the protest. The interview then

are dominated by students majoring in social sciences and humanities, Qinghua is mainly an engineering university. In comparison with People's University and Beijing University, Qinghua has tougher examination schedules and Qinghua students enjoy better job opportunities in China's booming high-tech and foreign business sectors. On the whole, Qinghua students are busier, less interested in politics and more content with the *status quo*. They are generally more amenable to being led by the government-sponsored student unions than students from People's University or Beijing University. As for the anti-US protest, few Qinghua students minded who had organized the event. Many of them might not even have joined the protest had it not been a weekend event. In contrast, because social sciences and humanities programmes in China teach predominantly Western culture in which liberalism and individualism figure importantly (Zhao 2001: ch. 3), students at People's University and Beijing University tend to be more individualistic and cynical. Many of them thus saw their student unions' efforts as a sign of government manipulation, and staged their own demonstrations. Second, not only were Qinghua students easier to discipline, Qinghua's student union cadres were also more motivated to cooperate with school authorities. At the time of the embassy bombing, 93 Qinghua graduates were in China's top political leadership. Next most favoured, Beijing University, had only 45 in that group (Li 2001: 107). Politically ambitious students at Qinghua competed hard for student union leadership positions, and once in those positions, worked hard to curry favour with the political leadership. These factors made Qinghua's student unions more effective organizations.

With the above discussion in mind, I hypothesize the following relationships:

Prior attitudes of the students and union leaders under larger context	→	Student union's capacity to mobilize students during the anti-US protest	→	Low →	Built-environment-based mobilization
				High →	Organization-based mobilization

I have argued that the student unions during the anti-US protest functioned similarly to movement organizations. If this is correct, then the following corollary for social movement mobilization, more generally, would apply:

---

proceeded variously according to the informant's narrative. Yet, I always probed into a few issues that I am particularly interested in: where the informant learnt the bombing-related news; the informant's reaction and personal activities after hearing the bombing and during the protest; the mobilization processes and the route of a protest in which the informant participated; the role played by the student unions in the protest that the informant was involved; the events or big-character posters by which the informant felt most impressed during the protest.

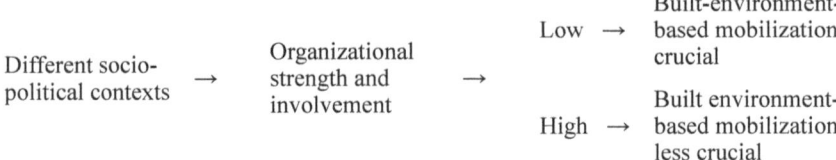

I replace 'organization-based mobilization' in the first diagram by 'built-environment-based mobilization less crucial' in the second diagram because I believe the empirical findings of this study should be interpreted cautiously.

These hypotheses suggest that each student union's mobilization capacity, and correspondingly the role of the built environment in the mobilization, is to a great extent rooted in the structural position of each university vis-à-vis the top political leadership of the state and, by implication, in prior political attitudes of students and student union leaders. It would therefore seem necessary to delve into the causes of the different structures of participant mobilization in the three universities. Such an approach, however, is not taken in this chapter. My aim here is very modest: to obtain a preliminary understanding of the relationship between a) organizational strength and involvement in a protest mobilization and b) the relative importance of built-environment-based and organization-based mobilization structures in actual mobilization processes. In other words, I examine only the relationship between the 'intermediate variable' (organizational strength and involvement in the mobilization) and the 'dependent variable' (relative predominance of built environment versus organization in mobilization). In social movement studies organizational strength and involvement is never considered an 'independent variable' in real-life mobilization processes because its roles in participant mobilization are always shaped by larger sociopolitical conditions. Nevertheless, studies on the relationships between organizational involvement in a mobilization and patterns of mobilization have been a popular topic in social movement research. This is because the relationship between organization and patterns of mobilization is not spurious in the sense that their association disappears when the larger sociopolitical conditions that shape the organizational strength and involvement are introduced into the analysis. Therefore, when social movement scholars study patterns of participant mobilization, they usually take the larger sociopolitical conditions as given by focusing on the relationships between organization and movement mobilization; this research simply follows that tradition.[12] In short, this research considers one more complication to the traditional analysis of mobilization structure: the relationships between built-environment-based and organization-based mobilization.

---

12   For example, when Gould (1991) examines the role of formal and informal organizations and networks during the mobilization of the 1871 Paris Commune, he takes as given all the factors that led to the Paris Commune's particular pattern of mobilization without even mentioning those factors.

The built environment and organization in mobilization

Let us now turn to the actual mobilization processes of the anti-US protest in each of the three universities.

Qinghua University

During the embassy-bombing incident, almost all of the demonstrations at Qinghua University were organized by the student unions. During the first demonstration, for example, the student union cadres simply asked students to get on the pre-arranged buses and afterwards informed them of the purpose of the trip. An informant recalled (no. 52):

> I heard about the bombing incident at noon on May 8. Two hours later we had a routine rehearsal [for a parade to celebrate Communist China's 50th anniversary]. About half an hour after the rehearsal started, [someone] came over and asked us to get on the buses. We did not know the purpose before boarding the bus, but as soon as we got on the bus an organizer told us that we were going to demonstrate at the US embassy. The organizer also mentioned that those who did not want to go were free to leave. Most of us stayed.

Qinghua students were less likely to be mobilized by spontaneously initiated demonstrations. On the evening of May 8, for example, several hundred students from Beijing University went to mobilize Qinghua students to join them in demonstrating at the US embassy. An informant from Beijing University recalled (no. 4):

> I saw the demonstration after I had heard some noises outside our dormitory. I followed them to Shaoyuan, the foreign students' compound. We shouted in front of Shaoyuan and asked the foreign students to follow us. After a while, some students suggested that we go to Qinghua to mobilize their students. Some university authorities that followed us, however, suggested that we return to the Triangle. They wanted us to limit the demonstration to the campus, but we ignored them and marched to Qinghua. ... When we were at Qinghua, however, very few of their students responded to our call.

It does not mean that students from Beijing University were unable to mobilize the Qinghua students at all. The problem is that such a bottom-up initiative was easily contained. A Qinghua student recalled the same event (no. 37):

> When the students from Beijing University arrived, we were holding a candlelight vigil for the victims outside the university stadium.[13] ... The students

---

13 The students from Beijing University entered Qinghua from its West Gate and walked directly to the university stadium located at the north-eastern side of the campus,

from Beijing University ... asked us to go with them and demonstrate at the US embassy right away. We almost followed when the president of our university's student unions stopped us. He said that if we went today, we had to walk. Besides, a demonstration off the campus needs to be approved by the Public Security Bureau. Since we only applied for a demonstration for tomorrow ... if we went right now, the police might stop us.

As a result, very few Qinghua students followed the Beijing University students.

The higher level of involvement of Qinghua's student unions was also revealed in other aspects of the incident. An informant from Beijing University, for example, told me how he was impressed by Qinghua students' actions during the demonstrations (no. 4):

I woke up late [on the morning of May 9th]. I went to the US embassy with several of my classmates by public transportation. After we arrived, however, we were for a long time unable to find the students of our university, but we were impressed by how well the Qinghua demonstrations were organized. Their students were led by a university banner. Behind the banner, the students were separated by department and each department had its own flag.

A Qinghua student also described with pride (no. 57):

Our demonstrations were very well organized. Our goals were achieved without resort to radical actions. Our slogans such as 'National revitalization' were also farsighted. The students in some other universities, such as Beijing University, perhaps because their demonstrations were spontaneously initiated, were poorly organized. They also tended to act more emotionally and radically.

More significantly, because of the effective control the student unions exercised over the anti-US demonstrations, the student mobilization in Qinghua never relied on everyday patterns of spatial interaction, structured by the built environment. My earlier example shows that Qinghua's student union leaders could simply round up students who were doing something else, wherever they were, and lead them to the US embassy. One Qinghua student (no. 27) told me that a student union cadre in her class was involved in organizing the protest. Another Qinghua student recalled how she had joined a protest (no. 30): 'A student union cadre in my class informed me that there would be an anti-US demonstration tomorrow

---

adjacent to student dormitories and dining halls. The vigil at Qinghua was organized by the student unions. When the students from Beijing University arrived, some Qinghua students wanted to answer their call and demonstrate outside the campus. While the response was a sign of spontaneity, this kind of induced/reactive spontaneity was very different from the proactive spontaneity that the students of the People's and Beijing University displayed during the anti-US protest.

morning [that is, May 9]. … She then told me when and where we should meet. [The next morning], the buses sent us from the tenth dining hall directly to the US embassy.' I would like to stress that unlike the mobilization at People's University and Beijing University, the major demonstrations by Qinghua students always moved directly from a central meeting point on campus, the 10th dining hall, to the US embassy. Everyday patterns of spatial interaction, structured by the built-environment, played an important role in mobilization at People's University and Beijing University, but not at Qinghua University.

People's University and Beijing University

In comparison with Qinghua students, the students in both People's University and Beijing University were more individualistic and less easily harnessed by the authorities. While Qinghua students accepted the student unions' leadership during the anti-US protests, many students at People's and Beijing Universities either refused to join the student union sponsored protests or staged their own demonstrations in a largely spontaneous manner. A Beijing University student recalled (no. 6): 'XX and his friends in our department all refused to participate in the protest. They reasoned that the demonstration must have been organized by the government. If it were a student-initiated event, it needed to be approved by the Public Security Bureau. How could it be possible for the Public Security Bureau to approve a demonstration that fast?' Another Beijing University student expressed similar feelings (no. 36): 'The government's manipulation was too obvious during the protest. I also did not like slogans such as 'Down with American Imperialism' shouted during the protest. I always kept a distance with events that involved the government. Consequently, I did not participate in any protest.'

Perhaps because of the existence of the above kind of mood among the students, the student unions at the two universities acted more cautiously. The student unions at Qinghua University could simply use university money to rent buses and send the students to the US embassy, but the student unions at both People's and Beijing University were unable to do so, even though the lack of the money was never the determining issue. At Beijing University, as I was told (no. 7), to avoid alienating the students, the student unions distanced themselves from the school authorities and raised money among the students for the protests.[14] At People's University, the student unions took a more managerial strategy. A top union leader at People's University (no. 21) told me that in those days they had to stay in the office day and night waiting for news from various locations on campus. Whenever they learned that some students intended to protest off campus, they sent union leaders along, bringing the university flags and loudspeakers with them. By controlling the university flags and loudspeakers, union leaders tried to gain control of the originally spontaneously initiated demonstration. The spontaneity of student

---

14   The fundraising does not suggest that the student unions at Beijing University have financial problems. Student unions in major Chinese universities are very well funded.

mobilization at People's University and Beijing University represented a very different mobilization process from that of Qinghua University. Indeed, the People's University and Beijing University protests were strikingly similar to those of the 1989 anti-government mobilization. Both episodes of mobilization were based in everyday routines of spatial interaction, structured by the built environment.

Here, I present two brief cases (both happened on the evening of May 8) to show how the built environment played its role in the 1999 anti-US protests.

People's University   'Come out if you are Chinese! Come out Chinese!' 'Come out those Chinese who have a backbone!' It was on the evening of May 8. An informant (no. 12) recalled that he went out of his dormitory room after being attracted by this kind of shouting outside the No. 8 dormitory (the largest dormitory on campus). As more and more students were gathered around that shouting crowd, the student union leaders also came over. They warned the students that such protest needed to be approved by the Public Security Bureau and persuaded the students not to hold the demonstration. Another informant (no. 58) recalled:

> The students insisted on going to protest in front of the US embassy, but the student union leaders said that we needed first to get permission from the Public Security Bureau. There was almost a physical fight. Eventually, no student union leaders dared to step forward because we claimed that we were going to dismiss all the leaders of the current student unions and elect new ones right away. After 9:00 p.m., we [started to march forward], disregarding objections from the student unions.

The student union leaders followed the students. When the students reached the university's sports ground, a union leader suggested that they march inside, but the students refused. Instead, the students marched further east first to the instruction quarters where students were studying, and then north of the east gate where several other dormitory buildings were located. Outside the instruction and dormitory buildings, the students shouted slogans and made noises. Eventually, the size of the student demonstration reached over a thousand.

Before the students marched out of the university's east gate, however, the student union leaders stopped the students again and warned them not to demonstrate off campus. At this moment, hundreds of students from Beijing University arrived at People's University. They shouted: 'Come out! Come out! We need your support!' The students rushed out and followed them. It was only at this point that the student union leaders gave up their attempt to control the demonstration.

Beijing University   Many students went to the Triangle that evening.[15] At the Triangle, one might see big-character posters condemning the embassy bombing.

---

15   The following narrative is based on interviews with three informants (nos. 8, 9 and 10).

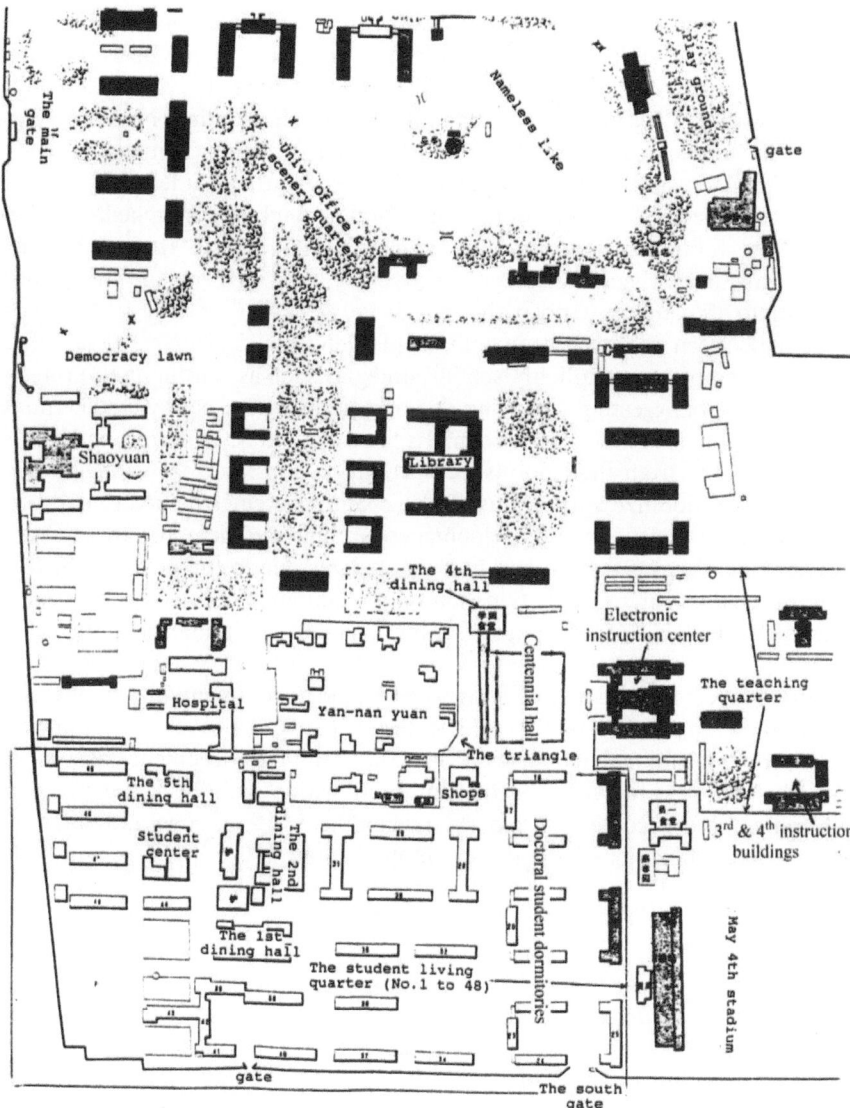

Figure 9.1 A partial map of Beijing University

One might also see students clustering in small groups, chatting and debating, and quite a few foreign students mingling and debating with the Chinese. Around 8:00 pm, a student proposed to demonstrate at the US embassy. A few students who supported the idea walked with that student to the Centennial Hall adjacent to the Triangle, where they started to call on people to join them. Eventually, the size of the formation reached between three and four hundred and the demonstration started. They first marched south to the doctoral student dormitories and then turned west to the undergraduate dormitories. After passing the dormitory area, they moved north to Shaoyuan, where the foreign students lived. Having stayed there for a while, they turned east first to the library, then to the Third and Fourth Instruction Buildings, and finally to the Electronic Instruction Centre (Figure 9.1). Whenever they reached a new place, they shouted to attract students. By the time they left the Electronic Instruction Centre, turned south and marched out of the campus, the formation had increased from a few hundred to several thousand. The student union leaders at Beijing University tried to persuade them not to demonstrate off campus, to no avail.

As can be seen, the students in both People's and Beijing University mobilized participants employing explicitly spatial strategies such as starting their protest at a central location, creating a moving demonstration whose path was determined by the built environment and students' spatial activities, shouting and yelling to attract people from their dorms and activity centres, and travelling to other universities to mobilize participants. These spatial strategies helped to overcome the organizational deficiencies of spontaneously initiated demonstrations. While such strategies were not used by Qinghua students due to the near ubiquity and organizational resources of the Qinghua student unions, they were routine for the protests at People's University and Beijing University. Significantly, such built-environment-based strategies of mobilization also figured very importantly during the 1989 Movement, which, totally opposed by the government, had an obviously different nature from the 1999 anti-US protest, but nevertheless also developed out of serious organizational deficiency (Zhao 2001). Built-environment-based spatial strategies are thus crucial for participant mobilization with weak organizational involvement, but become less important when there is the strong and ubiquitous presence of organizations in participant mobilization.

Conclusion

This study tries to understand the relationships between the built environment, everyday patterns of spatial interaction, and organization in participant mobilization during a protest. My initial hypothesis was that the built environment plays a more important role in participant mobilization for those protest activities with weak organizational involvement. To test this hypothesis, I examined the mobilization of the 1999 anti-US protest at three major universities in Beijing: Qinghua University, People's University and Beijing University.

My study indicates that, during the anti-US protest, the three universities exhibited very different student mobilization practices. At Qinghua University, most protests were organized by the student unions and the campus built environment played little role in the mobilization. By contrast, at both People's University and Beijing University, students resisted the manipulation by student unions and many protests were spontaneously initiated. A consequence of this was that the anti-US protests at the two universities exhibited a mobilization structure that was quite different from that of Qinghua University, but that closely resembled the built-environment-based student mobilization during the 1989 Prodemocracy Movement. In summary, this study suggests that built-environment-based mobilization is more important in weakly organized protests than in well-organized ones.

The built environment has always been at the centre of theorizing in American sociology due to the long-lasting influence of the Chicago School. The importance of the built environment in social movements and collective actions has been frequently noticed in a variety of studies (e.g., Feagin and Hahn 1973; Fogelson 1971; Heirich 1971; Lofland 1970; Tilly and Schweitzer 1982). It is somewhat puzzling that the role of the built environment was neglected in the early works of the resource mobilization and political process model traditions. This study shows that once an organization is able to develop apparatuses and resources that can be deployed in a nearly ubiquitous fashion, it may not adopt the built-environment-based mobilization strategies, even if the built environment itself is highly conducive to effective mobilization. In other words, when both strong organizational resources and the built environment are present, organization may have a higher order of importance in shaping participant mobilization. This result helps us to understand why the earlier studies in the resource mobilization and political process traditions neglected the role of built environment in movement mobilization. That is, with the process of democratization and the development of civil society since the 1960s (Rueschemeyer, Stephens, and Stephens 1992), social movements in the United States have become increasingly organized activities, many of which even come very close to interest group politics (Costain and McFarland 1998; McCarthy and Zald 1973; Meyer and Tarrow 1998; Polletta 2002). Poorly organized built-environment-dependent collective actions, such as riots, ceased to be a major focus of social movement scholars. Under such circumstances, as this study shows, the role of built environment in participant mobilization becomes less crucial. Understandably, sociological theories built under such an empirical setting would have neglected the role of the built environment.

This study, however, does not suggest that built environment plays no role in mobilization when a social movement is highly organized. Organizations frequently use built-environment-based strategies of mobilization, especially when they face various kinds of deficiencies in achieving mobilization. When movement leaders realize the mobilization potential of a built environment strategy, they often develop strategies to exploit such opportunities. But poorly organized protest activities may be compelled to rely strongly upon built-environment-based mobilization strategies, lacking the organizational resources to do otherwise. An effective organization is always freer to choose and alter its strategies than a poorly organized group, even though the benefits of organization do not come without costs (Piven and Cloward 1979).

References

Agnew, J. (1987) *Place and Politics: The Geographical Mediation of State and Society*. Boston: Allen and Unwin.

Bayat, A. (1997) *Street Politics: Poor People's Movements in Iran*. New York: Columbia University Press.

Costain, A. and A. McFarland (1998) *Social Movements and American Political Institutions*. Lanham: Rowman and Littlefield.

Cresswell, T. (2004) *Place: A Short Introduction*. Malden, MA: Blackwell Publication.

Earle, C. (1993) 'Divisions of Labor: The Splintered Geography of Labor Markets and Movements in Industrializing America, 1790-1930', *International Review of Social History*, 38 (1): 5-38.

Entrikin, N. (1991) *The Betweenness of Place: Towards a Geography of Modernity*. Baltimore: Johns Hopkins University Press.

Feagin, J. and H. Hahn (1973) *Ghetto Revolts: The Politics of Violence in American Cities*. New York: Macmillan Company.

Fernandez, R. and D. McAdam (1988) 'Social Networks and Social Movements: Multiorganizational Fields and Recruitment to Mississippi Freedom Summer', *Sociological Forum*, 3: 357-382.

Fewsmith, J. (2001) *China since Tiananmen: The Politics of Transition*. Cambridge: Cambridge University Press.

Fogelson, R. (1971) *Violence as Protest: A Study of Riots and Ghettos*. Garden City, N.Y.: Anchor Book.

Gieryn, T. (2000) 'A Space for Place in Sociology', *Annual Review of Sociology*, 26: 463-496.

Gould, R. (1991) 'Multiple Networks and Mobilization in the Paris Commune', *American Sociological Review*, 56: 716-729.

—— (1993) 'Collective Action and Network Structure', *American Sociological Review*, 58: 182-196.

—— (1995) *Insurgent Identities: Class, Community, and Protest in Paris from 1848 to the Commune*. Chicago: University of Chicago Press.

Gregory, D. and A. Pred (eds) (2007) *Violent Geographies: Fear, Terror, and Political Violence*. New York: Routledge.

Hedstrom, P. (1994) 'Contagious Collectivities: On the Spatial Diffusion of Swedish Trade Unions, 1890-1940', *American Journal of Sociology*, 99: 1157-1179.

Heirich, M. (1971) *The Spiral of Conflict: Berkeley 1964*. New York: Columbia University Press.

Klandermans, B. and D. Oegema (1987) 'Potentials, Networks, Motivations and Barriers: Steps toward Participation in Social Movements', *American Sociological Review*, 52: 519-531.

Li, C. (2001) *China's Leaders: The New Generation*. Lanham: Rowman and Littlefield.

Liu, Y. (2000) *Meiti Zhongguo* (Media's China). Chengdu: Sichuan Renmin Chubanshe.

Lofland, J. (1970) 'The Youth Ghetto', in E. Laumann, P. M. Siegel, and R. W. Hodge (eds) *The Logic of Social Hierarchies.* Chicago: Marrham Publishing Company, pp. 756-778.

McAdam, D. (1986) 'Recruitment to High-Risk Activism: The Case of Freedom Summer', *American Journal of Sociology*, 92: 64-90.

McCarthy, J. (1996) 'Constraints and Opportunities in Adopting, Adapting, and Inventing', in D. McAdam, J. McCarthy, and M. Zald (eds) *Comparative Perspectives on Social Movements.* Cambridge: Cambridge University Press, p. 141-151.

McCarthy, J. and M. Zald (1973) *The Trend of Social Movements in America: Professionalization and Resource Mobilization.* Morristown, N.J: General Learning Corporation.

Martin, D. and B. Miller (2003) 'Space and Contentious Politics', *Mobilization*, 8: 143-156.

Marwell, G., and P. Oliver (1993) *The Critical Mass in Collective Action: A Micro-Social Theory.* Cambridge, N.Y.: Cambridge University Press.

Meyer, D. and S. Tarrow (1998) *The Social Movement Society: Contentious Politics for a New Century.* Lanham: Rowman and Littlefield.

Miller, B. (2000) *Geography and Social Movements.* Minneapolis, MN.: University of Minnesota Press.

Piven, Frances Fox, and Richard A. Cloward (1979) *Poor People's Movement.* New York: Vintage Books.

Polletta, F. (2002) *Freedom Is an Endless Meeting: Democracy in American Social Movements.* Chicago: University of Chicago Press.

Routledge, P. (1997) 'A Spatiality of Resistances: Theory and Practice in Nepal's Revolution of 1990', in Steve Pile and Michael Keith (eds) *Geographies of Resistance.* London: Routledge, pp. 68-86.

Rueschemeyer, D., Stephens, E. and J. Stephens (1992) *Capitalist Development and Democracy.* Chicago: University of Chicago Press.

Sewell, W. (2001) 'Space in Contentious Politics', in R. Aminzade, J. Goldstone, D. McAdam, and E. Perry (eds) *Silence and Voice in the Study of Contentious Politics.* Cambridge: Cambridge University Press, pp. 51-88.

Snow, D., Zurcher, L. and S. Ekland-Olson (1980) 'Social Networks and Social Movements: A Microstructural Approach to Differential Recruitment', *American Sociological Review*, 45: 787-801.

Stillerman, J. (2003) 'Space, Strategies, and Alliances in Mobilization: The 1960 Metalworkers' and Coal Miners' Strikes in Chile', *Mobilization*, 8: 65-85.

Tilly, C. (2000) 'Spaces of Contention', *Mobilization*, 5: 135-159.

Tilly, C. and R. A. Schweitzer (1982) 'How London and its Conflicts Changed Shape: 1758-1834', *Historical Methods*, 15: 67-77.

Traugott, M. (1995) 'Capital Cities and Revolution', *Social Science History*, 19: 147-168.

Tuan, Y. (1977) *Space and Place: The Perspective of Experience*. Minneapolis: University of Minnesota Press.

Wang, J. (2000) 'Democratization and China's Nation Building', in E. Friedman and B. McCormick (eds) *What If China Doesn't Democratize? Implications for War and Peace*. Armonk: M. E. Sharpe, pp. 49-73.

Wolford, W. (2003) 'Families, Fields, and Fighting for Land: The Spatial Dynamics of Contention in Rural Brazil', *Mobilization*, 8:157-172.

Zhao, D. (1998) 'Ecologies of Social Movements: Student Mobilization during the 1989 Prodemocracy Movement in Beijing', *American Journal of Sociology*, 103: 1493-1529.

—— (2001) *Power of Tiananmen: State-society Relations and the 1989 Movement*. Chicago: University of Chicago Press.

—— (2002) 'An Angle on Nationalism in China today – Attitudes among Beijing Students after Belgrade 1999', *China Quarterly*, 172: 885-905.

—— (2003) 'Nationalism and Authoritarianism: Student-Government Conflicts During the 1999 Beijing Student Protests', *Asian Perspective*, 27: 5-34.

Zhao, S. (1998) 'A State-led Nationalism: The Patriotic Education Campaign in Post-Tiananmen China', *Communist and Post-Communist Studies*, 31: 287-302.

Zheng, Y. (1999) *Discovering Chinese Nationalism in China: Modernization, Identity, and International Relations*. Cambridge: Cambridge University Press.

Chapter 10

Energizing Environmental Concern in Portland, Oregon

Ted Rutland

Introduction

In 1979, the City of Portland, Oregon, passed a landmark energy policy. The policy was a response to diverse concerns about energy production and consumption in the Pacific Northwest region, and, more specifically, to the remarkably effective messaging and organizing of a broad environmental movement based in Portland. The major short-term effect of the policy was to establish energy conservation as the first-order response to any future increases in energy demand, both within the municipality's operations as well as across the city as a whole. Backing this general commitment were concrete conservation targets of 10 per cent in five categories of energy use: residential, commercial, industrial, land use, and transportation. To monitor energy consumption and co-ordinate the delivery of incentives and loans for conservation initiatives (to businesses and households), the policy mandated the creation of two new departments that would eventually be merged to form the Portland Energy Office. Over time, Portland's energy policy came to vastly exceed its initial objectives and sphere of application. The Energy Office's monitoring and management of local energy consumption provided the basic framework for reducing local greenhouse gas (GHG) emissions as well, which, in 1991, Portland became one of the first cities in the world to do (Rutland and Aylett 2008).

My primary purpose in this chapter is to examine how the movement responsible for the Portland energy policy was formed and structured. Drawing on secondary research as well as interviews with Portland-area activists I aim to disclose how a broad group of people came to join forces and advance energy conservation as a city-level solution to broader environmental concerns. My approach to these questions is informed by two literatures: social movement studies and science studies (particularly, the work of Bruno Latour). Informed by the first literature, I aim to document how the Portland-based movement relied on particular, pre-existing 'social networks' as well as how it deployed 'frames' as a means of defining political concerns, demands, and actors. Informed by Bruno Latour's actor-network theory, I aim in turn to situate relevant social networks and frames within a concatenation of human and nonhuman elements conjoined, or associated, across time and space, and thereby demonstrate how Latour's work can be used to enrich more conventional social movement analysis.

My argument unfolds across five substantive sections. In the first section, I develop an approach to the study of social movements by combining the concepts of social networks and frames with certain insights from Latour. Next, I turn to the Pacific Northwest and describe how the development of a hydro-based energy system in the middle half of the twentieth century helped to define a distinct regional environment and gradually bring the lives of people in cities like Portland into a relation with it. When, in the 1970s, a variety of energy-related environmental problems were made more apparent, Portland-based activists were able to frame their concerns in terms of collective responsibility for the regional environment. Across the next three sections, I examine three sites where movement-related social ties and knowledge were generated: the Oregon Office of Energetics, Rain House, and the Portland Energy Committee. It was in these three sites that energy conservation became a movement demand and that the City of Portland was framed as the relevant collectivity to implement this demand. My conclusion is threefold: first, the movement responsible for the Portland energy policy emerged largely from an existing social network; second, the movement's frames highlighted the regional environment (as its concern), energy conservation (as its demand), and the city (as the responsible collectivity); and third, this network and these frames were constituted within an assemblage that brought together activists, the Columbia River, Pacific salmon, a social centre, and a range of other material elements into a broad collective enterprise that had the effect of energizing environmental concern in Portland.

Networks, frames, and materiality

My approach to the study of social movements combines more conventional insights with those of the science studies scholar, Bruno Latour. In this section, I suggest how concepts like social networks and frames might be enriched by adopting a Latourian approach to human–nonhuman associations. The concept of social networks, of course, occupies a central place in social movement studies. Since the 1970s, it has been the most intensely discussed issue in the literature (Jasper and Poulsen 1995) and has played a major role in analyses of how 'latent' pre-movement or extra-movement social connections facilitate organized resistance (Melucci 1988: 248). Although movements themselves can be considered a social network, the latter term is generally reserved for the relatively distinct social connections that either precede, coincide with, or succeed a fully formed movement. Oft-cited examples of social networks include voluntary organizations, neighbourhoods, workplaces, churches, colleges, families, and friendship networks (McAdam, McCarthy and Zald 1996; Goodwin and Jasper 1999). In these diverse 'micromobilization locations', (McCarthy 1996: 141) information pertaining to movement campaigns and actions may be circulated, relationships from which organizers can recruit movement members may be formed and sustained, and shared practices and values that enable coordinated, collective action may be developed (Snow and Benford 1988; Mueller 1992; Diani 1995; Routledge 2003).

That social networks are relevant to mobilization is both intuitive and extensively documented. At the same time, perhaps because networks are approached as the cause or pre-condition for something else (i.e., a movement), less attention has been paid in the literature to explaining extra-movement networks themselves. Bosco (2006), for example, calls for closer attention to the means through which the 'internal cohesion' of networks is maintained over time and space. I will return to this issue below.

The concept of frames refers to the more discursive or symbolic conditions from which movements are formed. Though definitions vary, the concept typically refers to the movement messages that define the collectivity that is invited to act together, that diagnose the collectivity's major concerns, and that articulate a set of demands (Snow and Benford 1992; Kurtz 2003). Frame analysis has been particularly productive in geography; it has been used not just to highlight how movements formulate grievances and demands (Boudreau 2003; Gilbert 2004; Debanne and Keil 2004), but also to draw distinctions between parts of what might otherwise appear to be a single, undifferentiated movement (Uitermark 2004) and to evaluate how certain frames succeed and fail in enabling a broad group of people to work together toward common goals (Cinalli 2003).

While some accounts posit that frames enable social networks to be formed (Jasper and Poulsen 1995) or vice-versa (McAdam, McCarthy and Zald 1988; Pfaff 1996; Tindall 2002), it is more typical to conceive of the two elements as co-constituted (Diani 1995; Mische 2003). As Juris (2004: 342) suggests, networks offer 'arenas for the production, contestation, and dissemination of specific movement-related discourses [i.e., frames]', while they are also 'produced and transformed through the discourses and practices circulating through them'. In a similar vein, Martin (2003) demonstrates how the deployment of 'place-frames' helps to produce a social unit that is sometimes taken for granted: the neighbourhood. Living in the same area of a city, Martin suggests, may provide people with common experiences or common concerns, without necessarily leading to the formation of a collective identity. The role of place-frames, when deployed by neighbourhood organizations, is to highlight common experiences in an area (e.g., signs of disrepair, and other problems, in the material environment) and invite residents to join together as 'neighbours' to address their common concerns. What I find particularly instructive here is twofold: first, the idea that seemingly self-evident social networks may not exist automatically, but rather, if they are to exist, require some ongoing work to be performed; and second, the suggestion that people may be connected in certain ways (e.g., inhabiting the same area, encountering the same problems) that do not necessarily cause them to recognize themselves as part of a collectivity, but nevertheless provide a potential basis for collective action. I read Martin's account, in other words, as a partial description of how social networks are formed and sustained – in this case, on the basis of a shared material environment and the deployment of place-frames that strategically interpret this environment.

If Martin suggests one of the ways that collectivities are formed and sustained, the work of Bruno Latour suggests another. Since the late 1990s, Latour's work has been used by geographers primarily in order to conceptualize and describe how nonhuman animals and materials are involved in the making of 'social' worlds (cf. Murdoch 1997, 1998; Whatmore 1999; Castree 2002; Gandy 2005). For the same reason Latour has made an appearance in studies of social movements. Routledge (2008: 201), for example, suggests that 'associations between humans and nonhumans are crucial to the enactment of political action' and advocates drawing on Latour to enable these associations to be recognized (see also Lockie 2004). Particularly instructive for my purposes is Bickerstaff and Agyeman's (2010: 50) claim that attention to nonhuman elements can help explain how social forms 'endure in space and time' and become, effectively, 'structural'. Picking up these cues, I want to consider how concepts like social networks and frames might be enriched by investigating their relationship to various nonhuman elements. To do so, I will discuss two especially useful concepts from Latour's expansive (and expanding) repertoire.

Latour's main concept bears a misleadingly straightforward name: 'networks'. The concept refers, generally, to a concatenation of human and nonhuman elements (Latour 2005: 128). Relevant nonhumans may include animals, machines, texts, instruments, buildings – literally anything that performs some kind of work, or ordering, in conjunction with other elements. The network that enables a poster for an activist event or demonstration to mobilize participants, for example, might include the people, software, and machines involved in the poster's design; the paper and ink that allow it to be printed; the wall on which it is posted; and the roads, hallways, or sidewalks that carry people past its displayed location. The inclusion of nonhuman elements is not, however, Latour's only innovation in his conception of networks. For one thing, Latour's later work emphasizes that a 'network' is not a thing, but rather a tool for rendering describable the associations that constitute things, that enable 'things' to exist as such (Latour 2005: 129; compare 1998: 3). Latour's project, in other words, is not to define what exists in the world (e.g., networks), but rather to decompose anything that seems to exist, inherently or independently, into the elements and relations that underpin this apparent and practical existence. This is partly a political project, the details of which are beyond the scope of this chapter (see Latour 2004a, 2004b). For the present purposes, the major implication is that both social networks and frames – ostensible 'things' – must be analysed as networks (in Latour's sense) and therefore decomposed into their various constituent elements.

Latour's second innovation is to emphasize the unequal influence that some (usually human) elements exercise in their relations with affiliated others. To explain this inequality, Latour relies on the concept of 'oligoptica'. Emerging in his later work, oligoptica are defined as particular, well-connected sites from which a comprehensive, detailed, but very selective view of a dispersed 'whole' becomes possible (Latour 1999: 18-19; 2005: 175-183). Latour's examples include military command centres and securities trading rooms, but we might also imagine the laboratories of the Intergovernmental Panel on Climate Change (and its expansive view of the biosphere) or the World Social Forum (where a broad perspective on global justice movements may be garnered). Oligoptica, as we might expect, are constituted through associations. They depend upon the creation of links between dispersed elements that allow various 'local' knowledges to be gathered and delivered to a central site. Whereas peripheral 'delegates' in knowledge-making are confined to their own limited perspective (as well as what the 'centre' allows them to know), occupants of oligoptic sites are able to consolidate a mass of delivered information into an integrated and ostensibly comprehensive form of knowledge. The capacity to influence affiliated others is gained, partially, by creating and occupying oligoptica.

Combining insights from Latour with those of social movement studies provides, I suggest, an incisive approach to the analysis of contentious politics. Above all, it promises further purchase on the question of how social networks are formed and sustained. Adopting Latourian concepts like networks and oligoptica, however, modifies how frames are conceived. Confronted with any phenomenon, Latour urges us to ask: 'What is it made of? How is it held together?' Tracking down the manifold human and nonhuman elements that allow 'things' to exist should disclose how social networks, frames, and ultimately social movements are constituted in an extended concatenation of elements. This, in any case, is my contention. In what follows, I document some of the most important elements that gradually came into alignment in the Pacific Northwest and provided the conditions of possibility for a broad, effective environmental movement in Portland. I begin my account not in the 1970s, but in 1941, and not with a group of activists but with a hydroelectric dam.

Energizing the Pacific Northwest

At 1:25 pm on March 23, 1941, the turbines at the Grand Coulee Dam in central Washington State began to spin water into electricity for the first time. Dubbed the 'mightiest work of man', the new dam spanned a record 4,300 feet from one bank of the Columbia River to the other, towered 550 feet upward from its foundation, conducted a waterfall twice the height of the Niagara over its parapet, and formed a reservoir upstream just a few miles shorter than Lake Ontario. While the dam's most immediate effects were ecological – its reconfiguration of the nonhuman world was unparalleled – it was a social transformation that its proponents and planners sought primarily to bring about. Initially, the construction of the Grand Coulee had the primary social effect of providing employment at a time of mass unemployment. Eventually, however, the dam, and the broader regional energy system that developed from it, would provide some of the conditions for the flourishing of environmental activism in Portland in the 1970s.

From dams like the Grand Coulee, to the energy transmission lines gradually veining the Northwest landscape, the regional energy system involved an immense enlistment and re-creation of nonhuman nature. As the system expanded from a few small hydroelectric installations to a vast network of over 400 dams and other generating stations, ancient rock formations were blown apart, forests were uprooted, valleys were flooded, and a system of bounding, vibrant rivers became a docile series of lakes. Of the multifarious elements involved in the energy system, the most basic element of the system was the potent force of fast-moving rivers like the Columbia; it was this force that the dams tapped and converted into a more practical, transmittable form. Crucial, in addition, was Pacific Northwest limestone, which was quarried, transported, ground to a powder, burned, and added as a binding agent to water, sand, and other minerals to produce concrete – and thus, the limbs and sluices of dams. Finally, copper, a mineral recognized for its especially conductive electron band structure, was enlisted in the transmission of energy from its sites of production to its sites of consumption; mined and pressed into wire, copper allowed the force of distant rivers to energize the homes and businesses of the region.

People too were involved in the energy system, and in diverse roles. From the late 1880s, geologists periodically surveyed the Columbia River and assessed its potential energy output. So potent was the river, reported early surveys, that a single well-placed dam could be expected to provide more electricity than the region could use. During the Great Depression, economic planners advocated dam construction less as a means of providing energy than as a job-creation strategy. The National Industrial Recovery Act of 1933 provided federal funding to purchase materials, hire workers, and construct the Grand Coulee Dam primarily, as President Roosevelt put it, to put people to work and thereby avoid 'complete economic stagnation' (*New York Times*, 23 March, 1941). As the Grand Coulee and subsequent dams came online, the Pacific Northwest would possess an abundance of low-cost electricity that gradually allowed the region to attract and develop profitable new industries. People were thus enlisted in the energy system as investors and workers in several large, energy-intensive aluminium smelters established in the region, as well as in the similarly consumptive warship and aircraft factories built during World War II (which continued to operate long afterward). People were enlisted in the system, finally, as consumers of electricity. To generate a market for the energy system's near-perpetual surplus, programmes were initiated to stimulate expanded non-industrial consumption. People were urged to buy new home appliances and move into all-electric houses; pedestrians and drivers might find themselves travelling along streets suddenly illuminated by electric lighting; and new electric trains were introduced, providing yet another way for city-dwellers to be propelled by the far-off Columbia River.

One of the unexpected products of this human–nonhuman assemblage was a new perspective on the natural environment. As Mitchell (2002) explains, the general, everyday belief that 'nature' and 'society' are two ontologically distinct realms has been significantly nourished, if not inaugurated, by modern techno-scientific projects like hydro dams. Although such a nature/society split has an extended history in the American West (cf. Kollin 2003; Egan 1990), the construction of a dam like the Grand Coulee, with its massive human-engineered bulk visibly overpowering the mad rush of river water, would certainly help to reinforce the image of a separate natural world tamed by human ingenuity. Such a sight was available to thousands of visitors to the dam, as well as the readers of the *New York Times* and other publications that published photos of the marvel in the weeks following its completion. Perhaps more significant than the idea of a separate nature was the idea of a distinct regional nature. According to White (1995), it was the development of a system of dams that eventually encompassed the full extent of the Columbia River watershed – and thus spanned and linked together the neighbouring states of Washington, Oregon, Montana, and Idaho – that is most squarely responsible for the idea of the Pacific Northwest as a coherent region. 'The lines of the Bonneville Power Administration [energy system] marked the region's boundaries', explains White. 'Where interties with other transmission systems occurred, there the Pacific Northwest encountered other regions' (64). The idea of a separate, regional nature – an idea with a clear material basis – would become central to the framing of 1970s environmental activism in Portland.

Another outcome of the energy system was a set of interdependencies among people. The same transmission lines that linked the people of the Pacific Northwest to the Columbia River also connected them to each other: as investors, workers, consumers, and citizens, people had a common concern in the energy system. Like many interdependencies, these ones were most apparent in times of crisis. In the 1970s, for at least four reasons, a significant and escalating crisis emerged. First, as a result of more and more homes and businesses being heated with hydroelectricity, the pattern of electricity demand had become increasingly inconstant and seasonal (i.e., higher in the colder months). Such demands could not be easily accommodated by the Columbia River system and its almost opposite pattern of water flow (i.e., highest between April and September, lowest in the cold months between December and January). Second, the river's seasonal water flow proved, in 1973 and 1976, to be relatively unreliable. In those years, major droughts dramatically reduced the flow of water delivered to the dams and resulted in crippling electricity supply shortfalls. Third, there were, of course, only so many places that dams could be built, and thus only so many opportunities to expand electricity production. By the 1970s, with dams in place nearly everywhere feasible and the electricity surplus eliminated by spiralling consumer demand, the system faced a shortfall that it could not meet through more hydroelectric dams. Fourth, by the 1950s, the damaging effects of the dams on the Pacific salmon populations that used the river system to spawn had begun to be documented. Efforts to mitigate salmon population decline through the establishment of hatcheries largely failed. Major population declines in 1972 and 1974 were registered and the plight of the salmon was recognized and taken up by fledgling salmon activists and newly formed environmental organizations (Allen 2003; Lichatowich 1999).

The need for a new energy regime in the Pacific Northwest had become palpable. The first attempt to hold the system together – the development of a series of nuclear plants – ended up fracturing it further still. As with hydroelectricity, nuclear energy relies on the enlistment of nonhuman nature. In this case, uranium is mined, enriched, and 'split' in a reactor to release heat; the heat, in turn, is used to produce steam that spins a turbine. In the calculations of nuclear proponents, tapping uranium was a solution to the many problems of the tapped out river system: it would allow the electricity supply to be augmented virtually without bounds and would leave Pacific salmon populations undisturbed. Such optimism, however, proved unwarranted. In the 1970s, a consortium of publicly owned utilities organized under the name Washington Public Power Supply System (WPPSS) quickly announced plans to build five nuclear plants in the Pacific Northwest. To finance construction, it issued billions in municipal bonds that it imagined would be paid back out of the revenues of the forthcoming plants. As a result of miscalculations and mismanagement, construction expenses vastly exceeded expectations and, in the end, only one plant was built. WPPSS eventually had to write off $2.25 billion of its debt in what it still the largest bond default in US history. When the remaining portion of the so-called 'whoops debt' was passed along to utility customers, wholesale electricity prices climbed nearly eightfold in just a few months. Not only did WPPSS fail to expand electricity generating capacity significantly, it also increased the cost of existing electricity. Rather than providing a solution to the system's problems, it exacerbated them.

For the Pacific Northwest, the failure of the nuclear experiment was the last gasp of a certain trajectory of energy production and consumption. From that time on, it had to be admitted that energy production could go on increasing or low energy prices could be maintained, but not both. Decades of federally funded and relatively cost-effective investments in hydro power had bequeathed the region per-unit energy costs that could not be matched by other (non-hydro) means of energy production. Future increases in supply, therefore, would necessarily be more costly per unit and would put corresponding upward pressure on prices. This would be a concern for consumers, but especially for industries that had developed precisely on the basis of low-cost energy. Additional concerns arising at this point had to do with the effects of the energy system on the natural environment. People in cities like Portland were brought into relation with this environment principally as consumers: a separate, regional environment was constituted by the energy system and linked to towns and cities in the resource-form of kilowatt hours. Once involved, however, it would be possible for people to imagine other more responsible modes of involvement and, indeed, some people would. By the 1970s, a variety of concerns emerged on the basis of relationships brought into being by the development of the regional energy system. These concerns seemed to exceed the capacity of the system to address them. If a new approach to energy production and consumption was required, it was the task of an emerging social movement to define what this approach would be.

Oregon Office of Energetics

The Pacific Northwest energy system helped to delineate, and bring people into relation with, the regional environment. The struggle to define a new relationship with this environment was carried out, most significantly, in three particular sites: the Oregon Office of Energetics, Rain House, and the Portland Energy Committee. The first site, the State of Oregon's Office of Energetics, was established in 1972. The Office was the brainchild and purview of a technically-minded analyst named Joel Schatz, who had been recruited by Oregon governor Tom McCall to develop solutions to the unfolding energy crisis. Schatz was given a staff, a budget, and a broad mandate. His first move, upon taking up his new position, was to dispatch his staff to compile a mass of data on a range of energy sources that were then being considered as potential responses to still-increasing demand. The sources to be evaluated included nuclear, oil shale, and coal. The analysis was published in 1974 as The Transitions Report. The Report offered a sweeping view of the energy landscape and, owing both to the strength of its analysis as well as the ability of Schatz to put this analysis before the public eye, provided a framing of energy-related problems in the region that had enormous influence in Portland and much farther afield. The Report advanced conservation as the best available 'source' of energy and the best passage toward a viable future for the Pacific Northwest. It was precisely because of the Report, and Joel Schatz, that many Portland-based activists came to be persuaded of the merits of conservation as a political demand.

At the heart of Schatz's analysis were two kinds of consolidation, or translation. The first was analytical and relied on an idea developed by University of Florida polymath Howard Odum: the concept of 'net energy'. Extolled in Odum's book, Environment, Power, and Society – a book read by Schatz and, at the urging of Schatz, by many Portland activists – the concept compels an analysis of energy sources that takes into consideration both their expected output as well as their inputs. Inputs, here, refer to the typically overlooked quantities of energy consumed in the development and operation of generating facilities and production processes of different types. As the Transitions Report showed, once inputs are subtracted from output, energy sources like nuclear fission and oil shale seem rather miserly, while conservation, seldom regarded as a 'source' at all, appears much more generous. One of the benefits of net energy calculations is to allow a mass of information to be condensed into a few digestible figures. It was on the basis of this data, this work, and this consolidation that the Report provided a coherent analysis of a complex situation and a relatively straightforward argument for energy conservation.

The second consolidation was more explicitly political. Schatz chose energy as his primary focus in the hopes of bringing diverse political concerns into alignment. Energy production could be linked to a broad range of environmental problems in the region. 'Species elimination, deforestation, loss of topsoil, everything bad in the biosphere – all of these could be linked to energy', recalls Schatz. 'I used energy as a fulcrum to talk about all the other [environmental] ailments' (personal communication, November 2007). Perhaps more importantly, energy could also be placed at the centre of a set of problems that were not strictly environmental and that could potentially be addressed with or without environmental considerations. Rising energy costs were already raising costs of living and putting area businesses at risk. Large-scale investments in inefficient sources of energy, it could be argued, would make matters worse. 'Lots of people wanted a solution [to the energy crisis]', Schatz recalls, and one of the functions of net energy analysis was to frame energy policy debates in terms that would make conservation widely appealing, not just to salmon activists and anti-nuclear campaigners but also to a broader constituency more immediately disturbed by their monthly energy bills than by the plight of fish, rivers, and forests.

In the mid-1970s, Schatz and the Office of Energetics occupied an influential position in the struggle to define a response to mounting energy-related problems in the Pacific Northwest. Activists in Portland remember the appeal of Schatz's analysis. Energy activist Margie Harris, for example, suggests that net energy calculations provided a 'scientific basis for policy change' and allowed activists to 'combat the mainstream approaches of the day ... with statistical information and systemic analysis' (personal communication, December 2007). In addition to a comprehensive analysis, Schatz had also developed effective ways of getting his message out. He spoke regularly to the media about energy issues; he attended meetings, conferences, and talk shows across the country; and he consulted with a variety of organizations. By tapping into these communication circuits, his analysis could travel much more widely than he, as an individual, ever could. Schatz's position in communication circuits further enhanced his influence among Portland-based activists. 'We made alliances with all kinds of people and organizations', says Schatz. 'I had huge access to the media and did a ton of public speaking ... so people would come to me and we'd do collaborative work. We became the spokespeople for the environmental movement.' Drawing local knowledges into a central site, constructing a politically and epistemologically comprehensive perspective, and circulating this perspective to a broad constituency accounts, to a large extent, for the influence of the Office of Energetics and, indeed, for the widespread reception of conservation as a solution to energy-related problems in the Pacific Northwest.

Rain House

If Schatz and the Office of Energetics were especially effective in promoting energy conservation, it was nevertheless elsewhere, most significantly at Rain House, that conservation came to be articulated as an urban scale solution and that a social movement emerged to advocate for it. Rain House, located in Portland, was the physical headquarters of *Rain Magazine*. Founded in 1973 and published consistently until 1986, *Rain Magazine* was an inventive, paradox of a periodical: an 'appropriate technology' manual packaged in a Japanese-Zen aesthetic; a source of technical specifications and blueprints for environmentally beneficial projects, and an anthology of environmentalist poetry and visual art. This coming-together of diverse approaches and perspectives was reflected in *Rain*'s readership and, significantly, in the range of people who spent time at Rain House. A lively nexus of people and ideas, Rain House allowed Portland-based activists to forge connections across differences in philosophy and develop ways of working together toward common objectives. It was at Rain House that a broad group of activists came to support the establishment of conservation-focused energy policy in Portland.

One of the striking features of *Rain Magazine* was the array of perspectives that it brought together in a single volume. Its editors aimed explicitly to open a broad canopy under which, for example, spiritually oriented philosophies like bioregionalism could commingle with technical treatises on energy systems, architecture, and planning. As former editor Carlotta Collette recalls, the magazine saw its mission as addressing both the 'why' and the 'how' of environmental action:

> *Rain*'s really special contribution was that it provided a bridge between the spiritual/philosophical work of people like Gary Snyder and the practical/ political work of the Amory Lovinses of the world. It brought together the spiritual basis for why we should do something with the practical basis for how we might do it. (personal communication, December 2007)

Rain, in its affirmation of multiple environmental perspectives, suggested a way for renewable energy advocates, anti-nuclear activists, and wilderness protection campaigners – to name three prevalent categories of environmentalist inhabiting the Portland area – to advocate the same things without necessarily thinking or feeling the same thing.

What *Rain* brought together, in addition to multiple philosophies, was a mass of information on effective, replicable environmental projects and actions. From localities around the world, people involved in appropriate technology projects or activist campaigns contributed stories, pictures, blueprints, instructions, and other materials. The latter would accumulate, sort, and publish the best of this material. Readers of the magazine therefore gained a relatively comprehensive view of what was happening in the world and what seemed to be working. *Rain* made connections between dispersed projects, campaigns, and communities; it created circuits through which good ideas could spread; it even published the phone numbers and mailing addresses of the people involved in featured projects, enabling interested readers to contact the latter for advice. 'The idea with *Rain*', says Collette, 'was to survey who was doing good things and to describe these good things in a way that inspired and enabled [readers] to do them too.'

The capacity of *Rain* to enable social connections was most pronounced in Portland, where its physical headquarters – an old Victorian house in the Northwest of the city – became a home away from home for many activists. Most of *Rain*'s staff lived there, movements like Nuclear Freeze were given office space, and well-known radicals passing through Portland would often spend the night on a couch or the floor (Ivan Illich, Winona LaDuke, and Jerry Brown are a few examples). The house also became a repository for the many materials sent to the magazine (e.g., appropriate technology blueprints and instructions, zines, periodicals, and books from small and large publishers). When an archivist was eventually retained to organize these materials, which had been packed into boxes and stored in scattered nooks and closets, Rain House could henceforth boast a one-of-a-kind library. 'With no internet in those days', recalls Collette, 'people who wanted to learn about appropriate technology would physically have to come to Rain House; people would come from around the world to use the library.' As a dense point of connection for people, materials, and ideas, Rain House was a fertile site for political discussion and organizing. Important contacts were made, actions were planned, and conversations around the dinner table could last late into the night.

One idea that eventually came up at Rain House was that of establishing a strong city-level energy policy. The idea stemmed from a story that had been received by the magazine: Franklin County, Massachusetts, had become the first county in the US to adopt a conservation-focused energy policy. The idea was not entirely new to *Rain* editors; they had read the Transitions Report, for example, and had long supported the goal of energy conservation. What was novel about Franklin County was the focus on the local scale. While the effects of energy production and consumption were more spatially extensive in the Pacific Northwest – the problems of disappearing salmon, razed forests, dangerous nuclear plants, and so on, were largely regional in scope – Franklin County had demonstrated that local government was nevertheless a fruitful place to apply political pressure and make energy-related demands. Impressed by the precedent, *Rain* featured the story and proposed to the Rain House community that such an initiative could be pursued in Portland. Given that *Rain*'s purpose was to accumulate and disseminate knowledge about ground-breaking environmental projects and campaigns from around the world, its proposal carried weight. As a hub for Portland-based environmental activists, moreover, Rain House was a productive place to circulate an idea. In 1976, on *Rain*'s suggestion, a group of activists and radical city officials who were regulars at Rain House began to work toward the establishment of a conservation-focused municipal energy policy. Many activists would apply pressure from outside the political system. Many, however, worked within the system as members of the newly formed Portland Energy Committee.

Portland Energy Committee

The Portland Energy Committee was formed in 1977 to evaluate city-wide energy consumption (i.e., in the municipality's operations, as well as those of private businesses and households) and propose appropriate energy policies. Reporting to City Council and overseen by the City's manager of urban development (Mike Lindberg), the Committee was composed of a full-time project manager (Marion Hemphill), a 20-person unpaid steering committee, and five task forces of 10-15 members each. Unpaid committee members were drawn from an assortment of locations: the executive branch of local financial institutions and other corporations, public and private utilities, and the citizenry at large. The Committee's work was greatly facilitated by social connections forged, and knowledge produced, in other locations. The Transitions Report was read closely by most members of the Committee; Lon Topaz, former director of the Office of Energetics, was enlisted to carry out additional research for the Committee, which was formed partly on the recommendation of the Report. *Rain* was also a significant influence: many Energy Committee members – including Marion Hemphill and Mike Lindberg – were subscribers to the magazine, frequent visitors to Rain House, or both; such associations helped to establish a frame for the Committee's work. As activist and Committee member Margie Harris recalls, 'we were pretty unified on what needed to happen; the important pieces of the puzzle were already there' (personal communication, December 2007). That energy conservation was the primary objective was agreed upon. The question was how to achieve it.

Once assembled, the Committee set upon its major tasks: setting energy conservation targets and outlining effective mechanisms for achieving them. Local energy consumption was divided into five categories – industrial, commercial, residential, transportation, and land use – and a separate task force was assigned to each one. To win broad political support for its proposals, the Committee planned to set conservation targets that could be achieved cost-effectively; that is, every dollar spent on a retrofit or other energy-conserving initiative would be expected to produce an equivalent long-term reduction in energy consumption expenses. Setting targets would therefore require new, and more detailed, information. To that end, the Committee solicited funds from the Federal Department of Housing and Urban Development and the State Department of Transportation. With the funds, the Committee hired a local architecture and engineering firm to examine local energy consumption patterns and estimate the level of conservation that could be achieved cost-effectively (in each of the five categories). In addition, a group of students at Lewis and Clark Law School were enlisted to investigate the viability, from a legal perspective, of a variety of policy instruments. On the basis of this research, the Committee acquired an unparalleled and very detailed view of local energy consumption and the viable means of managing it. Indeed, before the Committee's work, the category 'local energy consumption' had no precise meaning; the line between 'local' and 'non-local' usage, after all, was not entirely obvious and certainly could not have been assigned a quantity.

But comprehensive knowledge was not, in itself, sufficient to achieve the Committee's goals. Although research indicated that certain levels of conservation could be achieved cost-effectively, not everyone was convinced. Particularly sceptical were the region's energy utilities, which saw conservation as a threat to their future profits and began to organize opposition to the Committee's efforts.

In 1978, the major utilities convened a series of secret meetings at an airport hotel outside Portland. Under consideration was a newly hatched plan to build several nuclear and coal-fired power plants in the region, assign the construction costs to the Bonneville Power Administration, and preclude Portland from receiving any of the new energy produced. News of the secret meetings eventually reached Portland City Hall, where Angus Duncan, an assistant to mayor Neil Goldschmidt, devised an intervention. Duncan was a long-time activist; as an aide to the mayor, he was involved with the Energy Committee and would be responsible for presenting its recommendations to City Council. Perceiving the utilities' scheme as an obstacle to the forthcoming energy policy, Duncan pressured the utilities to allow him to attend their meetings. He was successful. Relying on the Energy Committee's findings, Duncan sought to demonstrate to the utility executives that conservation would, in fact, be better for their profits than building new plants. 'I showed them figures comparing present retail [electricity] rates with the incremental costs [of providing additional electricity]', recalls Duncan. 'With a 5 cent per kilowatt retail rate and a 7 cent incremental rate, the costs of new supply were higher than the potential revenues' (personal communication, December 2007). After a brief discussion, Duncan left the research in the hands of Pacific Gas and Electric (PG&E) CEO Don Frisbee, and advised that he 'think it over'.

The following year, the Committee concluded its two years of research, discussions, and public consultations, and submitted a draft energy policy to Portland City Council. The policy established conservation as the first-order response to future increases in energy demand. Conservation targets of 10 per cent were applied to five categories of consumption. Included in these categories was the consumption not just of electricity, but also, significantly, fossil fuels (which would predominate, especially, in the categories of land use and transportation). The policy mandated the creation of two new municipal departments to monitor local energy use and co-ordinate the delivery of incentives and zero-interest loans for conservation projects carried out by local businesses and households. Overall, the policy committed Portland residents to certain modest changes. These changes would have their effects, of course, much farther afield. To the extent that Portland could conserve energy, for example, the need to build new generating facilities in the region would be reduced – and so, too, the effects of Portland residents upon their environment. When the draft policy went before Council there were two witnesses in attendance to speak in its favour: the director of the Portland Environmental Council, and PG&E's Don Frisbee. The latter had, apparently, heeded Duncan's advice. With the policy's foremost opponent now on board, it went ahead. With the policy's approval, a social movement whose conditions of possibility emerged in the development of the Grand Coulee Dam and the broader Pacific Northwest energy system, as well as in sites like the Office of Energetics and Rain House, had achieved a tangible victory.

Conclusion

Portland's municipal energy policy was a major achievement. Not only did it establish conservation as a priority and provide mechanisms for achieving it, it addressed the entire energy system: not just electricity, but also, for example, fossil fuels consumed for the purpose of ground transportation. Seizing on the relatively obvious problems with the Pacific Northwest energy system, activists succeeded in opening up a political terrain in which the broader impacts of human beings upon their environment could be discussed and addressed. Longer-term, the energy policy provided some of the conditions for taking 'local action' on other issues of broad concern. From the monitoring and managing of local energy consumption, it was not such a leap to address greenhouse gas emissions. Indeed, in 1991 Portland became one of the first cities in the world to commit to concrete emissions reductions (Rutland and Aylett 2008). Portland, along with a handful of other cities, has since become a world leader in this regard (Betsill and Bulkeley 2004; Slocum 2004). At the same time, the frame of energy conservation had major limitations. To conserve energy, after all, is still to approach nature primarily as a source of kilowatt hours, and activists who sought to bring about a more fundamental, spiritual-philosophical transformation in Portland did not move very far toward their objective. Similarly, conservation implies no necessary shift toward renewable forms of energy, something that some activists advocated.

My purpose in this chapter was to link the social movement responsible for Portland's energy policy to its major conditions of possibility. The development of the Pacific Northwest energy system, I argued, was particularly important. As the system expanded from a few major dams to a system of over 400 generating stations, it nourished the idea of a distinct regional environment and brought the lives of people in cities like Portland into relationship with it. Though it may not have been apparent without the crises of the 1970s and activists' framing of these crises, the energy system also brought people into relationships with each other, giving them a common concern. Also important to the 1970s environmental movement were a series of influential, 'oligoptic' sites. The Oregon Office of Energetics, Rain House, and the Portland Energy Committee provided the material basis for the creation of social networks among activists and helped to consolidate dispersed 'local' knowledges into more comprehensive perspectives that greatly influenced the articulation of activists' concerns and demands. The Portland-based movement's major frames highlighted problems in the regional environment (its concern), the responsibility of the residents of Portland (the primary collectivity), and energy conservation (its demand). This social network and these frames were constituted through a concatenation of human and nonhuman elements conjoined, or associated, across time and space. The mobilization of environmental concern in Portland was a genuinely collective achievement.

This analysis shows how certain insights from the work of Bruno Latour can be combined with more familiar concepts in social movement studies to address some of the limitations of the latter. While Latour's principal contribution may be to provide an incisive approach to the question of how social networks are created and sustained, it is ultimately an approach that applies equally to frames and movements themselves. Latour's work has its own limitations, however. Accordingly, I conclude by mentioning two major limitations and their implications for the account I have offered. First, critics have argued that Latour's approach, while compellingly demonstrating how associations are formed, largely ignores processes of disassociation and exclusion (Star 1991; Haraway 1997). My account, for example, makes no mention of Portland activists whose aims could not be satisfied by energy conservation and who, in various ways, contested this articulation of environmental concern. A second, related criticism is that Latour's approach attends to minute details at the cost of obscuring the broad, systemic conditions in which associations are formed (Castree 2002; Kirsch and Mitchell 2004; Wainwright 2005). Indeed, my account scarcely acknowledges how 'systems' such as capitalism, colonialism, and patriarchy, for example, were implicated in the environmental politics of Portland and the Pacific Northwest. While systems too might be analysed as networks, it is hard to imagine all of their important elements being enumerated in a single chapter. If the application of Latour's work in studies of social movements has its merits, it also has its limitations. I urge that both be taken into account.

Acknowledgements

Helpful comments on earlier drafts of this chapter were provided by Alex Aylett, Trevor Barnes, Pablo Mendez, Elvin Wyly, and the editors of this volume. Thank you all.

References

Allen, C. (2003) 'Replacing Salmon: Columbia River Indian Fishing Rights and the Geography of Fisheries Mitigation', *Oregon Historical Quarterly*, 104 (2): 196-224.

Betsill, M. and H. Bulkeley (2004) 'Transnational Networks and Global Environmental Governance: The Cities for Climate Protection Program', *International Studies Quarterly*, 48: 471-493.

Bickerstaff, K. and J. Agyeman (2010) 'Assembling Justice Spaces: The Scalar Politics of Environmental Justice in North-East England', in R. Holifield, M. Porter, and G. Walker (eds.) *Spaces of Environmental Justice*. Malden MA: Wiley-Blackwell.

Bosco, F. (2006) 'The Madres de Plaza de Mayo and Three Decades of Human Rights' Activism: Embeddedness, Emotions, and Social Movements', *Annals of the Association of American Geographers*, 96 (2): 342-365.

Boudreau, J. A. (2003) 'Questioning the Use of "Local Democracy"', *International Journal of Urban and Regional Research*, 27 (4): 793-810.

Castree, N. (2002) 'False Antitheses? Marxism, Nature, and Actor-Networks', *Antipode*, 32: 111-146.

Cinalli, M. (2003) 'Socio-politically Polarized Contexts, Urban Mobilization and the Environmental Movement: A Comparative Study of Two Campaigns of Protest in Northern Ireland', *International Journal of Urban and Regional Research*, 27 (1): 158-177.

Debanne, A., and R. Keil (2004) 'Multiple Disconnections: Environmental Justice and Urban Water in Canada and South Africa', *Space and Polity*, 8 (2): 209-225.

Diani, M. (1995) *Green Networks: A Structural Analysis of the Italian Environmental Movement*. Edinburgh: Edinburgh University Press.

Egan, T. (1990) *The Good Rain: Across Time and Terrain in the Pacific Northwest*. New York: Vintage Books.

Gandy, M. (2005) 'Cyborg Urbanization: Complexity and Monstrosity in the Contemporary City', *International Journal of Urban and Regional Research*, 2 (1): 26-49.

Gilbert, L. (2004) 'At the Core and on the Edge: Justice Discourses in Metropolitan Toronto', *Space and Polity*, 8 (2): 245-260.

Goodwin, J. and J. Jasper (1999) 'Caught in a Winding, Snarling Vine: The Structural Bias of Political Process Theory', *Sociological Forum*, 14 (1): 27-54.

Haraway, D. (1997) *Modest_Witness@Second_Millennium.FemaleMan_Meets_ OncoMouse*. New York: Routledge.

Jasper, J. and J. Poulsen (1995) 'Recruiting Strangers and Friends: Moral Shocks and Social Networks in Animal Rights and Anti-Nuclear Protests', *Social Problems*, 42 (4): 493-512.

Juris, J. (2004) 'Networked Social Movements: Global Movements for Global Justice', in Castells (ed.) *The Network Society: A Cross-Cultural Perspective*. Cheltenham: Edward Elgar Publishing.

Kirsch, S. and D. Mitchell (2004) 'The Nature of Things: Dead Labor, Nonhuman Actors, and the Persistence of Marxism,' *Antipode*, 36 (4): 687-705.

Kollin, S. (2003) 'North and Northwest: Theorizing the Regional Literatures of Alaska and the Pacific Northwest', in Crow (ed.) *A Companion to the Regional Literatures of America*. New York: Blackwell.

Kurtz, H. (2003) 'Scale Frames and Counter-Scale Frames: Constructing the Problem of Environmental Injustice', *Political Geography*, 22 (8): 887-916.

Latour, B. (1998) 'On Actor-Network Theory: A Few Clarifications', Centre for Social Theory and Technology, Keele University <http:/www.keele.ac.uk/depts/stt/staff/jl/pubs-jl2.htm>.

—— (1999) 'On Recalling ANT', in J. Law and J. Hassard (eds) *Actor Network Theory and After*. Oxford and Keele: Blackwell.

—— (2004a) 'Why has Critique Run Out of Steam? From Matters of Fact to Matters of Concern', *Critical Inquiry*, 30 (2): 225-248.

—— (2004b) 'The Promises of Constructivism', in Ihde and Selinger (eds) *Chasing Technoscience: Matrix for Materiality*. Bloomington: Indiana University Press.

—— (2005) *Reassembling the Social: An Introduction to Actor-Network Theory*. Oxford: Oxford University Press.

Lichatowich, J. (1999) *Salmon without Rivers: A History of the Pacific Salmon Crisis*. Washington: Island Press.

Lockie, S. (2004), 'Collective Agency, Non-Human Causality and Environmental Social Movements', *Journal of Sociology*, 4 (1): 41-57.

McAdam, D., McCarthy, J. and Zald, M. (1988) 'Social Movements', in N. Smelser (ed.) *Handbook of Sociology*. Beverly Hills: Sage.

—— (1996) 'Introduction: Opportunities, Mobilizing Structures, and Framing Processes: Toward a Synthetic, Comparative Perspective on Social Movements', in D. McAdam, J. McCarthy, and M. Zald (eds) *Comparative Perspectives on Social Movements: Political Opportunities, Mobilizing Structures, and Cultural Framings*. Cambridge: Cambridge University Press.

McCarthy, J. (1996) 'Constraints and Opportunities in Adopting, Adapting, and Inventing', in D. McAdam, J. McCarthy and M. Zald (eds) *Comparative Perspectives on Social Movements: Political Opportunities, Mobilizing Structures, and Cultural Framings*. Cambridge: Cambridge University Press.

Martin, D. (2003) '"Place-framing" as Place-making: Constituting a Neighborhood for Organizing and Activism', *Annals of the Association of American Geographers*, 93 (3): 730-750.

Melucci, A. (1988) 'Social Movements and the Democratization of Everyday Life', in Keane (ed.) *Civil Society and the State: New European Perspectives*. London: Verso.

Mische, A. (2003) 'Cross-talk in Movements: Reconceiving the Culture-network Link', in M. Diani and D. McAdam (eds) *Social Movements and Networks: Relational Approaches to Collective Action*. Oxford: Oxford University Press.

Mitchell, T. (2002) *Rule of Experts: Egypt, Techno-Politics, Modernity*. Berkeley: University of California Press.

Mueller, C. (1992) 'Building Social Movement Theory', in Morris and Mueller (eds) *Frontiers in Social Movement Theory*. New Haven: Yale University Press.

Murdoch, J. (1997) 'Towards a Geography of Heterogeneous Associations', *Progress in Human Geography*, 21 (3): 321-337.

—— (1998) 'The Spaces of Actor-Network Theory', *Progress in Human Geography*, 29 (4): 357-374.

Pfaff, S. (1996) 'Collective Identity and Informal Groups in Revolutionary Mobilization: East Germany in 1989', *Social Forces*, 75 (1): 91-117.

Routledge, P. (2003) 'Convergence Space: Process Geographies of Grassroots Globalization Networks', *Transactions of the Institution of British Geographers*, 28 (3): 333-349.

—— (2008) 'Acting in the Network: ANT and the Politics of Generating Associations', *Environment and Planning D: Society and Space*, 26 (2): 199-217.

Rutland, T. and A. Aylett (2008) 'The Work of Policy: Actor Networks, Governmentality, and Local Action on Climate Change in Portland, Oregon', *Environment and Planning D: Society and Space*, 26 (4): 627-646.

Slocum, R. (2004) 'Consumer Citizens and the Cities for Climate Protection Campaign', *Environment and Planning A*, 36 (5): 763-782.

Snow, D. and R. Benford (1988) 'Ideology, Frame Resonance, and Participant Mobilization', *International Social Movement Research*, 1: 197-217.

—— (1992) 'Master Frames and Cycles of Protest', in Morris and Mueller (eds) *Frontiers in Social Movement Theory*. New Haven: Yale University Press.

Star, S. (1991) 'Power, Technology and the Phenomenology of Conventions: On Being Allergic to Onions', in Law (ed.) *A Sociology of Monsters: Essays on Power, Technology, and Domination*. New York: Routledge.

Tindall, D. (2002) 'Social Networks, Identification and Participation in an Environmental Movement: Low-medium Cost Activism with the British Columbia Wilderness Preservation Movement', *Canadian Review of Sociology*, 39 (4): 413-452.

Uitermark, J. (2004) 'Framing Urban Injustices: The Case of the Amsterdam Squatter Movement', *Space and Polity*, 8(3): 227-244.

Wainwright, J. (2005) 'Politics of Nature: A Review of Three Recent Works by Bruno Latour', *Capitalism, Nature, Socialism*, 16: 115-121.

Whatmore, S. (1999) 'Hybrid Geographies: Rethinking the "Human" in Human Geography', in D. Massey and Sarre (eds) *Human Geography Today*. Cambridge: Polity Press.

White, R. (1995) *The Organic Machine: The Remaking of the Columbia River*. New York: Hill and Wang.

Chapter 11

Networking Resistances: The Contested Spatialities of Transnational Social Movement Organizing

Andrew D. Davies and David Featherstone

Introduction

The prominence and visibility of transnational forms of organizing is a defining feature of contemporary contentious politics. There are significant histories of transnational forms of contention and organizing, and they are by no means new despite being frequently depicted as such. The growing prominence and interest in such transnational forms of organizing has unsettled some of the key ways of understanding the geographies of contentious politics. This has opened up a challenge to the ways that both social movement theory and political geography have been structured by an implicit assumption that the national arena is the most obvious container for political activity.

Dislocating such nation-centred understandings of the political opens up important theoretical and methodological challenges that we explore here. We are concerned to foreground the practices through which social/political movements intervene in relations between different places. To do this we adopt an explicitly relational approach to understanding the geographies of contentious politics. We are concerned to understand the ways in which forms of contentious politics constitute diverse, multiple and contested geographies of connection. They are also defined by the practices through which they bring unequal geographies of power into contestation. This approach positions the networks that social movement organizing generates as the 'overlapping and contested material, cultural and political flows and circuits that bind different places together through differentiated relations of power' (Featherstone, Phillips and Waters 2007: 386).

The chapter develops these arguments through engagement with two different forms of transnationally networked politics. We dissent from those who argue that there is anything like a fully formed global civil society (e.g. Kaldor 2003). In contrast we argue that there is a much more diverse, multiple and contested set of transnational social movement actors (see also Routledge and Cumbers 2009). Further, such forms of mobilizing generate multiple and contested political identities. There is, for example, no singular opposition to neo-liberalism. To foreground this multiplicity we engage with transnational organizing defined by different forms of political identities. These are the transnational 'pro-Tibet Movement' and transnational resistance to the Coca-Cola Corporation.

The first part of the chapter engages critically with the influential work on transnational social movement activity associated with Sidney Tarrow. We argue that his engagement with the processes that constitute forms of transnational contention is significant, but ultimately tends to reduce such activity to a set of formalistic processes and mechanisms. We then outline an agenda for thinking relationally about the conduct of transnational social movement organizing that foregrounds its productive, contested and multiple characteristics. These arguments are demonstrated through a discussion of the ways political actions reshape and engage with power relations. We then engage with the contested and multiple practices through which articulations are formed between different places and sites through social movement activity. The chapter then explores the importance of the ongoing work of maintaining and generating connections to transnational organizing.

The chapter concludes by exploring issues in relation to the effectiveness of transnational political networks in facilitating and hindering political actors in shaping durable alliances and achieving defined political goals. It highlights the significance of solidarities in generating unintended outcomes and transforming both political identities and the terms on which contestation is generated. It also argues that foregrounding the work of maintaining and generating connections opens up important and neglected aspects of the forms of identity and agency constructed through transnational organizing.

Space, transnationalism and the constitution of political alliances

One of the most sustained attempts to think through the significance of recent forms of transnational politics from an explicitly social movement theory perspective is the work of Sidney Tarrow. Tarrow, both in his book *The New Transnational Activism* and in co-authored work with Douglas McAdam and Charles Tilly, has mobilized a set of 'robust concepts' to understand and interrogate emerging forms of transnational activism. Tarrow argues that 'there is no single core process leading to a global civil society or anything resembling one, but – as in politics in general – a set of identifiable processes and mechanisms that intersect with domestic politics to produce new and differentiated paths of political change' (Tarrow 2005: 9). This attention to the processes through which political activity and contentious politics is produced is useful. This is part of an attempt to move beyond 'static, variable-driven structural models' to engage more directly with the dynamic character of contentious politics (Tarrow and McAdam 2005: 124).

This attempt to foreground the dynamic processes through which contentious politics is generated is important and has opened up productive research agendas. The work of Tarrow et al., however, is structured by a particular approach to understanding the processes and mechanisms which constitute contentious politics. Tarrow et al. have adopted a mode of theorizing that seeks to develop a core set of concepts to explain common processes between radically different social movements (see especially McAdam, Tarrow and Tilly 2001). This produces accounts of social movement activity where such activity is reduced to identification with such concepts. It maps social movement activity on to processes and mechanisms which they see as common to many different forms of contentious politics (e.g. Tarrow and McAdam 2005: 126). Rather than allowing an engagement with the dynamic processual constitution of political activity, their approach reduces social movement activity to a set of generic processes and mechanisms.

A key consequence of such an approach is that it evades the generative character of social movement/political activity. This is very different from other approaches such as the work of Juris, which we will discuss below, which develops a more 'open' sense of the processual character of transnational social movement activity. It ignores the practices through which social movement/political activity can re-constitute the terrain of contestation and political identities, rather than working within the neat conceptual processes defined by social movement theorizing. The tensions involved with this approach can be illustrated by an engagement with their accounts of brokerage, diffusion and scale shift. These concepts have been central to Tarrow's recent work on transnational social movements, and are an attempt to think about the spatial constitution of contentious politics. They are therefore worthy of sustained discussion here.

Tarrow generates accounts of what he terms distinct processes such as brokerage, scale-shift, emulation and diffusion to explain how movements move and how alliances develop between movements in different places. Brokerage is a particularly central term in his work, in this respect. He defines brokerage as the 'linking of two or more previously unconnected social actors by a unit that mediates their relations with one another and/or with yet other sites' (Tarrow 2005: 190; see also Tarrow and McAdam 2005: 127). This develops a focus on some of the spatial practices through which transnational movements are constituted and generated. However, Tarrow (2005) and Tarrow and McAdam (2005) deal with such processes in rather reductive ways. Thus they argue that 'that successful brokerage promotes attribution of similarity, while unsuccessful brokerage promotes the recognition of difference' (Tarrow and McAdam 2005: 130).

Tarrow and McAdam's account of the term brokerage, then, rests on a problematic account of how forms of transnational political identification work. They suggest a very straightforward relationship between the ways in which brokerage works and the outcomes of political activity. Further, it is based on a reductive account of such processes which suggests they can be neatly categorized as either promoting similarity or difference. We contend that engaging with the conduct and following political activity disrupts such easy categorizations. As we argue below different movements can become articulated in opposition to shared enemies such as the Coca-Cola Corporation. Further, movements which may at first appear to be based around shared and rather fixed identities often produce more multiple forms of political practice. Below, for example, we demonstrate that diverse identities are constituted through transnational organizing in relation to the 'Tibet Issue' rather than the singular notion of Tibetan identity (see also Davies 2009a).

Tarrow et al. have deployed their analysis of forms of brokerage in conjunction with what they term 'scale shift' to engage explicitly with the spatial practices through which social movement activity is constituted. Tarrow argues that 'scale shift' is 'an essential element of all contentious politics, without which all contention that arises locally would remain at that level' (Tarrow 2005: 121). McAdam, Tarrow and Tilly define it 'as a change in the number and level of coordinated contentious actions to a different focal point, involving a new range of actors, different objects, and broadened claims' (McAdam, Tarrow and Tilly 2001: 331). They argue that:

> Scale shift is a moderately complex process within which the relative salience of diffusion and brokerage varies, but passage through attribution of similarity and emulation regularly produces a transition from localization to large-scale coordination of action. Like actor constitution and polarization, scale shift operates across the whole range of contentious politics in similar ways, yet in conjunction with other mechanisms and processes produces anything from strike waves to mass murder (McAdam, Tarrow and Tilly 2001: 339).

The stress on practices of emulation and similarity in their account of scale shift closes down a sense of the generative relations between different place-based movements which are constituted through transnational organizing. Rather than focusing on the formation of alliances as something productive and involving the dislocation and reworking of political identities their focus on practices of emulation and similarity gives a sense of these processes as being singular, rather than multiple. Further, their account of scale-shifting reduces it to a generic process, which happens in similar ways in radically different kinds of movements. This reduction of these spatial practices to the status of a generic process militates against engaging with the generative conduct of particular instances of political activity.

These tensions resulting from the reduction of social movement activity to the status of generic/identifiable processes structure Tarrow's account of the relations between processes such as 'diffusion' and 'scale shift'. There are a number of issues here. Diffusion is seen as essentially horizontal, and seemingly involves only two actors, in a relationship of 'initiator' and 'adopter'. We argue that the spaces of these engagements are not purely about operating across a smooth spatial 'plane' of diffusion. Instead, the often messy relationships formed through the movement of social movements means diffusion must be read as a more negotiated and seemingly unknowable process. The relations that bind or break between social movement actors alter the shape of the social movement itself.

Tarrow's account of diffusion then codifies the multiple arrangements between actors at a variety of different scales into a process whereby connections either work or fail, and gives little account to the important generative work that goes on through the practices of actually attempting to create and hold stable these relations. Further, while diffusion of movements provides a way of explaining the spatial extensity of a social movement, Tarrow argues that in order to move 'up' a hierarchical level of organization (i.e. from the local to the global) social movements need to enact a series of performances by agents. Thus, while transnational movements can spread through diffusion of their networks across space, they can also, by brokering and negotiating, incorporate new claims and claimants.

> Whereas diffusion is a traditional process that moves horizontally between one initiator and one adopter and has done so since long before the idea of globalization gained ground, scale shift would need to be the first process in the work of building a global social movement (Tarrow 2005: 139).

It is at this point that the lack of a truly transnational political action leads to a breakdown in the process of scale shifting. Tarrow argues that many actors remain rooted in local politics, unable to break away from the constraints of their uniquely local circumstances – they are unable to become the 'rootless cosmopolitans' who are the truly transnational actors. Those who do begin to act at a 'higher' scale, according to him, are more likely to transpose some of their work upwards, and still remain rooted in their particular circumstances.

This approach isolates local actors from ongoing engagement with spatially stretched power relations and relates to a final tension in the work of Tarrow and others. Their accounts of processes like brokerage or scale shift do not give a sense of these geographies as constitutive of the political practices and identities of these movements. Tarrow's account of scale-shift treats geography as a fixed backdrop to transnational struggles rather than something which is actively brought into contestation through the conduct of political activity. Further, there is little sense of how geographies of connection are made and re-made through social movement activity. Thus Tarrow and McAdam's account of 'relational diffusion' argues that this 'involves the transfer of information along established lines of interaction' (Tarrow and McAdam 2005: 129). As we will demonstrate below, transnational social movement activity, however, is often produced through the ongoing formation of networks linking different places. It involves the creation of connections, often fragile articulations between humans and non-humans, rather than merely following already constituted connections.

The work of Tarrow et al., then, opens up important aspects of the processual character of transnational contention. We have argued that this work, however, tends to reduce contention to forms of generic processes and mechanisms, rather than engaging with the productive and dynamic character of such contention. We contend that it is necessary to engage with a more open and productive account of the processual constitution of transnational social movement organizing. To develop such an account the next section turns to work which adopts a more generative engagement with the political and transnational social movement organizing.

Transnational politics as relationally networked

Jeffrey Juris in his account of transnational resistances to corporate globalization adopts what he terms a practice-based approach to the study of networks which links 'structure and practice to larger social, economic and technological forces' (Juris 2008: 11). This approach to the study of transnational organizing foregrounds its processual character, but in ways which are more generative and alive than the accounts of Tarrow et al. This allows a more open-ended and constitutive account of social movement organizing. This is also foregrounded through Juris's fine ethnographic work. This section draws on Juris and others to develop a more generative account of the activity of transnational organizing. This is part of a broader project to develop an explicitly relational approach to forms of transnational organizing. We contend that the following issues are central to such a theoretical project.

Firstly, it is necessary to assert that transnational political movements/alliances are not the products of a singular network, but are rather the coming together of different dynamic trajectories of political activity. This unsettles the tendency to counterpose place and space in accounts of such movements (see Miller 2004). It allows a focus on the relations between different place-based political movements, which are always already constituted through various interrelations. This ensures a focus on the situated practices and geographies that are constituted through transnational contentious politics, which often are occluded by a focus on processes such as scale-shifting or scale-jumping. Such theoretical imaginaries tend to treat resistances as primarily local unless movements actively 'jump scales' 'from local to regional, national or international' through their activity (Glassman 2002: 524; Smith 1993; for critiques see Amin 2002; Cox 1998: 2-3). This view of political activity fails to situate resistances as always already the product of different trajectories. Understanding how different trajectories of activity are combined and reworked through transnational political activity is a particularly important theoretical and methodological task.

Juris's work develops important accounts of the ways in which grievances are mobilized over space and time from particular 'local' issues through to large-scale resistances to corporate capitalism. These accounts produce differing trajectories and the utilization of network imaginaries by actors within dynamic systems of political action. This introduces a productive sense of the dynamism that contemporary activists are involved in as they mobilize. This results in an account of the political which stresses the importance of lived practice to the constitution of politics. David Graeber, in his engagement with anti-corporate globalization activists (2002, 2009) has also been important in understanding how democratic ideals are constituted and transformed through their enactment in the relations produced throughout the movement. A concern with how such trajectories are negotiated through political activity has been central to Davies's work on recent transnational protest events mobilized by the 'pro-Tibet Movement' in relation to the Beijing Olympics. He argues that it is necessary to understand the 'simultaneous nature of contemporary transnational politics' (Davies 2009a: 20). He notes that in 'various places, in a short period of time, events impacted upon one another, and the speed of communication between places allowed a degree of coordination to occur'. To understand the forms of connection constituted through such events necessitates an understanding of the co-production of such trajectories and their potentially productive effects.

Secondly, these networked forms of political activity are interventions in the relations between places in various different ways. This is significant in that understanding the geographies of connection produced, reworked and contested through transnational political activity offers important resources for understanding the forms of identity, solidarity and agency constituted through such activity. Engaging with the conduct of such activity necessitates dislocating the frequently human-centred accounts of the social and political mobilized through work on social movements. Thus social movement theory, including work on environmental politics, and much work on the geographies of resistance, remains firmly within a restrictively human-centred account of the political (see Miller 2004; Tarrow 2005). Understanding social/political activity as co-produced by a range of actants including both humans and non-humans opens up significant possibilities for accounts of transnational contentious politics. Below we note the importance of delegated forms of relations through post and paper in maintaining the ongoing construction of Tibetan transnational politics. Anti-Coca-Cola politics has been constituted through a range of materially heterogeneous politics which involves the contestation of the effects of water extraction on water-tables and the dumping of sludge in communities in India where Coca-Cola plants are located. Taking the non-human seriously in this way opens up hitherto ignored or marginalized aspects of social and political activity. This opens up important possibilities in thinking about the relations between place and geographies of connection.

Thirdly, situating transnational political networks as the coming together of dynamic political trajectories emphasises the multiple and contested character of their activity. Too much emphasis on emulation, similarity and common understandings closes down a focus on the radical plurality of political identities brought together through opposition to neo-liberal globalization. The constitution of transnational political networks can be much more fissiparous, contested and productive than this focus on common understandings suggests. The contested and power-filled geographies produced through such transnational networks have been foregrounded by the work of feminist scholars such as Janet Conway and Chandra Mohanty. Mohanty has argued that 'global justice movements' have tended not to make the unequal gender relations that are produced and reproduced through neoliberalism central to their analysis (Mohanty 2003: 249-250). Janet Conway has noted the very different forms of transnational feminism that have been constituted through processes like the *World March of Women* and at the *World Social Forum* (see Conway 2008). Foregrounding the multiple and contested geographies through which such transnational politics is constituted opens up both theoretical and political possibilities. Thus Juanita Sundberg has argued that understanding forms of 'mutual solidarity built from embodied experiences makes alliances between differently situated actors struggling against unequally constituted geometries of power more possible' (Sundberg 2007: 162). This allows a focus on the diverse forms of identity produced in relation to transnational organizing. We signal this here by engaging with the very different political identities generated through anti-Coca-Cola politics and that coalesce around the Tibet Issue.

Fourthly, forms of transnational organizing can have multiple effects. Understanding and accounting for such effects disrupts the very limited 'goal' oriented notions of political 'calculation' that have often structured accounts of transnational political alliances or movements. Tarrow, for example, draws on a definition of coalitions as '[c]ollaborative, means-oriented arrangements that permit distinctive organizational entities to pool resources in order to effect change' (Levi and Murphy, cited by Tarrow 2005: 164). The problems with such narrow ways of evaluating the effects of alliances and solidarities is that they close down a focus on the more open-ended effects of such political activity. This is not to argue that these goals are not important, or that the goals identified by particular movements don't matter in their own right. However, we contend that evaluating social movement's success and impact in narrow terms around such goals can be problematic and miss key effects and forms of agency that movements constitute. This involves problematizing the notion of the political that informs dominant work on social movement theorizing.

Finally, then, we contend that it is necessary to take the political seriously in engaging with transnational organizing. Social movement theory has tended to marginalize the political in various ways. Thus, Leitner et al. note that McAdam, Tarrow and Tilly's conception of contentions politics is state-centric and interest-oriented (Leitner et al. 2008: 157). The political has also been occluded in recent work on neo-liberalization in geography which configures the political and resistance as resolutely secondary to a spatial terrain already shaped by powerful and dominant political-economic processes (see Castree 2008; Peck and Tickell 2002). Interrogating the constitution of the political permits an understanding of how the terrain on which political activity is constituted and performed can be reshaped through such action. This is one of the reasons why the reduction of the activity of social movements to a set of 'identifiable processes and mechanisms' by Tarrow et al. is limiting to understandings of the forms of agency and identity produced through transnational social movement organizing. Their way of positioning such processes and mechanisms forecloses the generative possibilities shaped through such activity and to close it down around existing categories.

An alternative position for understanding the generative character of such activity is opened up by different understandings of the political associated with the work of post-structural theorists of the political such as Jacques Rancière. Rancière has drawn attention to the ways in which the political involves not the negotiation of already constituted, fixed interests, but 'the continual renewal of the actors and of the forms of their actions' (Rancière 2007: 61). The potential of such approaches has been signalled for the understanding of transnational forms of political mobilization by Benjamin Arditi. He argues that:

> Social movements, advocacy groups and NGOs are opening up a second tier of politics in civil society, and the new internationalists are building its supranational tier. Yet this expansion is not a mere arithmetic sum, for it also modifies what Rancière calls the partition of the sensible and therefore transforms the way the political field is coded; it creates a new condition of the sensible (Arditi 2007: 145).

Attending to the ways that political action can transform the way that the political field is coded and the way that the terrain of the political is constituted is central to engaging with the dynamic practices of transnational organizing. Thinking about these forms of politics in explicitly spatial terms necessitates understanding the spatial practices through which such processes are constituted and re-constituted. The following section engages with some of these practices through discussion of transnational mobilizations around Coca-Cola and the Tibet issue.

Power as (re)produced through engagement

We contend that in order to address the spatial construction of social movements/ alliances, it is necessary to recognize that the processes of transnational organizing are productive of new relations. These new relations shift dominant power geometries and can redraw maps of grievance previously held relatively stable. We use the term maps of grievance to signal the productive practices through which actors engage with and contest spatially stretched relations of power (see Featherstone 2008). It is through the processes of negotiation and mediation inherent in the movement of social movements that we see these processes at work. The politics of the Coca-Cola boycott and the 'Tibet Issue' all seek to shape and challenge particular orderings of the world. However, the coming together of the various actors within these movements necessarily shifts and alters the power relations at work. In this section, we use the example of the anti-Olympic Torch protests that erupted in early April 2008 to show how political actions reshape the power relations at work in them.

On a crude level, power could simply be codified as a type of mass. The increased numbers of people who turned out in the UK, France and North America to protest along the route of the Beijing 2008 Olympic Torch in London, Paris and

San Francisco represented an upwelling of support and the coming together of a mass of people. In London, while blocks of Tibet supporters challenged the state authorities policing the route, and came into contact with members of the Chinese community who had been mobilized in support of the Torch Relay, the mass of Tibet supporters would seem to be a homogenized reflection of a great upwelling in support of Tibet. In a time when, for many of the previous years, pro-Tibetan events and demonstrations in the UK were attended by at most a few hundred Tibetans and their supporters, these mass protests represented something of a pinnacle of protest within the pro-Tibet Movement. Thousands of people gathered along the parade route in central London, and with many of them waving brightly coloured Tibetan flags the seeming co-option of people into the Movement was smooth and unproblematic (see Figure 11.1). Indeed, this seemingly fits clearly into a diffusionist model of social action, with the core of the movement promoting a series of issues which the adoptees take up unproblematically.

Figure 11.1   Pro-Tibetan protesters on Whitehall, London, April 2008

This process becomes more complicated when comments at a similar demonstration a few weeks earlier are taken into account. Here, while waiting for the march to begin, a protester on the protest spoke to one of us (Davies) about his involvement with the Tibet Movement in the UK: 'Well, I've been a member of this [a UK Tibet Support Group] for years. But it's only now, when things have become really bad inside Tibet, that I've thought, "Right, enough is enough", and had to come down here [to London] and do something about it.' Here, we see that there are degrees of involvement within a particular movement. The seductive power of a Tibet Support Group was enough to draw in a member who was content simply to pay their membership fees but not make an outright bodily commitment to demonstrating against China's perceived injustices in Tibet. Thus, rather than being a wholesale adoptee of the Tibet Movement, this supporter occupied an indeterminate place, where he was neither co-opted wholesale into the movement, nor completely ambivalent towards it. It was only with the advent of a perceived series of injustices that this person became willing to commit in a more visible way (by being at a pro-Tibetan demonstration) to the movement. The upsurge in support for Tibet in spring 2008 could seemingly be read as representing a degree of diffusion, increasing the mass of the pro-Tibet Movement at one particular scale.

However, at the same time, these comings together challenged conceptions of protest events as homogeneous expressions/demonstrations of dissatisfaction. While the protests included many supporters who were usually content to sit in the background, as with the individual above, there were also a number of people who were drawn into the movement at this particular occasion for a variety of other reasons. These ranged from middle-aged, middle-class professionals who expressed themselves in terms of supporting Human Rights issues through to groups of teenagers who had to ask where Tibet was and what the purpose of the rally against the Beijing Olympics was. This is to say nothing of the variety of other campaigners such as Burmese pro-democracy campaigners, and practitioners of Falun Gong whose campaigns were relatively invisible under the weight of the Tibet supporters' mass protest. In the aftermath of the events, when speaking to Chinese associates about these events, most expressed a belief that the anti-Olympic protests were filled with predominantly Western people who had little or no understanding of the Tibet issue from a Chinese or a Tibetan perspective.

At the same time, activists within the Movement spoke of this time as an opening up, and a reworking, of the Tibet issue's political landscape. The upwelling of support was firstly seen as a justification of the ongoing struggle, with the Tibet issue's continuing unresolved status meaning that at times the movement has suffered from a lack of public visibility or momentum, particularly since the last significant wave of protests in the late 1980s and early 1990s. Secondly, however, this upwelling provided a key moment of contention for the Movement as a whole. Speaking to activists who had been engaged long term with the Movement, the Olympics and the protests in March/April 2008 were seen as important 'moments' within the struggle. The ability to shift attention in 'wider civil society' onto Tibet, and the ability to increase pressure on pro-Chinese lobbies meant that pro-Tibet activists saw these protests as particularly important. Indeed, the small number of dedicated Tibet activists engaged in the Movement on a day-to-day basis meant that the sheer number of working hours spent organizing the protests, reworking the Beijing Olympics into a political as well as a sporting occasion, had many activists approaching exhaustion.

Here, the exact details about who is right and who is wrong in this debate are not important, instead, we wish to emphasise that this series of events both reaffirmed and reworked a series of power relations. Through this coming together of a variety of actors who are more or less relatively well connected to particular aspects of the Tibet issue, the Movement alters and shifts. What legitimacy it may have had from the viewpoint of the Chinese state is diminished as uninformed actors become involved. Yet, at the same time, the growing media presence and the increasing number of people who are willing to 'do something' for Tibet seemingly increases both the mass and visibility of the movement. Crucially, these activists are also attempting to invest certain activities, in this case, the supposedly non-political Olympic Games, with an explicitly political edge. This reworking of a political landscape crucially alters and shifts the basis of any future negotiations of the Tibet issue. It is through the movement of these contested constellations of discourses, actors and materials that we argue that power is actualized. Rather than equating the Tibet protests as a simple upwelling of support, we argue that events like this, and the constantly changing parameters of most social movements, reshape and reorder power relations. New actors are drawn in, and others are made defunct. It is through these spatial entanglements of power (cf. Sharp et al. 2000) that we see transnational organization and resistance being constituted.

Contested articulations and spatial particularity

This understanding of the multiple geographies of power constituted and challenged through transnational organizing has implications for understanding the forms of articulation and solidarity generated between different sites of struggle through movements. One of the defining features of transnational mobilization against Coca-Cola has been the way it has networked different sites where Coca-Cola's actions have been brought into contestation producing a multi-faceted set of grievances in relation to Coca-Cola. Here we understand articulation, not merely as a purely discursive process, but as a set of spatially and materially heterogeneous practices through which different sites and movements become connected (see also Featherstone 2010). These practices, as this section argues, are generative and contested and are worked through militant-place-based activity.

Coca-Cola has been contested on various grounds and in various places. Coca-Cola has been the subject of mobilization around the rights of trade unionists because of its links to the assassination of trade unionists in Colombia. As with the record of other multinationals in Colombia, such as BP, there is strong evidence of both complicity and collusion by multinationals with attacks by paramilitaries on trade unionists (Gill 2007; Higginbottom 2007; Novelli 2004). SINALTRAINAL, the local branch of the food and drink workers union, has been active in bottling plants and initiated calls for a boycott in 2003. There have been nine assassinations of unionized workers from Coca-Cola plants (Higginbottom 2007: 2). These include the assassination of Isidro Segundo Gil who was murdered by paramilitary forces inside the Carepa plant of Bebidas y Alimentos' bottlers/ franchise for Coca-Cola (United Steelworkers Union and the International Labor Rights Fund 2001; Higginbottom 2007; War on Want 2006: 8). Coca-Cola's anti-trade-union practices have also been contested in Guatemala, Nicaragua, Peru, Russia and Turkey (War on Want 2006: 8-9). The boycott is part of a longstanding transnational mobilization against Coca-Cola, which included the concerted international organizing in support of workers in the Coca-Cola bottling plant in Guatemala City in 1979 and 1980 (Levenson-Estrada 1994: 190-192, 203).

Coca-Cola has also been contested due to its poor environmental record. There have been major campaigns against Coca-Cola in India due to the role of Coca-Cola plants in exacerbating water shortage. The most high profile of these has been the ongoing campaign in Plachimada, a tiny hamlet in the state of Kerala. Here a diverse coalition has campaigned against the effect of the company's plants on water supplies. The multinational has been accused of 'creating severe water shortage, of polluting its groundwater and soil, and also of distributing toxic waste as fertiliser to farmers in the area' (Ravi Raman 2005: 2481). Organizing in Plachimada brought together subaltern groups such as adivasis and dalits with human rights and left activists. The Coca-Cola Virudha Janakeeya Samara Smithi (Anti-Coca-Cola People's Strike Committee) was formed which comprised both Gandhian as well as moderate-radical groups such as the SUCI, Pottoram, and Ayyankali Pada, the latter Maoist outfits, in addition to the subalterns whose livelihood had come under direct threat (Ravi Raman 2008: 85).

This demonstrates how such militant particularisms, to use Raymond Williams's term for militant place-based politics, are constructed out of different connections and relations (see Williams 1989). As Ravi Raman argues in relation to the activists in Plachimada this generates alternative, subaltern articulations of cosmopolitanism (Ravi Raman 2008). These situated and antagonistic articulations of cosmopolitanism are constituted in very different ways from the elite spatial practices often associated with cosmopolitanism (see Ley 2004). The activists contesting Coca-Cola in Plachimada in Kerala, for example, would not generally be considered 'cosmopolitan' in the sense of leading the hyper-mobile lives that are associated with elite articulations of cosmopolitanism. Central to their politics, however, is an engagement with the unequal connections that stretch beyond Plachimada. Articulations between these different sites and militant particularist struggles have generated transnational forms of organizing against Coca-Cola.

Such organizing has been constituted through different practices and as responses to different tensions and local conditions. Thus SINALTRAINAL's leadership in response to the violent repression of organization has 'internationalized its conflict with the Coca-Cola Company as a human rights issue through the development of new relationships with trade unions, student organizations, lawyers, solidarity groups and intellectuals in various countries, and it has also forged alliances with various groups in Colombia' (Gill 2007: 238). This has generated different activities through which Coca-Cola has been brought into contestation. These include a law-suit in the US brought against Coca-Cola for its alleged links to the assassination of Isidro Segundo Gil by the United Steel Workers Union and the International Labor Rights Fund (United Steel Workers Union and the International Labor Rights Fund 2001).

There have also been significant campaigns on US campuses to divest themselves of Coca-Cola. A campaign at the University of Michigan lost Coca-Cola the lucrative contract for supplying soft drinks to the University of Michigan campus which was estimated to be worth $1.27 million (Hardikar 2006). Students have now 'thrown Coke products out of [at least] thirteen colleges and universities' in the US (Higginbottom 2007: 7). Student associations in Ireland, Italy and the UK have taken similar action. Most recently colleges and universities in Norway have decided not to renew an exclusive contract with Coca-Cola due to its 'mismanagement of water resources in India' (India Resource Center 2009). In India the *National Alliance of People's Movements* has been influential in networking opposition to Coca-Cola mobilizing through sites such as the World Social Forum where representatives from SINALTRAINAL have also spoken (Ravi Raman 2005: 2484; Gill 2007). The role of websites such as 'Killer Coke' has also been crucial in facilitating these forms of transnational organizing.

These forms of opposition to Coca-Cola are not just defined by practices of emulation. They are more productive than this suggests. They have brought together different forms of grievance and articulated together concerns frequently held apart, such as environmentalism, inequality and labour rights. This signals the productive spatial practices through which militant particularist actions like the struggles against Coca-Cola in Plachimada can be networked and constituted in relation to geographies of connection. In this sense these solidarities are actively produced through contesting the generation of transnational networks and the heterogeneous relations between 'commodities', labourers, water tables, trade unionists that they produce. In this sense the politics of the Coca-Cola boycott is both spatially and materially heterogeneous. Further, such solidarities and connections have effects on how militant particularist actions are envisioned. Thus Ananthakrishnan Aiyer has argued that 'Plachimada's struggle against the Coca-Cola Company' was and 'is important in having raised serious issues about the role of transnational corporations and globalization in India. It is now firmly tied in the hearts and minds of many activists to the struggles against the Coca-Cola Company in Colombia ... or the victory against Bechtel in Cochamba, Bolivia'. Further 'their struggle has been inspirational and played a significant role in generating opposition to The Coca-Cola Company and PepsiCo in other rural communities in India' (Aiyer 2007: 652). This emphasizes the productive character of political activity. The ways in which the Coca-Cola boycott has generated articulations of politics around inequality and environmental politics suggests how the terrain of the political can be re-shaped through the conduct of alliances.

This is not, however, to imply that the work of constructing such articulations is smooth, consensual or free of contested power-relations. Hardt and Negri's (2004) account of the 'multitude' have over-played the 'smoothness' through which diverse groups and collectives are constituted and act. Different tensions have emerged through the formation of such transnational forms of organizing which bear on the ways grievances are constructed and the conduct of solidarities. In a discussion of the relationship between the Coca-Cola boycott and anti-racist activism, for example, Ashwini Hardikar argues of the coalition she was involved in at the University of Michigan that:

> Although the coalition looks very 'representative' – that is, students of color and women are at the center of the coalition – the process through which the coalition was built over the period of a year has been a point of contention among some. I do believe that this campaign was initiated with the intention of synthesizing anti-racism and anti-globalization struggles, yet to some the outreach and coalition-building process seemed to put the racist aspects of Coke's violations as more of an afterthought rather than at the center of the movement (Hardikar 2006: n.p.).

Hardikar's analysis of the process of coalition building against Coca-Cola usefully indicates that even when coalitions look 'representative', practices of marginalization can still exert pressure on the practices through which maps of grievance are shaped and defined. Thus she argues that the coalition didn't make anti-racism central to the contestation of Coca-Cola, and that there was a disjuncture between the antiracist and antiglobalization aspects of the coalition. The formation of particular gender relations through such organizing has also been noted. Thus Gill argues that SINALTRAINAL has had 'little to say about the particular vulnerabilities of working-class women' (Gill 2007: 238). This raises important questions about the terms on which different groups are brought together through transnational organizing. It also emphasizes contra Hardt and Negri that differently placed political cultures have impacts on the ways in which such transnational connections are constituted. The next section develops these concerns with the constitution of transnational connections by interrogating some of the practices through which such connections are produced and maintained.

Production/maintenance of connections

In contrast to accounts of social movement activity which have focused primarily on their most visible and dramatic manifestations we also wish to emphasize that much social movement organization is done through the maintenance of national and transnational connections. Developing and fostering arrays of connections is something that is inherently difficult and we see within these networks a series of systems which are developed and organized in order to create stability within a given movement. Thus, rather than emphasize the greater speed and fluidity allowed by transnational organizing, we recognize that it is through a series of systems and structures that social movements are maintained and allowed to continue mobilizing their grievances. To illustrate, we examine a set of particular procedures that help to maintain one network within the pro-Tibet Movement.

> It's my first day working with [a London-based Tibet Support Group (TSG)], and they are struggling to think of a task that is suitable for me. It's mid-afternoon and the Office Manager is struggling with processing the latest campaign appeal slips that have been sent in by members and donors, so asks me to help out. The work consisted of processing cheques for one of the campaigns which they've just started – one for two Tibetan brothers who've just been imprisoned inside Tibet.
>
> This processing consisted of sorting out postal donations into different piles of documents. Firstly, the donation slip they had initially been sent and asked to fill in with their details. These are coded to record what type of supporter they were and were marked with a date they had been received. These were sorted into those who had donated by card or by cheque. Occasionally I found a second element – a completed postcard written in Chinese which the individual was supposed to send directly to China, but obviously people had often returned them to the TSG instead. The completed forms were then sent along to the database section, where their individual records were updated to show what they had contributed and to which campaign. The cheques were then processed and put into this specific campaign's account. The amount of money donated varied from coins taped to the form through to hundreds of pounds, and while many had simply sent a form and a cheque, occasionally people had attached a note expressing their anger or sorrow at the situation. Some would say how they were praying to God for a resolution for Tibet and Tibetans. Once a letter spoke of how the individual could not afford to donate any more than a small amount at this time in an attempt to justify their donation to the organisation.[1]

The above account attempts to highlight some of the deliberate systematization that formed one particular network of the UK-based TSG. It is worth going through this exact process in some detail as it highlights the complicated series of connections that a campaign produces and forms. At one level this system is created to ease the process of dealing with a campaign about a specific issue and its associated paperwork, and could simply be read as a process of good and accountable housekeeping by a political organization. However, there is a greater degree of both connection and separation going on here.

Having negotiated the intricacies of the postal system and its own set of networked relations, they are then (presumably) opened and read by these individuals, who then decide whether they want to donate to this particular campaign. Some will send the postcards, some will donate money, some will do both, some will do nothing, but each will act in their own particular way to these mailings. As such these mailings act as a means of connecting the members to a

---

1   Adapted from Research Diary – ResearchDocs/Diaries/UK/FTC/051006.

TSG, and form an important part of maintaining the bond between members and the TSG, however temporary and partial this bond turns out to be. As shown by the notes attached to some donations, these contacts allow a point of emotional, affective connection – those people who write in to explain why they cannot donate more obviously display a connection to both the Tibet Issue, and to this particular TSG and this particular campaign that is strong enough to provoke the need to justify their position. This could be down to a sense of guilt or another moment of affective contact, but what is clear is that these mailings provide a moment of relative intensity in the relational networks between TSG and supporters/members.

At the most basic level, these moments are a way of reaffirming and reproducing relations (both of connectivity and of separation) to the widest possible audience, and keeping people informed of the current struggles and campaigns helps to maintain a certain basic level of support. Members of TSGs can range in numbers from hundreds to tens of thousands, so mailshots like this serve to remind those who are less active that they are members of the organization and provide a stimulus to those who are more active. The regular production of new campaigns and associated materials is also an attempt to show the members that the TSG is active and that progress is being made on the Tibet issue, something that is not always readily apparent given the deadlock on negotiations. Thus the targeting of small, reasonably achievable goals like a campaign to release two political prisoners in Tibet is a way of reassuring members that their money is not going to waste.

In addition, the flows and connections created by the sending of this campaign leaflet also serve to highlight the immanent power relations built up within this particular TSG. The TSG in sending out these flyers is attempting to persuade people to conform to its campaign goals and commit to those goals. However, once these objects have left the office, their power to elicit a response from people emerges from the circumstances they arrive in. Those people who feel compelled to give an amount, no matter how small, are willingly complicit in allowing the TSG to inform them of the issue (in this case the two political prisoners) and the best way to react. However, those who do not react in such a conformist way are playing out their own relationship with the objects and, by extension, with the TSG. This power only arises in their particular space or place, and thus helps to determine the extent of the network. Thus, each mailshot will give a different impression of the TSG and its influence. Those campaigns that are more effective at getting people to donate will give the impression of a much wider network of connections, while those that are less fruitful will obviously give the impression of a smaller, more incoherent system.

However, the mailshot also becomes part of a system that produces connections to the Chinese and is therefore meant as a direct challenge to Chinese practices in Tibet. In this case, the inclusion of a postcard is also a way of connecting the individual to the struggle directly. Most campaign material asks people to write to the relevant Chinese authorities, the State Security Bureau, the Chinese Embassy etc. and to make their grievances known. This spreads the map of grievance allowing non-Tibetan people to legitimately call for change in what China perceives as its internal affairs. Indeed, much TSG literature is explicit in its attempts to show that letter writing can force changes in the cases of individual political prisoners in Tibet. Here though, the additional postcard is designed to simplify this process. The postcard removes the need for the individual to write a letter to the relevant authorities. Instead, all they have to do is attach a stamp and put the postcard in a post-box. By believing that they can influence and change the terms of struggle by writing a letter or sending the postcard, these individuals make a seemingly greater investment in the campaign than those who simply donate money. The deliberate use of Mandarin Chinese represents an attempt to ease communication with the Chinese authorities, opening up new ground for grievances to be aired. Presumably this allows everyday Chinese postal workers to read the postcards as well as the intended recipients, and thus creates a whole new set of connections for pro-Tibetan issues to be raised.

A final element of this particular network we want to highlight lies in the act of performance. By sending out regular updates and campaign requests, the TSG is *performing* a prescribed role – that of a campaigning TSG – and thus by doing this, it is proving to its membership that it is a valid organization that actually does things (whatever those things may be) and thus helps to 'shore up' people's belief in the TSG and its role as an important player in the Tibet issue. Thus, these micro-processes that occur within the organization become constitutive of the organization as a whole. The projection and performance of this constitution then helps to maintain the organization as a stable entity – as mentioned in an earlier example, during the time one of us (Davies) worked at a particular TSG there were lots of staff changes and a relative level of fluidity in the organizational structure of the group. However, the sending of the campaign material, which contained the TSG's logo and other standardized information such as addresses, meant that the TSG remained a stable organization to those who were not in immediate contact with it. Despite the movement of people and objects through the office, the perception and performance of the organization remains the same as certain key markers are deployed and redeployed to maintain a semblance of order.

This one example of mailing shows how the organizational networks that extend out from a TSG act as spaces of flows and movement that attempt to both extend and intensify relations within the TSG's membership. Thus, while seeming to give stability to the organization, in practice they are all about movement and change; the creation of rigid organizational structures is designed to help speed the flows of information between different networks. Each item of post constitutes its own assemblage that has different sets of relations as it moves through and around the TSG and its members. These relations are bound up with immanent regimes of power, as people dictate their compliance or resistance to the TSG's campaign according to their own set of spatial relations. Thus, the simple mailing of information to members becomes an infinitely more complex process of connection and negotiation and the seeming coherence of the organization becomes radically decentred and materially heterogeneous (after Law 1994). Rather than simply the

TSG and its staff contained by the walls of its offices, the various elements within and beyond this relatively intimate office space become complicit in the Tibetan Freedom Movement.

Such processes emphasize the day-to-day work and effort necessary to keep these processes of contestation and solidarity functioning. We argue that similar processes of connection and negotiation form a crucial part of any form of resistant organization, and it is through the workings of these networked systems that we can begin to understand the practice of political action. This, together with the previous empirically led sections, have shown what we believe to be the benefits of a spatially nuanced account of transnational organizing. The examples drawn from the pro-Tibet Movement and the transnational resistance to the Coca-Cola Corporation allow us to see that there is a good deal of insight that a more spatially sympathetic reading of politics can give us. However, to conclude this chapter, we move on to thinking about the effectiveness of political action.

Conclusions

This chapter has foregrounded the productive and contested spatialities of transnational social movement organizing. Here we conclude by exploring issues in relation to the effectiveness of transnational political networks in facilitating and hindering political actors in shaping durable alliances and achieving defined political goals. Both the pro-Tibet Movement and the movement against Coca-Cola have not been as successful as they would wish. In the case of Tibet, a campaign that effectively started with the move of the Dalai Lama into exile in 1959 has seemingly achieved little of its concrete aims of greater autonomy in ethnically Tibetan regions of the People's Republic of China. Similarly, while anti-Coca-Cola protesters have been relatively successful in making visible Coca-Cola's relation to human rights abuses and environmental degradation in different parts of the world, the power of the corporation and its ability to colonize legal and political processes has made organizing against it difficult.

We contend that making injustices visible, making unequal relations of power part of the terrain of contestation, is itself productive and a significant outcome of transnational of forms of organizing. The ability of an organization like the Tibet Movement – which, in terms of activists/supporters remains numerically small despite the upsurge of 2008 – to continue to promote its cause at any level can be seen as a degree of success in its own right. The fact that these issues continue to garner support despite a political process that seems to have stagnated means that the discursive arguments being deployed are effective to some extent.

We also wish to emphasize the unexpected and contested routes through which these solidarities and resistances must work. The variety of responses to a simple campaign leaflet detailed above show that politics do not play out as we would expect. Issues like this call into question Tarrow's and others' conceptions of diffusion and scale shift as part of a toolkit with which to study social movements. They also unsettle the very limited goal-oriented ways of assessing the effectiveness of political activity. We contend that following the activity of movements allows for more generous forms of calculation of what counts as effectiveness and can foreground significant aspects of social movement organizing that are frequently marginalized or ignored. Further, we argue that a more spatially nuanced understanding of the processes of social movement action foregrounds the multiple, dynamic and contestable character of such processes. In short, attention to spatial nuance opens up politics as a realm where the work of the everyday becomes important in shaping and reinventing political concerns, through and across space.

References

Aiyer, A. (2007) 'The allure of the transnational: notes on some aspects of the political economy of water in India', *Cultural Anthropology*, 22 (4): 640-658.

Amin, A. (2002) 'Spatialities of globalisation', *Environment and Planning A*, 34 (3): 385-399.

Arditi, B. (2007) *Politics on the Edges of Liberalism: Difference, Populism, Revolution, Agitation*. Edinburgh: Edinburgh University Press.

Castree, N. (2008) 'Neoliberalising nature 2: processes, outcomes and effects', *Environment and Planning A*, 40 (1): 153-173.

Conway, J. (2008) 'Geographies of transnational feminism: the politics of place and scale in the world march of women', *Social Politics*, 15 (2): 207-231.

Cox, K. R. (1998) 'Spaces of dependence, spaces of engagement and the politics of scale, or: looking for local politics', *Political Geography*, 17 (1): 1-23.

Davies, A. D. (2009a) 'Ethnography, space and politics: interrogating the process of protest in the Tibetan Freedom Movement', *Area*, 41 (1): 19-25.

—— (2009b) *Networks of Transnational Tibetan Politics*. PhD University of Liverpool.

Featherstone, D. J. (2008) *Resistance, Space and Political Identities: The Making of Counter-Global Networks*. Oxford: Wiley-Blackwell.

—— (2010) 'Contested relationalities of political activism: the democratic spatial practices of the London Corresponding Society', *Cultural Dynamic*, 22 (2): 87-104.

Featherstone D., R. Phillips and J. Waters (2007) 'Introduction: spatialities of transnational networks', *Global Networks*, 7 (4): 383-391.

Gill, L. (2006) 'Fighting for justice, dying for hope: on the protest line in Colombia', *North American Dialogue*, 9 (2): 9-13.

—— (2007) '"Right there with you": Coca-Cola, Labor Restructuring and Political Violence in Colombia', *Critique of Anthropology*, 27 (3): 235-260.

Glassman, J. (2002) 'From Seattle (and Ubon) to Bangkok: the scales of resistance to corporate globalization', *Environment and Planning D: Society and Space*, 20 (5): 513-533.

Graeber, D. (2002) 'The New Anarchists', *New Left Review*, 13, 61-73.
—— (2009) *Direct Action: An Ethnography*. Edinburgh: AK Press.
Hardikar, A. (2006) 'Declaring (incomplete) victory: the Coke campaign, corporate accountability & people of color movements', http://criticalmoment. org/issue15/coke, accessed 15th June 2007.
Hardt, M. and A. Negri (2004) *Multitude*. London: Hamish Hamilton.
Higginbottom, A. (2007) 'Killer Coke', in Dinan, W. and D. Miller (eds) *Thinker, Faker, Spinner, Spy: Corporate PR and the Assault on Democracy*. London: Pluto Press, pp. 278-294.
India Resource Center (2009) 'Norway campuses reject Coca-Cola contract not renewed, ethical concerns raised in decision' http://indiaresource.org/ news/2009/1036.html.
Juris, J. (2008) *Networking Futures*. Durham and London: Duke University Press.
Kaldor, M. (2003) *Global Civil Society*. Cambridge: Polity Press.
Law, J. (1994) *Organizing Modernity*. Blackwell: Oxford.
Leitner, H., Sheppard, E., and K. Sziarto (2008) 'The spatialities of contentious politics', *Transactions of the Institute of British Geographers*, 33 (2): 157-172.
Levenson-Estrada, D. (1994) *Trade Unionists Against Terror: Guatemala City 1954-1985*. Durham, NC: University of North Carolina Press.
Ley, D. (2004) 'Transnational spaces and everyday lives', *Transactions of the Institute of British Geographers*, 29 (2): 151-164.
McAdam, D., S. Tarrow and C. Tilly (2001) *Dynamics of Contention*. Cambridge: Cambridge University Press.
Miller, B. (2000) *Geography and Social Movements*. Minneapolis: University of Minnesota Press.
—— (2004) 'Spaces of mobilisation: transnational social movements', in Barnett, C. and M. Low (eds) *Spaces of Democracy: Geographical Perspectives on Citizenship, Participation and Representation*. London: Sage, pp. 224-246.
Mohanty, C. T. (2003) *Feminism Without Borders: Decolonizing Theory, Practicing Solidarity*. Durham, North Carolina: Duke University Press.
Nicholls, W. (2007) 'The geographies of social movements', *Geography Compass*, 1(3): 607–622.
Novelli, M. (2004) 'Globalisations, social movement unionism and new internationalisms: the role of strategic learning in the transformation of the Municipal Workers Union of EMCALI1', *Globalization, Societies and Education*, 2 (2): 161-190.
Peck, J. and Tickell, A. (2002) 'Neoliberalizing space', *Antipode*, 34 (3): 380-404.
Rancière, J. (2007) *On The Shores of Politics*. London: Verso.
Ravi Raman, K. (2005) 'Corporate violence, legal nuances and political ecology: Cola war in Plachimada', *Economic and Political Weekly*, June 18, 2005, 2481-2485.
—— (2008) 'Environmental ethics, livelihood and human rights: subaltern-driven cosmopolitanism?', *Nature and Culture*, 3 (1): 82-97.

Routledge, P. (1993) *Terrains of Resistance: Nonviolent Social Movements and the Contestation of Place in India.* Westport, CT: Praeger.

Routledge, P. and A. Cumbers (2009) *Global Justice Networks: Geographies of Transnational Solidarity.* Manchester: Manchester University Press.

Routledge, P., Nativel, C. and A. Cumbers (2006) 'Entangled logics and grassroots imaginaries of global justice networks', *Environmental Politics*, 15 (5): 839-859.

Sharp, J., Routledge, P., Philo, C. and R. Paddison (eds) (2000) *Entanglements of power: geographies of domination/resistance.* London: Routledge.

Smith, N. (1993) 'Homeless/global: scaling places', in J. Bird, B. Curtis, T. Putnam, G. Robertson, and L. Tickner (eds) *Mapping the Futures: Local Cultures, Global Change.* London: Routledge, pp. 87-119.

Sundberg, J. (2007) 'Reconfiguring North–South solidarity: critical reflections on experiences of transnational resistance', *Antipode*, 39 (1): 144-166.

Tarrow, S. (2005) *The New Transnational Activism.* Cambridge: Cambridge University Press.

Tarrow, S. and D. McAdam (2005) 'Scale Shift in Transnational Contention,' in D. Della Porta and S. Tarrow (eds) *Transnational Protest and Global Activism.* New York: Rowman and Littlefield, pp. 121-147.

United Steelworkers Union and the International Labor Rights Fund (2001) Campaign to Stop Killer Coke. http://killercoke.org.

War on Want (2006) *Coca Cola: The Alternative Report.* London: War on Want.

Williams, R. (1989) *Resources of Hope.* London: Verso.

Chapter 12

Global Justice Networks: Operational Logics, Imagineers and Grassrooting Vectors

Paul Routledge, Andrew Cumbers and Corinne Nativel[1]

Introduction: Global Justice Networks: Diversity and Operational Logics

Over the past twenty years, social movements have been increasing their spatial reach in terms of constructing multi-scalar networks of support and solidarity for their particular struggles, and also by participating with other movements in broader campaign networks (e.g. to resist neoliberal globalization), in what have been termed the 'movement of movements' (Mertes 2004; Tormey 2004a). These have often come to public attention through heavily mediatized 'global days of action' whereby international summits of key neoliberal actors such as the G8, WTO and IMF have been met with collective protests. Rather than a coherent 'global justice movement', our findings lead us to support a conception of a series of overlapping, interacting, competing, and differentially placed and resourced networks (Juris 2004a), or what we term Global Justice Networks (GJNs). Through such networks, different place-based movements are becoming linked up to much more spatially extensive coalitions of interest.

We term such formations Global Justice Networks (GJNs) because such networks, and the movements that comprise them, articulate demands for social, economic and environmental justice. Although precise conceptions of justice will differ between movements and unions in different cultural and political-economic circumstances, the participants in GJNs share common claims to broadly defined notions of justice pertaining to issues of redistribution (e.g. class) and recognition (e.g. identity), seeking to address and correct inequitable outcomes (e.g. concerning economic development) and the underlying processes that give rise to them (see Fraser 1997). Such notions of justice within GJNs act as a master frame enabling different themes to be interconnected and convincing different political actors from different struggles and cultural contexts to join together in common struggle (della Porta et al. 2006). These notions of justice require knowledge of processes of inequality and injustice in the world and activists' personal involvement in attempting to transform them, and incorporate all scales of transformative action

---

1   An earlier version of this chapter appeared in *Environmental Politics* 2006, 15, 5.

– from the personal, the community, the state – into international arenas and institutions. Therefore, GJNs represent the ability of different movements to be able to work together without attempting to develop universalistic and centralizing solutions that deny the diversity of interests and identities that are confronted with neoliberal globalization processes.

In addition to their diversity, GJNs are frequently characterized by: creativity (e.g. the global days of action have involved a range of creative and productive activities such as counter-summits and convergence camps); the embodiment of a political vision and practice of autonomy (i.e. self-organized, cooperative, and interactive spaces and activities that seek to remain autonomous of the state, capital and organized political parties); convergence (such as when activists join together in targeting global institutions during global days of action); spatially extensive politics; the attempt to create spaces for participatory democracy; and the attempt to forge solidarities through the making of connections grounded in place- and face-to-face based moments of articulation that occur at conferences, social forums, and joint campaigns and protests (Routledge and Cumbers 2009).

However, the diversity inherent in GJNs will inevitably give rise to conflicting goals, ideologies, and strategies, and as a result, conflictual geographies of power. Such concerns are not only tied up with considerations of place and the ability to act politically in coalitions across diverse geographical scales (see Routledge 2003a), they are also associated with the (place-specific) operational logics of GJNs' participant movements. Due to the diversity of their participants, GJNs contain political, operational, and geographical 'faultlines'. These include differences between ideological and post-ideological positionalities; reformist and radical political agendas; the resource and power differences between movements from the global North and the global South; and different types of activism associated with NGOs, political parties, and direct action formations (Juris 2004a; Tormey 2004a).

Juris (2004a, 2004b, 2005) and Tormey (2004a, 2005) argue that two principal political imaginaries – vertical and horizontal – are at work within the operational logics of political formations.[2] Tormey (2005) argues that the vertical imaginary sustains an operational logic dependent upon hierarchical structures, a recognized leadership, and elections, viewing a political party or social movement as the key vehicle of social transformation. By contrast social movements that are autonomous of political parties and eschew representative political structures (e.g. direct action formations such as Reclaim the Streets and Italy's *Ya Basta*) articulate a horizontal imaginary and are not concerned with generating a distinct *place* that can be seized or captured from a ruling elite, but instead are interested in generating *spaces* that resist overcoding or incorporation into a governing ideology (Tormey 2005).

---

2   The tension between verticalism and horizontalism in revolutionary politics goes back to the antagonisms between anarchists, syndicalists and orthodox Marxists in the nineteenth century and resurfaced in the 1968 uprisings between new social movements and a 'new left' and the more established communist and socialist parties and trade unions.

Autonomous spaces are in this sense relational, open, contingent, and immanent (Juris 2004a, 2005). The World Social Forum was constituted as providing such spaces – spaces of discussion, comparison, affinity and affiliation. The Social Forums – at least theoretically – facilitate ways in which networks can coalesce, develop, multiply and re-multiply (Juris 2004a, 2005; McLeish 2004; Sen 2004; Tormey 2004b).

In reality, both operational logics are present within GJNs and the movements that comprise them (Routledge et al. 2006, 2007; Cumbers et al. 2008; Routledge and Cumbers 2009). Networks can be both topological (relational) and topographical (territorial) – the relative balance of these factors giving rise to different spatialized forms. Understanding the dynamics of network operational logics entails thinking about power relations across space and not necessarily in any one place. This relational perspective focuses on the effectiveness of connections (e.g. communication and operational links) between network actors (Massey 1994; Braun and Disch 2002; Castree 2002). In order to do this, we consider how connections are fostered and sustained within GJNs. Sidney Tarrow (2005) has noted how social movement scholars studying transnational contention need to inquire into the relationships between movement 'insiders' and 'outsiders' in forging transnational connections. In this chapter, we argue that the work of key organizers (those we term the 'imagineers') and key events constitute important 'grassrooting vectors' which literally 'produce' the network enabling it to act. In so doing, we highlight the power relations at work within GJNs, and argue that the work of the imaginers, while crucial to transnational organizing, also entangles verticalist and horizontalist operational logics within such networks. Hence, the binaries often used to conceptualize GJNs are somewhat ill-suited as they exclude a more complex analysis of actors' spatial positionings and interactions.

In order to analyse how such operational logics become entangled within the workings of GJNs, we focus upon two particular networks, People's Global Action, an international network of grassroots peasant movements, and the International Federation for Chemical Energy Mine and General Workers' Unions (ICEM), a global union federation (GUF) which brings together around 400 affiliate trade unions. Our research focused upon the Asian part of PGA which has been the most active in recent years and the European affiliates within the ICEM network, which are the source of most of the funds and decision-making power.[3]

---

3 Participant observation research within PGA was conducted in South and Southeast Asia during 2002-2004 by Paul Routledge. This consisted of working on and attending a PGA Asia conference (in Dhaka, Bangladesh, 2004), conducting workshops with activists and attending movement meetings, and interviewing activists in the participant movements of PGA Asia. Andy Cumbers and Corinne Nativel conducted a series of semi-structured interviews with representatives of ICEM in Brussels and with union affiliates in four case study countries (France, United Kingdom, Norway and Germany) in 2004-2005.

People's Global Action (Asia)

People's Global Action (Asia) (PGA Asia) represents a network for communication and co-ordination between diverse social movements, whose membership cuts across differences in gender, ethnicity, language, nationality, age, class and caste. The broad objectives of the network are to offer an instrument for co-ordination and mutual support at the global level for those resisting corporate rule and the neoliberal capitalist development paradigm, to provide international projection to their struggles, and to inspire people to resist corporate domination through civil disobedience and people-oriented constructive actions.

PGA Asia is concerned with five principal processes of facilitation and interaction between movements. It acts as a facilitating space for communication, information-sharing, solidarity, coordination and resource mobilization (Routledge 2003a). The network articulates certain unifying values – what we would term collective visions – to provide common ground for movements from which to coordinate collective struggles.[4] PGA has also established regional networks – e.g. PGA Europe and PGA Asia – to decentralize the everyday workings of the network. The principal means of materializing the network have been through the internet (PGA has established its own website (www.agp.org) and email list in order to facilitate network communication); and performative events such as conferences;[5] activist caravans (organized in order for activists from different struggles and countries to communicate with one another, exchange information, and participate in various solidarity actions); and global days of action (Routledge 2003a; see also Juris 2004a).

The PGA network is facilitated by a Convenors' Committee which comprises social movements within the network but much of the organizational work –

---

4   The collective visions of PGA, are as follows:

1. A very clear rejection of capitalism, imperialism and feudalism; and all trade agreements, institutions and governments that promote destructive globalization.
2. We reject all forms and systems of domination and discrimination including, but not limited to, patriarchy, racism and religious fundamentalism of all creeds. We embrace the full dignity of all human beings.
3. A confrontational attitude, since we do not think that lobbying can have a major impact in such biased and undemocratic organizations, in which transnational capital is the only real policy-maker.
4. A call to direct action and civil disobedience, support for social movements' struggles, advocating forms of resistance which maximize respect for life and oppressed peoples' rights, as well as the construction of local alternatives to global capitalism.
5. An organizational philosophy based on decentralization and autonomy

(Taken from the PGA website: www. agp.org).

5   PGA Asia held a regional conference in Dhaka, Bangladesh in May 2004.

preparing, organizing, and participating in discussions, meetings, conferences and caravans – has been conducted by 'free radical' activists and key movement contacts (usually movement leaders or general secretaries) who have helped organize conferences, mobilize resources (e.g. funds), and facilitate communication and information flows between movements and between movement offices and grassroots communities. These free radicals and key contacts – who must possess English language skills (the *lingua franca* of PGA Asia), and be computer literate (see also Juris 2004a) – represent what Sidney Tarrow (2005) terms 'rooted cosmopolitans', i.e. transnational activists who move physically and cognitively outside their origins; draw on, and are constrained by, domestic and international resources networks, and opportunities of the societies/places in which they live, and advance claims on behalf of external actors, against external opponents, or in favour of goals they hold in common with transnational allies. However, we use the term 'imagineers' because these activists attempt to translate the concept or imaginary of the GJN (e.g. what it is, and/or how it works, and/or what it is attempting to achieve in terms of campaigns and network goals) within the broader constituency that comprises the network participants. In PGA Asia there is an Asian convenor (All Nepal Peasant's Association, ANPA), a South Asia sub-regional convenor (Indian Farmer's Union, [BKU]), a Southeast Asia sub-regional convenor (Assembly of the Poor [AoP], Thailand), and a 'free radical' group of five activists.

PGA Asia favours non-hierarchical horizontal networking practices. Such transnational networking processes generate the communicative infrastructure necessary for the emergence of 'transnational counterpublics' (Olesen 2005: 94; see also Juris 2004a): open spaces for the self-organized production and circulation of oppositional identities, discourses, and practices. These refer to the larger social contexts in which networks are embedded and 'are constituted by a range of geographically dispersed actors, and are often centred around local or national issues considered to be of relevance to people outside of the geographical location, or around issues with a cross border nature' (Olesen 2005: 94). Transnational counterpublics are not exclusively virtual phenomena; they are also physically manifest within performative events: mass actions, global activist conferences, and regional network gatherings (see Routledge 2003a). This process of networking is always dynamic and constantly changing: it can involve conflict, expansion, convergence and dissolution (Juris 2004a).

Although PGA Asia is a grassroots-based, decentralized network, it nevertheless involves both newer and more traditional political formations, including NGOs, unions, and leftist parties. PGA Asia operates as a network wherein the participant grassroots movements (at both local and national scales) tend to be organized through more verticalist structures and logics: hierarchy, elections, delegation, and, in some cases political party structures too. For example, among the movements involved in PGA Asia are the Bangladesh *Krishok*[6] Federation (BKF) which

---

6   Signifies 'peasant'.

holds internal elections for a series of hierarchical functional positions within the movement; the All Nepal Peasants Association (ANPA) which operates similarly, and is also affiliated with the Communist Party of Nepal (Marxist-Leninist); the *Narmada Bachao Andolan*, India (NBA) which has a powerful core group activists leaders (Routledge 2003b); the Borneo Indigenous Peoples' and Peasants Union (Panggau) which has an elected secretariat and a mobile 'core catalyst' group of 20-30 people, who organize local communities throughout Sarawak, Malaysia; and the Assembly of the Poor, Thailand (AoP) which comprises a network of anti-dam, peasant, student and labour movements, with their own differing modes of operation.

This can create certain instabilities within the network, for many activists are unaware of the difference in operational logics between their own movements and the networking logic of PGA. For example, many activists who were interviewed expressed the need for more familiar operational logics and structures in the PGA Asia process. Hence an activist in the Bangladesh Kisani Sabha (peasant women's assembly) thought that the network was too loose and needed tangible structures to facilitate coordination, communication and contact:

> The network is too loose. We need an operational secretariat. Kisani Sabha is interested in coordinating a national PGA process in Bangladesh, but we need a tangible structure in Bangladesh for coordination, communication and contact. The Dhaka conference was the start of this (interview, Dhaka, Bangladesh, 2004).

There are also different notions of power manifested in the PGA Asia network. Many of the constituent movements within PGA Asia articulate more traditional forms of organizational logic, predicated upon taking political power. For example, while the BKF operates autonomously from Bangladesh's principal political parties (the BNP and Awami league), the Karnatakan State Farmer's Union (KRRS) participates in electoral politics in order to draw attention to rural, grassroots issues, and ANPA is affiliated to the CPN(M-L), which during 2004 was participating in a coalition government in Nepal. Mirroring local level dynamics, conflicts between networking and more vertical command logics generate constantly shifting alliances as activists alternatively participate within, abandon, or create autonomous spaces with respect to broader social networks. Hence within PGA Asia, an earlier participant in the network, the Federation of Indonesian Peasant Unions (FSPI) has left the convergence to operate as the Asia secretariat for *La Via Campesina*, (network of peasant farmers).

The International Federation for Chemical Energy Mine and General Workers' Unions (ICEM) and the European Mine Chemical and Energy Workers Federation (EMCEF)

ICEM is one of the 10 Global Union Federations (GUFs), formally known as International Trade Secretariats, set up to represent and coordinate worldwide labour interests.[7] Like other global unions, ICEM is industry-based and dedicated to practical solidarity. It unites 408 trade unions in its sectors (essentially chemicals, oil, energy and mining) on all continents. All together ICEM represents some 20 million workers in 125 countries. Its head office is located in Brussels and until late 2004, ICEM had five further regional offices in Latin America, Asia, Eastern Europe, Africa and North America.[8,9] In contrast to PGA Asia, ICEM is significantly more formalized, whose paid-up membership creates a principal-agent relationship that does not exist in the more horizontal 'open-access' PGA Asia network.

Like PGA, ICEM operates mainly through five processes: The most obvious is its World Congress which mirrors the traditional verticalist logic described earlier. ICEM world congresses take place every four years – so far three such Congresses have been held, in 1995, 1999 and 2003 in Stavanger, Norway. In addition, its Presidium and Executive Committee holds a general assembly every year and further regular formal meetings. Secondly, ICEM produces newsletters and information briefs diffused through its website (www.icem.org) and via electronic correspondence. Thirdly, the Brussels office acts as an intermediary to facilitate the establishment of bilateral relationships, unions in countries such as Iraq or China, and joint campaigns between affiliate members. It also intervenes directly towards particular struggles in selected countries such as Colombia. Fourth, global company networks provide a decentralized communication and information exchange mechanism for shop stewards from different countries. Since ICEM does not have the capacity to administer more than a very small number of networks, the Executive Committee decided that it should be the responsibility of affiliates to take on the role of administering any network prior to it being established. To date, ICEM has created six global company networks including Bridgestone/Firestone in Japan or Rio Tinto in the UK and Australia (see Sadler 2004 for an overview of the campaign against Rio Tinto). Finally, Global Framework Agreements (GFAs) are signed with multinational companies which pledge to respect a set of principles on trade union rights, health, safety and environmental practices.

---

7   Other GUFs include Education International (EI) and the International Metalworkers Federation (IMF) which both have 25 million members.

8   The five regional offices were closed down in late 2004 because of corruption. The decision was taken to conduct regional activities with international regional officers from the Brussels office.

9   ICEM's founding declaration can be downloaded at http://www.icem.org/media/foundec.html.

There are important variations in the number of affiliate union members across countries reflecting the organizational cultures that prevail in each country. For example, in Europe, countries such as Sweden, Finland and France have around eight or more unions that affiliate to ICEM when Switzerland or Germany only have one. However, this is not a true reflection of the relative weight and dominance of each country within the network: the level of affiliate fees depends on the density of union membership which grants more or less voting power to each of its members. In fact, the powerful German *Industriegewerkschaft Bergbau, Chemie, Energie* (IG-BCE) is the largest contributor to the ICEM, although its contribution in mining has been declining in recent years.[10] As pinpointed by several union officials we spoke to, it is not rare for individual unions to declare more members than they have on their books and subsequently pay higher fees to the federation as a strategy to obtain more voting power, although conversely some may declare less to reduce their fee. Thus the relative status and power of each member varies tremendously and creates tensions as reflected by an official arguing that the payment of high affiliation fees does not entail that these particular unions will be more active:

> The Norwegians are loaded with money so they can feed in. Even if they do nothing, they give the money and it's the others who act. So as a result, it is possible to implement projects that are financed by others (interview, FCE-CFDT, Paris, 12 April 2005).[11]

According to ICEM's 2003 Congress report, due to several factors – such as the impact of globalization on workers in both the developed and developing world, the continuing struggle of workers in ex-CIS countries, the tragic effects of war, conflict and terrorism, and the global consequences of disease, including HIV/AIDS – 'there has been as expected a gradual increase in demands on the ICEM from our affiliates' (page 4). Yet its action is constrained by the declining affiliation fees, itself linked to the worldwide decline in union membership. There is hence a significant interdependency between the international federation and national unions: declining resources have meant that the head office recently had to undergo some restructuring and reduce its team of permanent employees.

Networking processes

Despite the different cultures and organizational logics of the two networks studied here, both face common problems in developing and sustaining genuine

---

10   Unfortunately, access to the affiliation fees of ICEM and EMCEF members was denied to us, which does not allow for a precise mapping of affiliates positions within the network.

11   Quotes from interviews with French union officials have been translated.

international solidarity and collective action. Sustaining collective action over time requires the development of strong interpersonal ties that provide the basis for the construction of collective identities (Bosco 2001). As noted earlier, PGA and ICEM have periodic international and regional conferences that provide material spaces within which representatives of participant movements and affiliate unions can converge, and discuss issues that pertain to the functioning of the network. At the conferences, the hosts (a social movement or movements in the case of PGA, the presidium or a national affiliate in the case of ICEM) explain the specific problems and ongoing history of their campaigns.

In PGA Asia, caravans are organized to enable cross-movement exchanges and to encourage new movements into the convergence. The emphasis on such processes is the two-way communication regarding struggles, strategies, visions of society, and the construction of economic and political alternatives to neoliberalism. Such conferences, caravans, and meetings also enable strategies to be developed in secure sequestered sites, beyond the surveillance that accompanies any communicative technology in the public realm. Moreover, such gatherings enable deeper interpersonal ties to be established between different activists from different cultural spaces and struggles.

While an important aspect of the work is to build interpersonal relationships with contacts within these movements, another is to coordinate joint actions across space, for example against particular neoliberal institutions such as the WTO. Such joint actions, when embodied in collective experiences such as conferences, enable connections and exchanges between activists to be made, and such interrelations can build trust between activists and shape collective political identities and imaginaries. For example, through the re-drafting of the network's collective visions (or 'hallmarks', see footnote 4) at the international conferences in Bangalore, India, 1999, and Cochabamba, Bolivia 2001, the network was able to define who it was (in terms of its political beliefs and practices, see hallmark 2 and 5), what constituted its common problems (e.g. poverty), who constituted its opponents (see hallmark 1) and how the network operated through the creation of common political strategies (see hallmark 4) (Featherstone 2005; Wood 2005). Gatherings (such as the Dhaka conference for PGA or the union congresses and assemblies for ICEM) provide performative spaces that play a vital role in face-to-face communication and exchange of experiences, strategies and ideas (see also Juris 2004a).

These events enable alternative social movement networks to become embodied and where transnational counterpublics are able to be produced and reproduced (Juris 2004a). For example, one PGA activist from an indigenous people and farmers' movement in Panggau, Borneo, explained his reasons for attending the Dhaka conference:

> We wanted to share our experiences of the struggle. We don't have many linkages to other movements or the space to speak. For example at the World Social Forum time is too limited. The Dhaka conference provided us with an

opportunity and the space to speak (interview, Kuching, Borneo, Malaysia, 2004).

Moreover, people's positionality in relation to others can be re-assessed, as an activist in the slum dwellers network (within Thailand's AoP) noted:

> There was a real chance for exchange between activists. We usually stereotype people by nation but when we meet face to face it breaks down the borders between us, and generates collective strength to make change (interview, Bangkok, 2004).

However, networks are fragile entities which can be disrupted by a range of different issues. First, there is the constant problem of securing funding to provide key resources. For example, the PGA Asia conference in Dhaka was delayed by almost two years due to the problems associated with fund-raising, and was further hampered by the late availability of some of those funds. Here also, there is resonance with ICEM since the GUFs are currently facing the difficulty of transferring resources from the national to the international level:

> Again with the GUFs, we'd like them all to be more active and I think it's a question of resources. Collectively, compared to conversations three or four years ago that we had internally with the GUFs, we're thinking 'look, issues are moving internationally' so the question is how you transfer resources then internationally to be able to handle them at an international level. That hasn't happened and I think again the resources have been cut back when the unions have had a rough time domestically that they're having at the moment and that means even priority work gets under-sourced and that problem has become more acute not less acute over the last three or four years (interview with the general secretary of TUAC, Paris, 5 October 2004).

Second, there is the problem of movements potentially leaving the network or joining other networks which leads to less involvement with the original network. One example was provided by the British engineering union Prospect who left ICEM in 2004 after establishing that the focus of the federation was not the most relevant to its own activities. For the PGA, the formation of the World Social Forum (and the regional Asian Social Forum) has seen both the participation of several of the movements also involved with PGA Asia (such as BKF, ANPA), and the loss of movements within the global PGA network such as the FSPI as they have directed their energies towards the World Social Forum. In addition several movements involved in PGA Asia, such as ANPA, the BKF, and KRRS, are also participants in *La Via Campesina* network. In the case of unions, such networks include human rights NGOs such as the French 'right to energy' (*Droit à l'énergie SOS Futur*) in which the vice-president of the CFTC-CMTE union is also acting as a vice-president. This particular individual argued that he was 'wearing multiple hats'.

Third, a network can be compromised by the ineffectiveness of the communication and operational links between its participant movements. ICEM for example has recently restructured its activities to improve the communication between highly specialized and hence fragmented unions, by moving away from sector committees which have been replaced with four transversal thematic committees on collective bargaining, industrial policy, social dialogue and European Works Councils.

A further point relates to issues of communication and interpretation. In interviews with participants at the PGA Dhaka Conference, many people felt that the conference interpretation process was inadequate – there were many different languages and regional dialects present at the conference. Delegate participation at the conference was uneven because there were few interpreters (particularly women interpreters) and thus they missed much of the discussions. As an activist in the AoP noted: 'language was a real obstacle, because the full experience of activists was unable to be communicated' (interview Chiang Mai 2004). Few activists in grassroots communities have English language skills, and thus there is a danger that when they participate in PGA conferences, interpretation ghettos can emerge where folks barely communicate outside of their language group. An associated problem is that interpreters themselves accrue power and influence by virtue of their language skills. For one Thai activist, the operational logic of the network is underpinned by 'literate' and conceptual communicational forms (e.g. the writing of emails and documents, the analysis of how networks function), whereas the operational logic of most grassroots movements is based upon oral communication:

> There is a real limitation to the capacity of grassroots movements to take ownership of the process. Movements do not know each other very well, and some SE Asia movements do not really know the PGA process at all. Thus participation is limited and language affects this too. Most movements are based on oral communication, whereas the PGA process is more literate and concept-based, thus it is difficult for grassroots movements to understand (interview, Bangkok, Thailand, 2004).

The linguistic dimension is echoed by affiliates of ICEM, particularly the French who, whilst aware that they are lagging behind in terms of language training, complain that the hegemony of the English language in conferences and meetings gives rise not only to communication problems, but also to asymmetric power relationships (see also Stirling and Tully 2004 for an overview of a similar debate in European Works Councils). Several officials we spoke to argued that language differences can be used to the advantage of certain affiliates and to the detriment of others. Moreover whilst there are interpretation facilities in international Congresses, the lack of linguistic abilities can hamper the establishment of bilateral links. As a result, partnerships tend to be forged according to linguistic affinities. Thus the Nordics are more likely to cooperate with the British and the French with the Belgians. Yet the ability to master foreign languages seems less valued than the shared language of class consciousness:

> If it is to have around the table a plethora of people who speak dozens of languages but don't have any class consciousness, I'm not interested (interview, Force Ouvrière, Paris, 21 June 2005).

This view resonates with Wagner's (2004) analysis of the tensions inherent in the construction of what she terms 'activist capital' between an older generation of trade unionists with considerable experience of militant action at the local and national level and the newer generation of well-travelled 'union technicians' with university degrees and multi-linguistic skills. There is no doubt that this generational change will considerably impact upon relational networking processes within GJNs. Some actors have far more capacity to direct the course of relations than others, which partly stems from their ability to collect 'power' and condense it within networks (Castree 2002). This is related to the process of network relays of communication and information.

Imagineers and information/communication relays

Understanding the dynamics of network operational logics entails thinking about power relations across space and not necessarily in any one place. This is a relational perspective that focuses on the effectiveness of connections (e.g. communication and operational links) between the actors in the network (Massey 1994). Horizontal relations within networks necessitate the free flow of information between all participants in all directions. However, the reality of network agency is invariably compromised by various factors which give rise to more vertical network relations. PGA is organized primarily through the Internet which acts as a communicative and coordinating thread in the network. Whist the Internet is a significant resource to ICEM affiliates, members often point to the problem of 'information overload' and try make a selective and focused use of ICT tools. Moreover, a recurrent theme is that the Internet cannot replace face-to-face contacts. Large union federations rely on international officers and smaller ones on their general secretaries to play the role of imagineers. A similar tendency to rely on imagineers for communication, i.e. individuals who have various international contacts owing to their participation in various networks also occurs in PGA Asia but for diametrically opposed reasons. In the global South, grassroots movements have varying and often limited access to electricity, let alone computer technologies. Hence participant movements in PGA Asia effect communication and information relays via the imagineers.

Imagineers work relationally across space, they may be working in relatively more (or less) local, national or international spatial contexts depending upon the operational logics of a particular GJN; its internal power relations; levels of resourcing; and specific spatial dynamics etc. They are the points where information accumulates: the movement offices in Kathmandu, Dhaka, Bangkok etc. and the free radicals' laptops and office computers. Imagineers serve to embody the networks in which they work (see Olesen 2005). Key networking tasks – such as fundraising and planning for conferences – are delegated to the imagineers. They also work to enrol other movements into the network by visiting them, and holding meetings and discussions. The imagineers represent the connective tissue across geographic space working as activators, brokers, and advocates for domestic and international claims (Tarrow 2005). In the spaces of GJN 'performance' such as conferences, workshops, activist exchanges, or protest events, imagineers are able to promote all aspects of coalition-building (Juris 2004a). Certain imagineers are more powerful than others, and usually more powerful than the grassroots members of particular movements. Hence within the diversity of a network such as PGA Asia there is 'controlled heterogeneity' (Riles 2001: 120; see also Freeman 1970).

Agency is a relational effect generated by interaction and connectivity within the network. Power becomes the ability to enrol others on terms that allow key actors to 'represent' the others (Castree 2002). Owing to differential access to (financial, temporal) resources and network flows differential material and discursive power relations exist (see Routledge 2003a). Within the two GJNs, the imagineers – because of their structural positions, communication skills and experience in activism and meeting facilitation – tend to wield disproportionate power and influence within the network. Globally mobile (both physically – in that they have the time and resources to travel outside of their home countries – and through their access to distance-shrinking technologies), they perform much of the routine work that sustains the network. They possess the cultural capital of (usually) higher education, and the social capital inherent in their transnational connections and access to resources and knowledge (Missingham 2003).

Within PGA Asia, it is the interpersonal relations between key movement contacts and the free radicals that enable the network to function. However, in the case of PGA, the existence of an 'informal elite' can be partly due to the attitudes of grassroots activists themselves, who at times tend to defer authority to key movement contacts and let them get on with the work of international networking. Hence, an activist in the KRRS, explained:

> We need to involve the grassroots in the PGA process, but the attitude of local activists can be a barrier. They are often happy to depend upon me to organise the international side of the movement. When I report back about PGA events such as Dhaka, no-one really takes it very seriously, because they do not see a link between their movement and their daily lives, and PGA. This is

also accentuated by the fact that few people in the movement speak English (interview, Kathmandu, Nepal, 2004).

Similar issues arose in discussion with ICEM affiliates where there was a difficulty experienced in getting workers at the local level to connect up their everyday struggles to more global concerns:

> That is the dilemma. International work is becoming more and more important for trade union movements and I think that our union has to be involved more and more in international work than we do today but when membership is declining, the ordinary member or shop steward says, 'The issue is here and now. It's in our plant. It's the Norwegian problems with the industry. It is here that we have to use the resources and the people.' And to get the understanding of the bigger picture, and to see these things not as separate issues, I think it's quite a challenge. (Interview Vice President, Norwegian ICEM Affiliate, 16th August 2005)

The result is that international work gets deferred to an elite cadre made up increasingly of university educated 'bureaucrats' from different social backgrounds to those at the grassroots of union movements, and in many cases with a background in international development/NGO work (authors' interviews). Thus, there is the danger of a growing social distance, measured in terms of culture, politics and ideology between those who network globally and those engaged in day-to-day struggles in the workplace. For both PGA and ICEM, therefore, a key issue interwoven with the problem of differential activist powers (and the vertical social relations that this implies), is how the network's imaginary is visualized at the grassroots and how this is played out across material and virtual space.

Spatial dynamics of GJNs

Networks are a means of acting upon space – through mobilizations, connections, and combinations that link subjects, objects and locales. Networks can expand as well as contract, and the geographic fluidity of a network can contribute to its resilience to external threats and thus to its sustainability. A network can be spatially decentralized but power may remain highly centralized, and vice versa. In PGA Asia, a spatially decentralized network contains a concentration of power within certain individuals (i.e. the imagineers) although these individuals are themselves spatially dispersed. In ICEM, as argued above, several unions, such as the German IG-BCE, the South African National Union of Mineworkers (NUM) or the American Paper, Allied Industrial, Chemical and Energy Workers' International Union (PACE) are dominant.

A focus on processes necessitates an understanding of network growth and contraction strategies. This must take into account the fact that while networks of resistance operate transnationally, the struggles and the identities of resistance are often born locally through activists' sense and experience of place (Pile and Keith 1997). What also gets diffused and organized across space is the 'common ground' shared by different groups – often the result of groups' entangled interests (Routledge 2003a). Differential involvement occurs at different scales since participants within GJNs come from different cultures and political beliefs. PGA participants tend to hold a common vision while ICEM affiliates attach different meanings and expectations from global labour solidarity:

> Some of our affiliates are very Marxist oriented in various parts of the world. In the developed countries, some are fairly conservative. There's no bearing on that in our minds (ICEM office, Brussels, 18 November 2005).

Networks operate unevenly in space owing to differently 'placed' movements (in terms of their available material and discursive resources), and the uneven character of their connectedness within networks. Because space is bound into local to global networks, which act to configure particular places, places such as Dhaka (the recent conference site for PGA) or Stavanger (venue for the last ICEM Congress) become the focus of a distinct *mixture* of wider and more local social relations' (Massey 1994: 156), and hence can be imagined as 'articulated moments' (ibid.: 154) in networked social relations. These represent a network's common meeting and recruitment grounds (e.g. PGA conferences, where new movements from Bangladesh, Vietnam, Thailand and Malaysia all participated in the Dhaka conference for the first time; ICEM headquarters in Brussels or Conference meeting places); and thus their capacity for growth and contraction (Juris 2004a).

However, the technology for reaching out cannot extend networks where the notion of extension fails to capture the imagination (Riles 2001: 26). Some grassroots activists interviewed in Nepal and Dhaka articulated the need for more traditional, tangible (verticalist) organizational structures than the (horizontal) notion of a 'coordination tool' implied. In part, this may be attributed to the entangled character of organizational logics, discussed earlier. This leads us to a discussion of how GJNs are 'grassrooted' in the countries of their participant movements.

Grounding GJNs: imagineers as 'grassrooting vectors'

In order to 'ground' the idea of GJNs in the communities that comprise the membership of such networks' participant movements, it is essential to have 'grassrooting vectors' which work to intervene in the work by which networks are formed and developed, acting to further the process of communication, information sharing and interaction within grassroots communities. Such vectors include conferences, feedback by conference delegates to their grassroots, and the imagineers themselves.

However, the PGA Asia imaginary remains abstract to many grassroots activists, for whom the networking logic of many direct action groups (and PGA) is unfamiliar. As one BKF activist remarked:

> We have to disseminate information to people in rural areas, but so far they have not been able to visualise what the network is. We need a national conference to begin the process of visualization of the PGA process in Bangladesh (interview Dhaka, Bangladesh, 2004).

The idea of grassrooting the imaginaries of global labour struggles is felt as being a major challenge by the union officials interviewed in Europe. The rejection of the European constitution project by the French on 29 May 2005 has acted as a powerful reminder of the threats felt by many destitute citizens and workers and the difficulty of creating fraternal bonds and shared imaginaries with distant fellow workers. Imagineers from national unions see it as part of their remit to pay visits to their fellow colleagues and enlighten them. Indeed these visits not only help them build unions in their country, but also increase their awareness as to the negative effects of delocalization in the mature economies of the EU. For example, an official from the FNEM CGT-Force Ouvrière recalls a recent visit to Romania:

> I said to them, 'You see I'm here with you. We're trying to organize and create a union for you to have better conditions but at the same time, in France, some people have just lost their jobs. And we'll have to organise them too ...' This helped raise awareness in the discussion. He said to me, 'Explain that again,' so I said: 'The fact that we're fighting here for you to get better wages, when I go back I have to fight for those who've lost their job because the work they've lost is in your country, it is here. You've had subsidies from Europe to rehabilitate the factory, so the French employer settles here and on top of that he's gonna exploit you.' And then he really understood. (interview, FNEM CGT-FO, Paris 21 April 2005).

Likewise, within the PGA network, regional and international conferences have enabled grassroots activists to (a) learn about other struggles in other countries and decrease their sense of isolation; (b) communicate with other activists from other countries; (c) share tactics and strategies; and (d) generate a sense of common ground as a precursor to solidarity between movements. Hence a BKF activist noted:

> PGA can provide information about struggles around the world; about how economic globalization works, and how this is affecting grassroots communities, and agriculture in Bangladesh. This increases consciousness of international issues and struggles and thus we are able to identify common enemies. With PGA

support, the BKF struggle will be more effective. Information on other struggles also provides information about how others struggle and also knowledge about the history of different struggles. This inspires other movements when they know they are part of an international process (interview, Dhaka, Bangladesh 2004).

The role of the imagineers has also been important in grassrooting the PGA imaginary. For poor peasant communities which comprise PGA Asia's participant movements, their only source of connection to the network is primarily through the activist organizers who operate from the movements' offices, and who visit the communities as part of their organizing practices. 'Free radical' activists (accompanying activist organizers) have also often travelled to visit social movements in Asia before PGA events such as conferences to discuss with them the PGA process, conduct workshops, and invite them to participate in forthcoming events. The imagineers act as 'grassrooting vectors' furthering the process of communication, information sharing and interaction within grassroots communities. They frequently act to displace the network's collective visions (including the shared experiences of oppression, problems, opponents) from one context to another in order to further the processes of connectivity and affinity amongst peasant communities who comprise the 'grassroots' of GJNs' participant movements.

For the poor of grassroots movements such relational dynamics can constitute an expansion of their geographical imagination and practical political knowledge. The presence of imagineers in grassroots communities embodies the network, and can constitute proof of sorts of the international character of the network – a tangible example that peasants are part of something wider and larger. They help put human faces on what otherwise may be abstracted differences among distant organizations, allowing for greater interpersonal trust and intercultural education (Bandy and Smith 2005). It also enables the concept of PGA to begin to take root in people's imaginations.

The imagineers tend to act as the driving force of the network imaginary coordinating and controlling the majority of informational traffic. Certain decision-making power accrues to them by virtue of their access to resources (time, money, technology, language skills etc.), as well as personal qualities like commitment and charisma (Juris 2004a; King 2004). Social capital accrues to these imagineers by virtue of their networking capacities. Disproportionate power accrues to networking vectors as a result of their capacity to enrol others into the network, to travel, and to act as channels of communication between activists located in different places who are not as 'mobile'. They may also possess differential access to resources and mobility compared to others in the network (see Routledge 2003a).[12]

---

12 There has also been debate in the international PGA network about the disproportionate participation at international conferences by activists from the global North and the role of some of those activists (as imagineers) in logistical coordination, fundraising, and organizing newsletters and the PGA web page (Routledge 2003a; Wood 2005).

This belies the decentralized horizontal coordination that supposedly operates within PGA Asia, since grassrooting vectors constitute a hierarchy of communicational, informational, and decision-making 'powers' within the operational and relational dynamics of the network. However, grassrooting vectors are essential (at least at present) because they conduct the primary work that organizes the network.

In contrast, for most of the grassroots activists of PGA Asia's participant movements, their most immediate source of self-recognition and autonomous organization is their locality: they mobilize to protect their community, their land, and their environment (Castells 1997). However, these immediate issues of survival and livelihood nevertheless can act as motivations for people to participate (as social movement members) within transnational networks, in order to meet activists in other movements, to learn from them, and increase their understanding of the issues that affect them. They can also form the basis for common grievances between movements, as a prelude to forging mutual solidarity.

However, many activists believed that an important step in bringing the PGA imaginary to the grassroots, lay not only in having local post-conference debriefing meetings, or meetings where imagineers spoke, but also to create a national PGA process within their respective countries (which would also involve caravan activities such as meetings between activists from different countries):

> We need to bring the PGA process to the national level and then down to the grassroots workers, we need a national PGA process to which the grassroots are linked, via conferences, workshops, discussions, trainings. I have begun to talk to the grassroots communities in my district (Saptari) and in my union about my Dhaka experiences. But this has to be a collective process of growth. We also need to bring other international activists to the grassroots communities. The problem with the grassroots process is that we do not talk in depth, we need a national action plan for PGA (ANPA activist, interview Kathmandu, Nepal, 2004).

Again, as argued above, the idea of grassrooting network imaginaries which amounts to promoting a renewed labour internationalist movement is much less evident (albeit not less relevant) within ICEM and EMCEF and its constituent members than within PGA. We have found major differences of views between advocates of a reformist view (the dominant actors within ICEM/EMCEF) willing to bargain and compromise with employers, and more radicalist positions. In this sense the geopolitics of PGA are closer to the ideal type we have conceived for GJNs compared to the international and European union federations whose members have not yet established a consensus on resistance to neoliberalism and on the most effective mechanisms to effect it. For PGA, the challenge is thus less aspirational than practical as it seeks to establish more ongoing grassroots programmes, whereby some of the experiences that activists would normally only get at conferences (such as learning about the dynamics of globalization, and the struggles of other movements) could be provided.

Hence grassrooting vectors, while important, are insufficient for the construction of durable networks. The vertical power relations that they embody require replacement with more horizontal relations between movements within countries and between movements from different countries. Grounding network imaginaries necessitates forging mutual solidarity.

Forging mutual solidarity within GJNs

An important aspect of network dynamics entails deepening the process of network imagination within grassroots communities for whom digital technologies remain relatively inaccessible. Network imaginaries at the grassroots level remain uneven and potentially 'biodegradeable' (Plows 2004: 104), i.e. they may dissipate without sufficient and constant nurturing. Moreover, while networking as a process constitutes much of the vitality of GJNs, the question remains as to whether such networking is sufficient to enable transformative political projects to be realized.

Protest itself may mean little for the social and participatory rights of groups at the bottom of social hierarchies, whose specific interests remain un-represented. Social change is not about events (e.g. strikes) but the processes that are entailed in and result from events (Ettlinger 2002). An over-emphasis on resistance can ignore the lives of a variety of people with diverse relationships to globalization, including unorganized workers, undocumented immigrants, and those not involved in political movements. As one Panggau activist noted:

> We need to develop consciousness (e.g. through educational trainings) about legal rights, and how to develop sustainable economies and sustainable forms of resistance. We need to discuss ways that movements can meaningfully support one another (interview Kuching, Borneo, Malaysia, 2004).

This view reflects broader concerns about the establishment of lasting alternatives to neoliberalism. Such concerns have led to the emergence of certain projects from within the relationships generated through the PGA Asia convergence. For example, activists from the KRRS and Bangladesh Kisani Sabha have begun a long term project in southern India to establish an agro-ecological community for women's empowerment (personal communication, Kathmandu, Nepal, 2004). Likewise, the ICEM secretariat and some individual affiliates such as AMICUS in the UK have set up long-term HIV/AIDs projects in Africa.

Global Justice Networks can facilitate multiple localized oppositions which articulate diverse critiques, approaches, and styles in various places of action (Schlosberg 1999). In particular, what can get transnationalized in the network imaginary are notions of mutual solidarity – constructing the grievances and aspirations of geographically, culturally, economically and at times politically different and distant peoples as interlinked (Olesen 2005). Mutual solidarity across place-based movements enables connections to be drawn that extend beyond the local and particular. Such mutual solidarity recognizes differences between actors within networks while at the same time recognizing similarities (for example, in people's aspirations).

However, the construction of mutual solidarities is not a smooth process: they involve antagonisms (often born out of the differences between collaborators) as well as agreements: they are always multiple and contested, fraught with political determinations (Featherstone 2005). Nevertheless, network imaginaries may help to reconfigure distance in different ways – which emphasizes commonalities rather than differences. As Olesen (2005) argues, mutual solidarity 'entails a constant mediation between particularity and universality – that is, an invocation of global consciousness resting on recognition of the other' (2005: 111). A network imaginary that can invoke interconnections, opens up potentials for mutual solidarity that enables a diversity of struggles to articulate their particularities while simultaneously asserting collective identities (Holloway and Pelaez 1998). Such solidarity takes place in the form of changing and overlapping circuits of relations that are enacted both virtually and materially in particular forums and face-to-face meetings such as conferences.

Towards Grassroots Imaginaries

Networking processes help constitute alternative counterpublics, which form the communicational basis for transnational social movements, understood as highly complex and contradictory spaces of convergence rather than unified collective actors. As Juris (2004a) notes:

> Beyond creating open spaces for reflection and debate, forums and conferences also provide 'temporary terrains of construction' where activists generate and exchange innovative ideas, resources, and practices, and within which alternative social movement networks are physically mapped and embodied ... activist gatherings provide alternative mechanisms for generating affective attachments ... [that allow] ... movements to continue reaching out to a broader audience (466-467).

Tarrow (2005) is correct to argue that enduring GJN coalitions need to equalize resource allocations, overcome cultural differences, and bridge the differences in opportunities and constraints that participant movements and imagineers face in their home societies. However, the sustainability of such networks and their ability to develop a credible alternative politics to neoliberal globalization will also depend in part, on the extent to which network imaginaries are grounded successfully, and meaningfully, in grassroots communities and workplaces. As one Indian PGA Asia activist commented:

Movements in South Asia have a limited resource capacity to fully engage in global solidarity, things like time, money, language skills and computer skills. Hence most Indian movements are not really ready to fully participate in a global movement, to commit to it full time, or to fully involve and engage the grassroots in it. Most movements in India are leader based and many of these leaders have neither computer skills nor English language skills and thus they profess to be uninterested in global organizing since they do not possess the necessary skills for it. Most folk who do global organizing primarily like to travel and enjoy the benefits of conference hotels – they aren't serious about global solidarity. The language of many movement leaders is influenced by NGO discourse and not by the language of the grassroots. We need to return to the grassroots since most global work is too much in the air (interview, Kathmandu, Nepal 2006).

Network imaginaries must thus be grounded in the geopoetics of resistance, i.e. the cultural and ideological expressions of social movement agency – e.g. drawn from place-specific knowledges, cultural practices and vernacular languages – which inspire, empower, and motivate people to resist (Routledge 2000). While GJNs entail hybrid mixings of 'horizontalist' and 'verticalist' imaginaries, they need to be grounded in sustainable forms of material resistance to prevent the performative events of the network – the conferences, caravans, union meetings, days of action – from becoming only memories in the imaginations of grassroots peasant communities and workers. Moreover, GJNs involve ongoing antagonisms related to power, language, authority, and the GJNs entangled operational logics are thus constantly in process.

Acknowledgements

This research was made possible by a research grant from the UK's Economic and Social Research Council (award ref: ESRC RES-000-23-0528).

References

Bandy, J. and J. Smith (eds) (2005) *Coalitions Across Borders*. Oxford: Rowman and Littlefield.

Bosco, F. (2001) 'Place, Space, Networks, and the Sustainability of Collective Action: The *Madres de Plaza de Mayo*,' *Global Networks*, 1 (4): 307-329.

Braun, B. and L. Disch (2002) 'Radical Democracy's "Modern Constitution"', *Environment and Planning D: Society and Space*, 20: 505-511.

Castells, M. (1997) *The Power of Identity*. Blackwell: Oxford.

Castree, N. (2002) 'False Antitheses? Marxism, Nature and Actor-Networks', *Antipode*, 34 (1): 111-146.

Cumbers, A., Routledge, P. and C. Nativel, C. (2008) 'The Entangled Geographies of Global Justice Networks', *Progress in Human Geography*, 32 (2): 183-201.

della Porta, D., M. Andretta, L. Mosca and H. Reiter (2006) *Globalization from Below: Transnational Activists and Protest Networks*. Minneapolis: University of Minnesota Press.

Ettlinger, N. (2002) 'The Difference that Difference Makes in the Mobilization of Workers' International', *Journal of Urban and Regional Research*, 26 (4): 834-843.

Ettlinger, N. and F. Bosco (2004) 'Thinking Through Networks and Their Spatiality: A Critique of the US (Public) War on Terrorism and its Geographic Discourse,' *Antipode*, 36 (2): 249-271.

Featherstone, D. (2005) 'Towards the Relational Construction of Militant Particularisms: Or Why the Geographies of Past Struggles Matter for Resistance to Neoliberal Globalisation', *Antipode*, 37(2): 250-271.

Fraser, N. (1997) *Justice Interruptus: Critical Reflections on the 'Postsocialist' Condition*. New York: Routledge.

Freeman, J. (1970) 'The Tyranny of Structurelessness,' retrieved from http://flag.blackened.net/revolt/hist_texts/structurelessness.html on June 18, 2004.

Holloway, J. and E. Pelaez (1998) 'Introduction: Reinventing Revolution', in J. Holloway and E. Pelaez (eds) *Zapatista! Reinventing Revolution in Mexico*. London: Pluto Press.

Juris, J. (2004a) *Digital Age Activism: Anti-Corporate Globalization and the Cultural Politics of Transnational Networking*, Unpublished Ph.D. Dissertation, University of California, Berkeley, Department of Anthropology.

—— (2004b) 'Networked Social Movements: Global Movements for Global Justice,' in M. Castells (ed.) *The Network Society: a Cross-Cultural Perspective*. Cheltenham: Edward Elgar, pp. 341-362.

—— (2005) 'Social Forums and their Margins: Networking Logics and the Cultural Politics of Autonomous Space', *Ephemera*, 5 (2): 253-272.

King, J. (2004) 'The Packet Gang', *Metamute*, 27, www.metamute.com, site accessed June 2004.

McLeish, P. (2004) 'The Promise of the European Social Forum', *The Commoner*, 8, available at: www.commoner.org.uk/01-12groundzero.htm.

Massey, D. (1994) *Space, Place and Gender*. Minneapolis: University of Minnesota Press.

Mayo, M. (2005) *Global Citizens: Social Movements and the Challenge of Globalization*. London: Zed Books.

Melucci, A. (1996) *Challenging Codes*. Cambridge University Press, London.

Mertes, T. (2004) *A Movement of Movements: Is Another World Really Possible?*. London: Verso.

Missingham B.D. (2003) *The Assembly of the Poor in Thailand*. Chiang Mai: Silkworm Books.

Olesen, T. (2005) *International Zapatismo*. London: Zed Books.

Pile, S. and M. Keith (eds) (1997) *Geographies of Resistance*. London: Routledge.

Plows, A. (2004) 'Activist Networks in the UK: Mapping the Build Up to the Anti-Globalization Movement', in J. Carter and D. Morland (eds) *Anti-Capitalist Britain*. Cheltenham: New Clarion Press, pp. 95-113.

Riles, A. (2001) *The Network Inside Out*. Michigan: University of Michigan Press.

Routledge, P. (2000) 'Geopoetics of Resistance: India's Baliapal Movement', *Alternatives* 25: 375-389.

—— (2003a) 'Convergence Space: Process Geographies of Grassroots Globalisation Networks,' *Transactions of the Institute of British Geographers*, 28 (3): 333-349.

—— (2003b) 'Voices of the Dammed: Discursive Resistance Amidst Erasure in the Narmada Valley, India', *Political Geography*, 22 (3): 243-270.

Routledge, P. and A. Cumbers (2009) *Global Justice Networks: Geographies of Transnational Solidarity*. Manchester: Manchester University Press.

Routledge, P., Cumbers, A. and C. Nativel (2006) 'Entangled Logics and Grassroots Imaginaries of Global Justice Networks', *Environmental Politics*, 15 (5): 839-859.

—— (2007) 'Grassrooting Network Imaginaries: Relationality, Power, and Mutual Solidarity in Global Justice Networks', *Environment and Planning A*, 39: 2575-2592.

Sadler, D. (2004) 'Trade Unions, Coalitions and Communities: Australia's Construction, Forestry Mining and Energy Union and the international stakeholder campaign against Rio Tinto', *Geoforum*, 35: 35-46.

Schlosberg, D. (1999) 'Networks and Mobile Arrangements: Political Innovation in the U.S. Environmental Justice Movement', *Environmental Politics*, 6 (1): 122-148.

Sen, J. (2004) 'How Open?', in J. Sen, A. Anand, A. Escobar, and P. Waterman (eds) *World Social Forum: Challenging Empires*. New Delhi: The Viveka Foundation, pp. 210-227.

Stirling, J. and B. Tully (2004) 'Power, Process and Practice: Communications in European Works Councils', *European Journal of Industrial Relations*, 10 (1): 73-89.

Tarrow, S. (2005) *The New Transnational Activism*. Cambridge: Cambridge University Press.

Tormey, S. (2004a) *Anti-capitalism: a beginner's guide*. Oneworld: Oxford.

—— (2004b) 'The 2003 European Social Forum: Where Next for the Anticapitalist Movement?', *Capital and Class*, 84: 151-160.

—— (2005) 'After the Party's Over: The Horizontalist Critique of Representation and Majoritarian Democracy – Lessons from the Alter-Globalisation Movement (AGM)', paper presented at the European Consortium on Political Research, Granada.

Wagner, A (2004) 'Syndicalistes européens. Les conditions sociales et institutionnelles de l'internationalisation des militants syndicaux', *Actes de la Recherches en Sciences Sociales*, 155: 13-33.

Wood, L. (2005) 'Bridging the Chasms: The Case of People's Global Action', in
    J. Bandy and J. Smith (eds) *Coalitions Across Borders*. Oxford: Rowman and
    Littlefield.

Conclusion

Spatialities of Mobilization: Building and Breaking Relationships

Byron Miller

One might be tempted to say that each of the 12 empirical studies in this collection highlights a particular spatiality critical to understanding the mobilization of a social movement or other contentious political action. But this would be far too simplistic. In fact, while each study begins with a particular spatial focus, each goes on to examine multiple spatialities through which extremely complex mobilization processes are constituted, contested, and re-constituted. Social movement mobilization, and contentious politics more generally, is inherently spatial, as are all social and political processes. The spatial constitution of social and political processes almost always involves multiple spatialities. However, the roles of specific spatialities – e.g., place, space, territory, region, scale, networks – are always contingent and subject to change. Indeed changing the spatial constitution of social and political struggle is central to the course it takes.

Social movement studies have been bedevilled by debates over the appropriate way to conceptualize the spatial constitution of social movements. Place, space, territory, region, scale and networks have each been placed front and centre by a variety of social movement scholars and each spatiality-specific approach has yielded valuable insights, yet there has been little progress toward a more integrative approach. Increasingly scholars concerned with sociospatial relations, and contentious politics specifically, acknowledge this lacuna. As Leitner, Sheppard and Sziarto (2008) summarize in their influential article 'The Spatialities of Contentious Politics',

> The scholarly literature on geographies of resistance and social movements has produced valuable insights into each of these various spatialities (scale, place, networking, socio-spatial positionality and mobility), showing how they have shaped both political mobilisation and the trajectories of contentious politics (e.g. Knopp 1997; Moore 1997; Rose 1997; Slater 1997; Miller 2000; Wainwright et al. 2000; Rose 2002; Featherstone 2003, 2005; Routledge 2003; Wainwright 2007). Yet the co-implication of these diverse spatialities remains at times under-exposed, in face of the tendency in contemporary geographic scholarship either to privilege one particular spatiality, or to subsume diverse spatialities under a single master concept. (Leitner et al. 2008: 165).

Moreover, Leitner et al. assert that 'No single spatiality should be privileged since they are co-implicated in complex ways, often with unexpected consequences for contentious politics (2008: 169). This is not a claim that every spatiality is always equally important, but rather that different spatialities may be of greater or lesser significance in the articulation of specific processes.

But why is it that spatialities matter at all? Space matters because it is relational. It is the medium through which all social relations are made or broken – and making and breaking relationships is at the core of all questions of collective action. This very simple idea belies a tremendous complexity, as different spatialities vary in their implications for making and breaking relationships. The construction of relationships has enormous implications for social and political power, so it should come as no surprise that the construction of spatialities is born of struggle, and social and political struggle is bound up in the construction of spatialities.

There have been two major interventions in the spatialities debate since the call of Leitner et al. to examine the co-implication of spatialities. In 2008 Jessop, Brenner and Jones published 'Theorizing Sociospatial Relations', calling on scholars to recognize the polymorphic '... organization of sociospatial relations in multiple forms ... Territories (T), places (P), scales (S), and networks (N) must be viewed as mutually constitutive and relationally intertwined dimensions of sociospatial relations' (2008: 389). Like Leitner et al., they lament the focus of a series of 'spatial turns' on a particular spatiality, only to neglect other spatialities. After laying out the problems caused by one-dimensional methodological territorialism, place-centrism, scale-centrism, and network-centrism, Jessop et al. advocate adoption of a 'TPSN framework' that identifies principles of sociospatial structuration, as well as patterns of sociospatial relations, associated with each of the four spatialities they identify. They argue that the problems of one-dimensional frameworks 'can be avoided through more systematic, reflexive investigations of the interconnections among the aforementioned spatial dimensions of social relations – that is, the mutually constitutive relations among their respective structuring principles and the specific practices associated with each of the latter' (Jessop et al. 2008: 393). The TPSN framework is rooted in what Jessop calls the strategic-relational approach (SRA), an epistemological perspective grounded in critical realism and related to regulation theory. From the perspective of an SRA approach, Jessop et al. assert that sociospatial relations can be understood as a 'path-dependent, path-shaping dialectic of strategically selective structural constraints and structurally attuned strategic action' (2008: 395). Moreover, they suggest that the TPSN framework can 'fruitfully inform the field of "contentious politics", which examines different forms of contestation, resistance, mobilization, and struggle "from below"' (2008: 397).

In contrast to the structurally and more materially oriented TPSN approach, another body of literature addressing 'assemblage' offers a poststructural take on many of the same concerns addressed by Jessop, Brenner, and Jones. Assemblage has a long and diverse pedigree, tracing back to the work of Deleuze and Guattari (1987/1980 original French), Latour (2005), DeLanda (2006), and others.

Conceptions of assemblage are so diverse and varied that in a 2011 collection on assemblage published in *Area*, McFarlane and Anderson (2011: 162) explicitly state that they wish to 'avoid legislating for what assemblage is'. Nonetheless, they offer an impression of the commonalities of the various assemblage approaches:

> ... assemblage functions as a name for unity across difference, i.e. for describing alignments or wholes between different actors without losing sight of the specific agencies that form assemblages. Assemblage appears as a specific form of relational thinking that attends to the agency of wholes *and* parts ... thinking with assemblage is also in part about the play between stability and change, order and disruption (McFarlane and Anderson 2011: 162).

Assemblage thinking is clearly relational with an emphasis on agency, contingency, emergence, and process. It foregrounds the importance of space-time in understandings of social and political processes.

Recent attempts to apply the concept of assemblage to the analysis of social movements are of greatest interest here. McFarlane (2009), in his study of Mumbai's Slum/Shack Dwellers International, focuses on different spatial imaginaries and practices, describing their 'translocal' qualities as an alternative to accounts of place, scale, and networks. McFarlane's analysis clearly emphasizes 'how actors construct and move between different spatialities ... [and that we] should be attentive to how scale or network, as particular spatial imaginaries, become key devices used by actors as they attempt to structure or narrate assemblages' (McFarlane 2009: 564). Highlighting the perspectives of individual and collective agents, McFarlane argues that it is through spatial imaginaries and narratives that actors construct and reconstruct particular spatial practices. Davies (2011) also draws on the concept of assemblage, in his case to analyse the activities of Tibet Support Groups. Davies offers an account that draws more directly from the social movements literature and, instead of seeking to move 'away from particular spatial master concepts which often structure the discussion of space in relation to social movements' (McFarlane 2009), seeks to reconceptualize the relationships among specific spatialities. Davies concludes that:

> ... space is open to being structured and altered continually by those actors who are drawn into a specific social movement organisation, and therefore into a wider political movement. Territorial and relational structures are important in these actions, but are always present and always being reworked and negotiated, and are thus more or less visible at any one time. The relationship between territoriality and relationality then is something of a false dichotomy as we see elements of both with the same organisation/event ... The key for understanding political action is to uncover when and (more crucially) why either of these spatialities becomes pre-eminent (Davies 2011: 11).

Davies recognizes that the forces shaping social movement mobilization cannot be reduced to the spatial imaginaries and narratives held by agents, important though they may be. And while he is correct to point to the false dichotomy of networked relationships and territoriality, one might add that conceptualizing territory, region and scale as non-relational is itself highly problematic, given that these spatialities only exist in relationship to other territories, regions, and scales, and that their relationships are always contingent and subject to change through political struggle. Different spatialities, moreover, are frequently relationally entwined.

To borrow the terminology of Leitner et al., networks and territories may be co-implicated, each with their own particular effects, enabling and constraining collective action in particular ways. For example, building networked relationships creates the capacity for the communication of information, sharing of resources, strengthening of understandings, and coordination of action by individuals and institutions, while the construction of territory may serve to define available resources, create territorial identities that may underlie collective action, and set parameters of inclusion and exclusion – all of which may serve to strengthen or weaken, expand or sever, networked relationships.

Both the TPSN and assemblage approaches spur us to think about the relationships among diverse spatialities and how they are related to social movement mobilization. Both stress the relational nature of spatialities, both consider material practices as well as discursive constructions of the social world, and both see processes of struggle and resistance as constituted through diverse, articulating, and shifting spatialities. Yet there are important differences too. The TPSN approach, grounded in critical realism, explicitly recognizes structural properties and processes – particularly of capital and the state – and their relationships to agency. The TPSN approach also tends to emphasize the materiality of spatial practices, although certainly recognizing the importance of discursive practices and hegemonic projects. Assemblage approaches, in contrast, heavily emphasize agency with little attention to structure and, while considering material practices, conceive of material practices largely as the product of discursive practices and imaginaries.

There is clearly a gulf between the TPSN and assemblage approaches, rooted in their differing epistemologies, ontologies, and theories. Indeed, as Mayer (2008) has pointed out, scholars focusing on different spatialities often work from

> ... divergent theories of power and society. Thus most network theorists operate with a poststructuralist concept of society, viewing social power as diffused throughout topological networks and downplaying centralized, hierarchical, and class power. Such sociotheoretical assumptions mesh neither with the more regulationist paradigms implicit in most scale theories nor with the materialist perspectives of the international division of labor or state theory approaches that influenced the earlier territorial and place theories. Before embarking on a project of synthesizing such different sociospatial perspectives into one (TPSN) framework, their divergent and contradictory sociotheoretical assumptions, their

methodological and political differences would need to be sorted out and made congruent (Mayer 2008: 418).

Mayer identifies the central problems of developing an integrated approach to understanding the relationships among diverse spatialities. Yet a thread to guide us out of the impasse is contained in her observations. There are indeed many theories of power and society, but these need not be considered mutually exclusive, at least not always. The recognition of networked micro-circuits of power, following Foucault and Latour, need not be taken as a denial of centralized and hierarchical state power (Driver 1985; Jessop 2007). The recognition of exclusionary, authoritative, and allocative powers associated with scale and territory need not be seen as denial of networks (Cumbers et al. 2008; Miller 2009; Nicholls 2009). And certainly the study of material spatial practices need not entail the denial of spatial discourses and imaginaries. The primary barrier to developing an integrated approach to co-implicated spatialities would seem to lie in the relatively recent focus on ontology and epistemology with their a priori commitments to what does and does not exist, what can and cannot be known. Rather than operating on the plane of ontology and epistemology, greater insight may come from attending to practices and the multiple modalities through which power is exercised. Rather than choosing a singular conception of power and attempting to interpret the world through it, an open and pluralist stance toward power may be more illuminating.

To the point, we may find a way beyond the current impasse by conceiving of spatialities, in a Foucauldian sense, as *technologies of power*. Such technologies, as Rose (1999) summarizes, are

> ... imbued with aspirations for the shaping of conduct in the hope of producing certain desired effects and averting certain undesired events. I term these 'human technologies' in that, within these assemblages, it is human capacities that are to be understood and acted upon by technical means. A technology of government, then, is an assemblage of forms of practical knowledge, with modes of perception, practices of calculation, vocabularies, types of authority, forms of judgement, architectural forms, human capacities, non-human objects and devices, inscription techniques and so forth, traversed and transected by aspirations to achieve certain outcomes in terms of the conduct of the governed (which also requires certain forms of conduct on the part of those who would govern) (Rose 1999: 52).

Spatial technologies of power are particular types of technologies that shape the *formation and breaking of relationships* – technologies that are employed, counter-deployed, and altered in processes of social struggle. Such technologies may be employed by any actor – individual or collective, civil society or state – in efforts to mobilize or repress collective action, although specific types of actors may have greater capacity to employ particular spatial technologies, and less capacity to employ others, e.g., the capacity of the state to define and regulate territory – but for a fascinating counter-example see Agnew and Oslender (Chapter 6).

Numerous social theorists have observed that all social relations are simultaneously spatial relations, e.g. Lefebvre's (1991/1974) concept of the social production of space, Soja's (1989, 1996) idea of the socio-spatial dialectic, and Harvey's (1989, 2006) grid (or matrix) of spatial practices. This conceptualization of the social as spatial is clear, if implicit, in Deleuze and Guattari's (1987/1980) poststructural treatise *A Thousand Plateaus*. In all of these seminal works social struggle is simultaneously spatial struggle. It should come as no surprise, then, that contentious political actors adopt a variety of spatial strategies and tactics as they attempt to mobilize and gain power. As Neil Smith famously observed with regard to the production of scale, 'the scale of struggle and the struggle over scale are two sides of the same coin' (Smith 1992: 74). In other words, scale is not a fixed and inert container of social struggle, but rather actively produced as part and parcel of social struggle, as a means of altering power relations and conduct. The same is true of every spatiality. Places, scales, territories, regions and networks are produced, altered, and in some cases dismantled as part of the process of social struggle, to advance the interests of particular actors. These spatialities, moreover, exist materially in the form of the amalgamation, distribution and mobility of resources – including people, money, skills, equipment, information, etc. They exist as well in the form of narratives and imaginaries that may frame and motivate particular courses of action or inaction.

Originating in the work of Lefebvre and elaborated upon by Soja and Harvey, a tripartite spatial schema consisting of material space, conceptual space, and lived space has become commonplace in many theoretical and epistemological treatments of space. This tripartite schema, which should be regarded as a heuristic guide through simultaneously overlapping spatialities, can be adapted to address the core concerns of social movement and contentious politics theory. Each of the specific spatialities we have considered can be viewed as a spatial technology of power, with associated material and representational practices that shape sociospatial power relations (Table Conc.1). From a contentious politics perspective, *material space* primarily addresses the distribution, amalgamation, claiming, and mobility of resources. *Conceptual space* primarily addresses signification, particularly with regard to the construction of common understandings, values, discourses, and identities necessary for collective action. *Lived space* is the realm in which the material world and the conceptual frameworks through which we interpret and shape it come together. Here our tangible experiences collide with our conceptions and interpretations, giving rise to imaginaries of what the world is, what it can be, and what it should be. With respect to contentious politics, lived space has direct bearing on the legitimacy of existing social orders, and the legitimacy of proposals for other possible worlds. It is through lived space that the 'order of things' becomes 'common sense' and dominant norms are transmitted, challenged, and defended. Just as lived space is where power relations become legitimate, it is also where people discover the precariousness of legitimacy. People discover the

Table Conc.1 Spatialities and socio-spatial power relations in social movement mobilization (drawn from Lefebvre 1991; Harvey 2006; and Jessop et al. 2008)

Spatiality	Spatial technologies of power	Socio-Spatial Power Relations		
		Material space (material resources)	Conceptual space (signification)	Lived space (legitimation)
Place and Space	Social condensation and amalgamation through relations of co-presence; areal differentiation and segregation	Create condensed patterns of co-present interaction through which to mobilize people, wealth, income, skills, etc.	Create conditions conducive to the construction of strong ties, common understandings, shared values, shared identities	Define grievances, diagnoses, legitimate motivations for mobilization based on perceived shared place-based values and interests
Territory and Region	Bordering, enclosure, claiming on an areal basis	Lay claim to and/or controlling people, wealth, income, etc.	Create shared flows of information, common understandings, shared identities; Imagined communities	Define grievances, diagnoses, legitimate motivations for mobilization based on perceived shared territorial/regional values and interests; Determine which political actors have standing in political contests
Scale	Hierarchical ordering of territorial institutions; Horizontal nesting of territory and regionalized processes	Differentiate institutional and state powers to authorize transformation and/or allocation of resources	Construct hierarchical territorial identities and imagined communities; Prioritize particular hierarchical identities, e.g., national vs. regional	Define appropriate institutional/state arena for political contention – scale jumping
Networks	Connectivity among individual actors, institutional actors, and non-human actants	Generate topological connections and associations to mobilize people, wealth, income, skills, etc.	Create conditions conducive to construction of mostly weak ties, information sharing, common understandings, common associational identities	Define who may join the network; Determine who has power and influence within the network; Conform to or challenge dominant network discourses

contradictions of the existing order – the constant disconnects between practice and what is supposed to be – opening up interstitial spaces where people develop alternative norms and values from which wrongs are diagnosed and alternative possibilities and prescriptions are crafted.

Struggles over resources, understandings, values, and legitimacy can be waged, and the distribution of power altered, through the deployment of the spatial technologies of power – of place and space, territory and region, scale, and networks. It is important to stress that these spatial technologies of power are not to be viewed as static states but rather as practices that exist simultaneously with, and often being countered by, other practices. They are critical to the evolving processes of social and political struggle, as spatial technologies of power are deployed and counter-deployed by multiple contentious political actors, e.g., social movements, state actors, corporations.

Rather than focus exclusively on the practices of particular movement organizations or networked political actors, it is imperative to recognize the relational characteristics of social and political struggle. Social movements and other forms of contentious political activity almost always face opposition, i.e. other contentious political actors that attempt to repress mobilization through their own spatial strategies and tactics. Opposition and repression most commonly come from the state, but can come from almost any source including corporations, community groups, religious groups, property owners, or others whose interests are threatened or are ideologically opposed. For example, when a movement attempts to mobilize more resources by expanding its transnational networks, the state may tighten border controls on financial transfers and deny entry visas to foreign activists, cutting off crucial resources. Or as a movement gains support and legitimacy, an opposition group may attempt to portray the movement as linked to groups outside the territory or region, undermining its legitimacy.

It is through the back and forth struggle of opposing contentious collective actors that movements blossom, fizzle, or are crushed. Not uncommonly, the actions of one contentious collective actor may unwittingly serve to delegitimize one side of the struggle and arouse support for the other, as a variety of spatial technologies are employed and counter-deployed. This was the case on September 24, 2011, when New York City police corralled Occupy Wall Street protestors on a public sidewalk and pepper sprayed them. The action was undoubtedly intended to repress dissent by spatially delimiting and punishing public protest, but had the unintended consequence of delegitimizing state repression and dramatically increasing support for the movement as images of the incident were circulated through social media networks and mainstream media. The spatialities at play in this incident and its aftermath were numerous: the enforced confinement (regionalization) of protest in a space coded for public interaction, inclusion, and speech; the development of shared understandings of the Occupy protest and the causes of the financial crisis facing the world as discussion spread through expanding topological network space and face-to-face interactions in thousands of places around the world; the convergence of people and resources at places of protest, particularly Zuccotti Park. In these and many other ways spatial technologies of power, affecting the mobilization of material resources, the creation or disruption of common understandings, and the legitimation or delegitimation of contentious collective actors, were at play.

A comprehensive discussion of all conceivable spatial technologies of power, and how they may shape sociospatial relations in material, conceived, and lived space is beyond the scope of this chapter. The preceding discussion only scratches the surface of how spatial technologies may be deployed in processes of political contention. Indeed, considering the implications of deploying and counter-deploying spatial technologies in material versus conceived space, conceived versus lived space, and lived versus material space would add another massive level of complexity to any attempt at comprehensive analysis. Rather than trying to account for every possible permutation of sociospatial political contention, this extremely complex matter can perhaps be simplified by observing that all collective action involves ongoing tensions between building and breaking relationships. Collective action requires the building of relationships among individual actors, relationships that enable them to act collectively. Successful collective action – in most cases – involves building and shaping relationships not only among significant numbers of like-minded activists, but also with apparatuses of states, corporations, or other powerful institutions or groups in positions of authority in order to make meaningful claims upon them. Spatial technologies of power are at their root about building, as well as breaking, relationships. Expanding networks is about expanding relationships; reducing the permeability of territorial borders is about breaking cross-border relationships. Constructing a narrative of shared regional identity is about building relationships; claiming that political activists are illegitimate outsiders is about breaking relationships. Spatialities matter because they are how relationships are built or broken.

* * *

Each of the empirical chapters of this book deals with the struggles of social movements or other political actors as they attempt to build relationships that enable them to act collectively to challenge existing systems of authority and practice. Many of the chapters also address the actions of their opponents as they attempt to undermine insurgent collective action, breaking the relationships that make movements powerful and threatening to those in dominant positions of power. While each of the chapters begins with a particular spatial theme, as each narrative unfolds it becomes clear that multiple spatialities are at play.

The first four empirical chapters revolve around the theme of place and space. In 'Putting Protest in Place: Contested and Liberated Spaces in Three Campaigns', Donatella della Porta, Maria Fabbri and Gianni Piazza make a powerful case that protest in place is critical to building common understandings, common identities, and strong senses of solidarity. The place-based spatialities common to the No TAV, No Bridge, and No Dal Molin movements in southern Italy facilitated the negotiation of meaning and the building of strong bonds of trust. They also entailed a reconceptualization of relationships between the local and the global, between citizens and their territorial allegiances. As well, the three cases illustrate how redefining territory can be part and parcel of a strategy of greater inclusiveness and mobilization. In 'The Liberalization of Free Speech: Or, How Protest in Public Space is Silenced', Don Mitchell focuses on the role the state, through the courts, has played in repressing political speech in the United States. Beginning from the obvious but typically overlooked observation that speech always takes place somewhere, Mitchell makes a compelling case that 'the regulation of location, or place [of speech], becomes the surrogate for the regulation of content'. By regulating where speech is allowed, the effectiveness of political speech can be regulated as well. A series of Supreme Court rulings have progressively privileged the exclusionary rights of private property owners over the rights of citizens to be included, and free to speak, in public space. In 'Struggling to Belong: Social Movements and the Fight to Feel at Home', Jan Willem Duyvendak and Loes Verplanke, drawing on cases from San Francisco and the Netherlands, focus on the different ways in which home, as a particular type of place, can be constructed. They distinguish between places as 'heavens' where affective bonds are built, symbolic meaning constructed, and outward links are forged, and 'havens' which are primarily defensive territories based on social exclusion to ensure security and privacy. The latter form of place may not lend itself well to collective action, but may nonetheless be critical for survival in the face of a variety of threats. In 'Place Frames: Analysing Practice and the Production of Place in Contentious Politics', Deborah Martin examines how neighbourhood actors in Athens, Georgia, USA, construct conceptions of place and how these meaningful constructions set the stage for the collective diagnosis of, and motivations for dealing with, grievances. Martin's analysis of the discursive construction of place also points to important scalar issues, as the neighbourhood activists adopted prognostic, diagnostic, and motivational frames that exhibited scalar disjunctures. Places are themselves scaled in processes that are open to negotiation and contestation.

The second four empirical chapters revolve around the themes of scale, region, and territory. In 'Polymorphic Spatial Politics: Tales from a Grassroots Regional Movement', Martin Jones applies the TPSN framework that he helped develop to the analysis of a regional movement. The centrepiece of his analysis focuses on the development of the Movement for Middle England (MFME) which evolved into another movement known as *Devolve!* MFME and *Devolve!* contested a state-driven form of regional identity through the creation of a bottom-up networked civil society regionalism. His example demonstrates the role of territory, place, scale and networks in the shifting socio-spatial construction of regionalism. In 'Overlapping Territorialities, Sovereignty in Dispute: Empirical Lessons from Latin America', John Agnew and Ulrich Oslender analyse the challenge FARC (the Revolutionary Armed Forces of Colombia) has posed to the established territorial framework of the Colombian state. FARC's strategy has been based on claiming territory for the peasant communities it represents, creating a situation in which FARC and the Colombian state make overlapping territorial claims, which mean overlapping claims to resources, identity, and legitimacy. A central strategy for dealing with this conflict has been to rescale state functions. In 'LimiteLimite: Cracks in the City, Brokering Scales and Pioneering a New Urbanity', Johan Moyersoen and Erik Swyngedouw examine the dysfunctional governance of one of the most deprived neighbourhoods in Brussels, the Brabant neighbourhood. This neighbourhood has been squeezed between two different realms: one associated with major globally oriented businesses; the other associated with the everyday living spaces of its poor residents. To bridge this scalar disjuncture neighbourhood activists built the LimiteLimite tower as a focus for neighbourhood interaction, essentially serving as a place that could broker scales, allowing for the creation of relationships, common understandings and mutually agreeable courses of action. Moyersoen and Swyngedouw's complex analysis weaves together place, territory, scale and networks in material, conceptual, and lived space. In 'Multiscalar Mobilization for the Just City: New Spatial Politics of Urban Movements', Margit Mayer looks at the fields of social and power relations that operate at every scale, and the scalar cross-fertilization of these relations. She specifically examines the localization and urbanization of global movements, focusing on their networked relationships and scalar and territorial strategies. The co-implication of a variety of spatialities is a central theme.

The final four empirical chapters focus on networks. In 'The Built Environment and Organization in Anti-US Protest Mobilization after the 1999 Belgrade Embassy Bombing', Dingxin Zhao shows how mobilization of student protests against the 1999 US bombing of the Chinese embassy in Belgrade was mediated not only by different socio-political contexts and network characteristics at three different Chinese universities, but also by the physical layout of the campuses – the latter observation providing a very useful reminder that topological network characteristics cannot be neatly separated from the Euclidean spaces of the built environment. In 'Energizing Environmental Concern in Portland, Oregon', Ted Rutland provides a fascinating narrative of the mobilization of the renewable energy focused environmental movement in the Pacific Northwest of the United States. Rutland adopts Latour's actor-network theory to trace the relationships among a wide range of human actors and non-human actants, demonstrating the critical importance of non-human relationships in social movement mobilization. Rutland's account also stresses the importance of how issues are discursively framed, how these framings affect the growth of, and circulation of information through, networks, and how multi-scalar governance structures can influence the material basis of network formation. In 'Networking Resistances: The Contested Spatialities of Transnational Social Movement Organizing', Andrew Davies and David Featherstone examine ways in which transnational organizing around the Coca-Cola boycott and the Tibet issue produce new relationships. Networked relationships, in other words, are not encountered by social movement organizers fully formed and ready to be harnessed through strategies of brokerage and scale shift, but are also produced through the process of making injustices and unequal power relations visible, in turn constituting new identities, alliances, and spatial practices. Finally, in 'Global Justice Networks: Operational Logics, Imagineers and Grassrooting Vectors', Paul Routledge and Andrew Cumbers lay out a compelling account of how global justice networks, led by key organizers or 'imagineers', are able to build network relations by employing both vertical and horizontal political imaginaries. Their account stresses the need to link networks at different scales, translate global justice concerns into the language of everyday place-based lived experiences, and recognize both the topological and territorial constitution of networks.

In all of the empirical chapters we see the production of multiple spatialities as political actors seek to build and expand relationships. A variety of spatial technologies are employed as contentious collective actors seek to gain advantage vis-à-vis their opponents. In different contexts different spatial strategies will be more or less effective. In some cases mobilizing more material resources will be critical. In other cases blocking the resources of opponents will be paramount. In some instances building new understandings, or more broadly based identities, will be of utmost importance. In other cases gaining legitimacy, or delegitimizing an opponent, will be critical. There is no teleology to the mobilization of social movements or other forms of contentious politics. Rather, mobilization is more like a chess game as actors match strategic move with strategic move. Central to this game is the deployment of spatial technologies of power by activists attempting to build the relationships that will advance their cause, and by opponents attempting to dismantle the relationships that form the basis of any form of resistance. In the ever-changing field of struggle technologies of place and space, region, territory, scale, and networks all play their roles, albeit in different ways at different times and in different contexts. Explicit recognition of spatial practices in material, conceptual, and lived space helps us to better understand the dynamics of contentious politics. It may also help us to build more effective movements for a better world.

References

Cumbers, A., Routledge, P. and C. Navitel (2008) 'The Entangled Geographies of Global Justice Networks', *Progress in Human Geography*, 32, 183-202.

Davies, A. (2011) 'Assemblage and Social Movements: Tibet Support Groups and the Spatialities of Political Organisation', *Transactions of the Institute of British Geographers*, 36, 1-14.

DeLanda, M. (2006) *A New Philosophy of Society*. London: Continuum.

Deleuze, G. and F. Guattari (1987/1980) *A Thousand Plateaus*. Minneapolis: University of Minnesota Press.

Driver, F. (1985) 'Power, Space and the Body: A Critical Assessment of Foucault's Discipline and Punish', *Environment and Planning D: Society and Space*, 3, 425-446.

Featherstone, D. (2003) 'Spatialities of Transnational Resistance to Globalization: Maps of Grievance of the Inter-Continental Caravan', *Transactions of the Institute of British Geographers*, 28, 404-421.

—— (2005) 'Towards the Relational Construction of Militant Particularisms: Or Why the Geographies of Past Struggle Matter', *Antipode*, 37, 250-271.

Harvey, D. (1989) *The Condition of Postmodernity*. Cambridge: Wiley-Blackwell.

—— (2006) *Spaces of Global Capitalism*. New York: Verso.

Jessop, B. (2007) 'From Micro-Powers to Governmentality: Foucault's Work on Statehood, State Formation, Statecraft and State Power', *Political Geography*, 26, 34-40.

Jessop, B., N. Brenner and M. Jones (2008) 'Theorizing Sociospatial Relations', *Environment and Planning D: Society and Space*, 26, 389-401.

Knopp, L. (1997) 'Sexuality and Urban Space: Gay Male Identities, Communities and Cultures in the US, UK and Australia', in S. Pile and M. Keith (eds) *Geographies of Resistance*. London: Routledge, pp. 149-176.

Latour, B. (2005) *Reassembling the Social: an Introduction to Actor-Network Theory*. Oxford: Oxford University Press.

Lefebvre, H. (1991/1974) *The Production of Space*. Oxford, UK and Cambridge, USA: Blackwell.

Leitner, H. Sheppard, E., and K. Sziarto (2008) 'The Spatialities of Contentious Politics', *Transactions of the Institute of British Geographers*, 33: 157-172.

McFarlane, C. (2009) 'Translocal Assemblages: Space, Power and Social Movements', *Geoforum*, 40, 561-567.

McFarlane, C. and B. Anderson (2011) 'Thinking with Assemblage', *Area*, 43 (2), 162-164.

Mayer, M. (2008) 'To What End Do We Theorize Sociospatial Relations?', *Environment and Planning D: Society and Space*, 26, 414-419.

Miller, B. (2000) *Geography and Social Movements*. Minneapolis: University of Minnesota Press.

—— (2009) 'Is Scale a Chaotic Concept? Notes on Processes of Scale Production', in R. Keil and R. Mahon (eds) *Leviathan Undone? Towards a Political Economy of Scale*. Vancouver: UBC Press, pp. 51-66.

Moore, D. (1997) 'Remapping Resistance: Ground for Struggle and the Politics of Place', in S. Pile and M. Keith (eds) *Geographies of Resistance*. London: Routledge, pp. 87-106.

Nicholls, W. (2009) 'Place, Relations, Networks: The Geographical Foundations of Social Movements', *Transactions of the Institute of British Geographers*, 34 (1): 78–93.

Rose, G. (1997) 'Performing Inoperative Community – the Space and the Resistance of Some Community Arts Projects', in S. Pile and M. Keith (eds) *Geographies of Resistance*. London: Routledge, pp. 184-202.

Rose, M. (2002) 'The Seductions of Resistance: Power, Politics, and a Performative Style of Systems', *Environment and Planning D: Society and Space*, 20, 383-400.

Rose, N. (1999) *Powers of Freedom*. Cambridge: Cambridge University Press.

Routledge, P. (2003) 'Convergence Space: Process Geographies of Grassroots Globalisation Networks', *Transactions of the Institute of British Geographers*, 28, 333-349.

Slater, D. (1997) 'Spatial Politics/Social Movements: Questions of (B)orders and Resistance in Global Times', in S. Pile and M. Keith (eds) *Geographies of Resistance*. London: Routledge, pp. 258-276.

Smith, N. (1992) 'Geography, Difference, and the Politics of Scale', in J. Doherty, E. Graham and M. Malek (eds) *Postmodernism and the Social Sciences*. London: Macmillan, pp. 57-79.

Soja, E. (1989) *Postmodern Geographies*. New York: Verso.

—— (1996) *Thirdspace*. Malden, MA and Oxford: Wiley-Blackwell.

Wainwright, J. (2007) 'Spaces of Resistance in Seattle and Cancun', in H. Leitner, J. Peck and E. Sheppard (eds) *Contesting Neoliberalism*. New York: Guilford, pp. 179-203.

Wainwright, J., Prudham, S. and J. Glassman (2000) 'The Battles in Seattle: Microgeographies of Resistance and the Challenge of Building Alternative Futures', *Environment and Planning D: Society and Space*, 18, 5-13.

Index